Engineering Systems

Modelling and Control

Essential Maths for Students

Engineering Systems

Modelling and Control

Martin Hargreaves

Longman

Longman
Addison Wesley Longman Limited
Edinburgh Gate, Harlow
Essex CM20 2JE, England
and Associated Companies throughout the world.

First published 1996

British Library Cataloguing in Publication Data
A catalogue entry for this title is available from the British Library.

ISBN 0-582-23419-0

Library of Congress Cataloging-in-Publication data
A catalog entry for this title is available from the Library of Congress.

Produced through Longman Malaysia, GPS

To my mother, Jean Hargreaves, and sister, Janet Platek, with love

Contents

Preface

This book was written to serve the needs of undergraduate engineers embarking on a first course in engineering systems. The study of engineering systems requires a multidisciplinary approach which crosses the traditional boundaries of engineering. As such, many different types of engineers can benefit from the study of engineering systems. Of course, control engineers must engage in such study as the ability to model a system is often a necessary prerequisite to being able to control it. Most textbooks on control engineering only devote a small amount of space to the topic of system modelling. This is understandable in view of the large amount of material that has to be presented. Unfortunately, modelling is the part of control engineering that presents the most difficulties when first encountered and so a sympathetic approach, with plenty of practical examples, is especially valuable. This is the approach adopted in this book. The topic of control system design is also introduced but the decision was taken not to discuss frequency domain methods as this would have made the book too long. From my experience, students prefer to stick to time domain methods when first encountering control system design. More advanced books on control engineering are available which discuss frequency domain methods in some detail but these tend to be used for more advanced courses, usually taken in the final year of a degree course.

Martin Hargreaves

Acknowledgements

I would like to thank the series editors, Anthony Croft and Robert Davison, for their support and encouragement during the writing of this book. Also, a special thanks to Roger Lawrence and Ian Mann for their thorough appraisal of the manuscript. Finally, the comments of unknown reviewers are gratefully acknowledged.

1 Introduction to engineering systems

Objectives

This chapter:

- explains the basic concepts associated with the study of engineering systems
- describes the relationship between signals and system components
- identifies the elements of a block diagram
- describes the role of a mathematical model of an engineering system
- discusses the difference between linear and nonlinear models
- describes the SI system of units
- explains how to balance units in equations involving physical quantities

1.1 Introduction

It is useful to be able to see the links between the various areas of engineering. This leads to a greater depth of understanding which in turn results in better engineering designs.

The study of engineering systems allows a broad view of engineering to be taken. Barriers can be broken down between traditional engineering disciplines such as electrical, mechanical and chemical engineering, thus allowing links to be formed between a wide variety of engineering processes and products. The advantage of this approach is that what at first sight appear to be entirely different systems are suddenly seen to be very similar.

This saves a lot of the duplication of effort that occurs if each system is analysed in isolation. In addition, a systems approach makes it much easier to analyse and design systems that are a close mix of components from different engineering disciplines. Systems engineers have always been present in engineering, charged with the role of integrating the design of complex systems. Today, however, their importance is increasing as engineering systems become more and more complex. There is now a drive to break down the traditional barriers of engineering. For example, the new discipline of mechatronics has emerged because of the need to break down the barrier between electronic and mechanical engineering when designing many of today's products. A good example of such a product is a camcorder. Here the mechanics and the electronics are closely related and design of each requires knowledge of the other if a successful product is to emerge.

1.2 Basic concepts

KEY POINT

An **engineering system** is a set of **components** connected together to accomplish a useful task.

Example

1.1 Suggest possible components of a coal-fired power station.

Solution Typical components are the boiler, the turbine and the electrical generator.

The choice of appropriate system components is a task for the engineer designing a particular system.

KEY POINT

Engineers like to have clear divisions between the components of a system. It makes it easier to understand the system as a whole. The engineer's motto is, 'keep it simple'.

Guidelines for choosing system components:
(i) If possible, choose components that are easily identifiable as separate entities.
(ii) Try to choose components that have a simple and clearly defined interaction with other components.
(iii) Endeavour to keep the number of system components to the minimum necessary for the analysis being carried out. In other words, avoid too much detail as this leads to unnecessary complication.

To illustrate these guidelines consider the case of a single engine jet aircraft. A good choice for one of the system components is the jet engine. It is easily identifiable as a separate entity. It interacts with the rest of the system in a relatively simple manner in that, for a given input of fuel, it produces a certain thrust which in turn is transmitted to the airframe.

Now suppose that the jet engine is broken down to produce several new system components of which two are the compressor and the combustion chamber. This is not such a good choice because the interaction between these two components is complicated and so not easily defined. Also, it introduces too much detail for the purposes of analysing the aircraft as a whole. As a point of interest, the design and manufacture of jet engines is usually carried out by different companies from those that design and manufacture aircraft. Thus, aircraft manufacturers order engines to produce a suitable thrust characteristic for their needs and are not concerned with the internal components of the engine.

Example

1.2 Consider a high fidelity music system. Suggest a good set of system components. Suggest a more detailed set that may not be so suitable.

Solution A good set of system components are the loudspeakers, the amplifier, the record player, the tape deck and the CD player. If more detail is required then the amplifier could be broken down into various circuit boards and the record player into a motor drive, a cartridge, a turntable and so on. However, for this more detailed breakdown the interaction between the system components is more complicated.

Many of today's engineering products are so complicated that they can only be understood if a systems approach is adopted.

When analysing an engineering system an engineer is much more interested in the interactions between the various system components than in the internal workings of the components themselves. This is sometimes termed the **systems approach**. It provides a convenient way of reducing a complex system to something that is more easily understood. The system components are described in terms of their **inputs** and **outputs**. Consider again the power station of Example 1.1. The boiler has fuel, air and water as inputs and steam, ash, smoke and waste heat as outputs.

Example

1.3 Recall the music system of Example 1.2. Suggest possible inputs and outputs for the amplifier and loudspeakers.

Solution The amplifier has electrical power from the mains as an input together with various low power signals from the record player, the tape deck and the CD player. Outputs are two high power signals which are fed to the loudspeakers, a medium power signal which is fed to the headphones, and waste heat. You may be able to suggest other inputs and outputs. The loudspeakers each have an electrical power signal as input and sound energy and heat energy as outputs.

It is convenient to think of the components of a system as being interconnected by means of their inputs and outputs. For example, for the hi-fi system of Example 1.2 one amplifier output is a high power signal which provides an input signal to a loudspeaker. It must be stressed that this is just a convention that makes the analysis of systems easier. A layman would argue that the amplifier and loudspeaker were connected together by a wire!

The concept of a general signal is extremely powerful. It allows comparisons to be made between many different engineering products.

It is common to refer to all of these interconnecting inputs and outputs as **signals**. However, this does not mean that they are all electrical signals. For example, the signals may be mechanical, fluid or thermal quantities.

Example

1.4 Suppose that an electric train is a system with two components: the electric motor and the rest of the train. State the main input and output signals for these components together with any interconnections.

Solution The electric motor has an input signal of electrical power and an output signal of rotational mechanical power. The rest of the train has an input signal of rotational mechanical power which it converts to an output signal of translational motion. The two components are connected together by means of the rotational mechanical power signal.

It is useful to think of a system as being surrounded by its **environment**. An imaginary boundary is considered to divide the system from its environment and is known as the **system boundary**. Usually the system interacts with its environment and this gives rise to various signals crossing the system boundary. The signals which pass from the environment to the system are known as **system inputs** while those that pass from the system to the environment are known as **system outputs**. These concepts are illustrated in Figure 1.1.

Example

1.5 Consider again the power station of Example 1.1. Determine the system inputs and outputs.

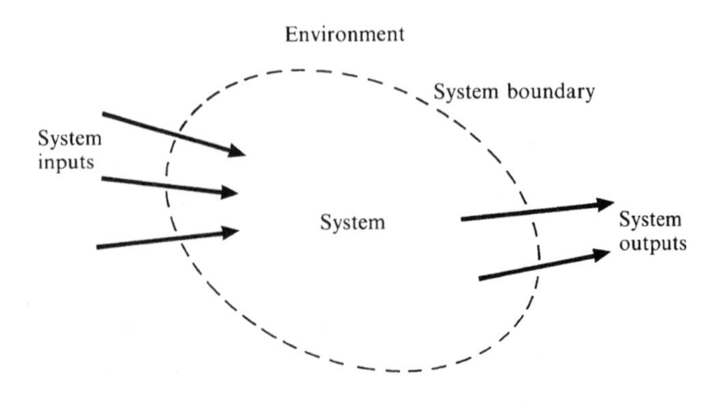

Figure 1.1
A general engineering system

Solution The main system inputs are fuel, air and cooling water. The main system outputs are electricity, waste heat, fuel ash, noise and pollutants. Note there are many other possible inputs and outputs. For example, the materials needed to maintain the power station can be regarded as inputs and waste paper from the offices is a possible output. The choice of appropriate inputs and outputs is dependent on what type of analysis is being carried out by an engineer.

A power station is a very complex system. The systems we shall analyse are much simpler. However, many of the ideas we develop carry over to larger systems.

When analysing a large system it is sometimes convenient to think in terms of a **systems hierarchy**. Items that are considered to be a component of a system at one level become systems themselves at a lower level. For example, a car may be thought of as a system with one component being the engine. However, the engine may also be thought of as a system with components such as the pistons and the crankshaft. Several levels may be needed for some systems. This approach allows an engineer to handle complexity more conveniently.

Self-assessment questions 1.2

1. State the definition of an engineering system.
2. Describe the guidelines needed when choosing a suitable set of components for an engineering system.
3. Explain what is meant by the systems approach to engineering.
4. Explain what a signal is in an engineering system.
5. What is a system input and what is a system output?
6. Explain what is meant by the term 'system boundary'.

Exercises 1.2

1. Suggest a suitable set of system components for a bus.
2. Suggest a suitable set of system components for an electronic calculator.
3. Suggest a suitable set of system components for a windmill producing flour.
4. Suggest possible system inputs and system outputs for a house.
5. Consider the music system of Example 1.2. Suggest possible system inputs and outputs.
6. Determine suitable signals and interconnections for the components of the bus of Exercise 1.2.1.
7. Determine suitable signals and interconnections for the components of the electronic calculator of Exercise 1.2.2.
8. Determine suitable signals and interconnections for the components of the windmill of Exercise 1.2.3.

1.3 Block diagrams

It is much easier to see the interactions between the various components of a system if they are presented visually. **Block diagrams** provide a convenient mechanism for doing this. The components of a system are represented by rectangular boxes and the signals that interconnect the various components are

Most engineers have very good spatial abilities and so they prefer pictures to symbols.

represented by lines with arrows. When creating a block diagram, components are limited to one input and one output. We shall see later that this restriction is used to enable block diagrams to be analysed mathematically. In practice this restriction is not too limiting and there are ways to overcome it should it be necessary to do so. The three main elements of a block diagram are shown in Figure 1.2.

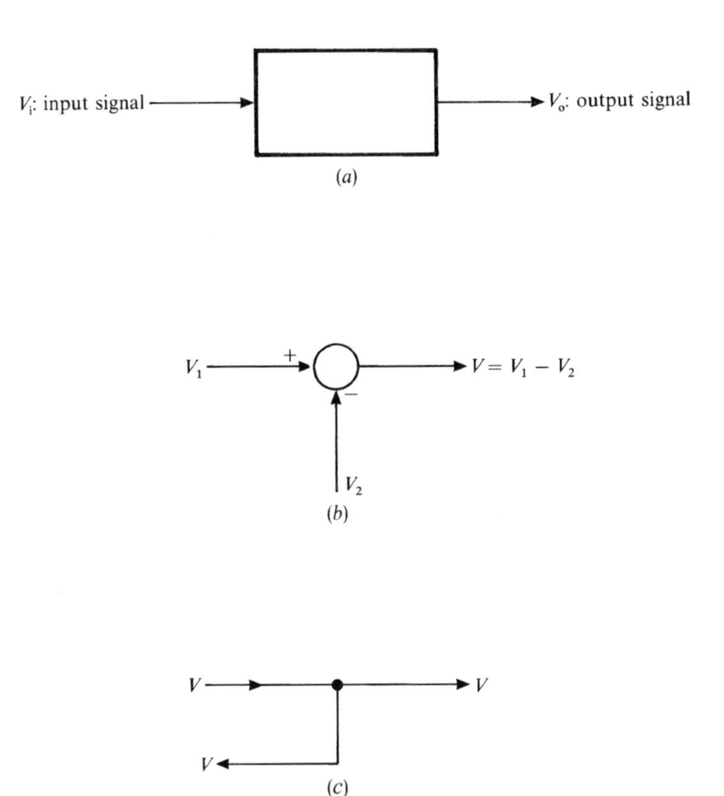

Figure 1.2
The three main elements of a block diagram: (a) a basic block; (b) a summing junction; (c) a take-off point

A **basic block** represents a system component and has an **input signal** and an **output signal**. It is convenient to think of the component **operating** on the input signal to produce an output signal. This is a very similar concept to that of a mathematical function and we shall explore this connection later.

A **summing junction** allows signals to be added together. It is depicted by a circle with one arrow coming out and one or more arrows going in. The output signal is the sum of all the input signals. Either a plus or a minus sign is placed against an input arrow head depending on whether the signal is to be added or subtracted.

A **take-off point** allows a signal to be 'tapped' and used elsewhere. An assumption is made that tapping into the signal does not **load** the signal. This means that the signal is not affected by introducing the take-off point.

Example

1.6 Consider again the electric train of Example 1.4. Draw a block diagram for this system.

Solution The block diagram is shown in Figure 1.3. There are two components, namely, the train motor and the rest of the train. There are three signals, namely, the electrical power to the motor, the rotational mechanical power from the motor to the train and the translational motion of the train.

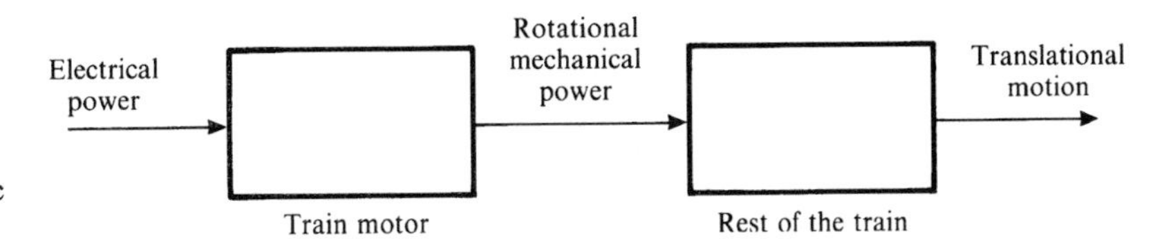

Figure 1.3
Block diagram of an electric train

It is common practice to label the signals within a block diagram as well as the various components, thus making it easier to understand the various interactions taking place within the system. We see that this has been done in Figure 1.3 for the electric train.

Example

1.7 Consider again the power station of Example 1.1. Draw a block diagram to represent the relationship between the fuel input rate to the boiler and the electrical power produced by the electrical generator which is connected to the steam turbine.

Solution For simplicity we will assume the signal connecting the boiler and the turbine is the steam pressure. Also, there is a rotational mechanical power signal connecting the turbine to the generator. The block diagram is shown in Figure 1.4.

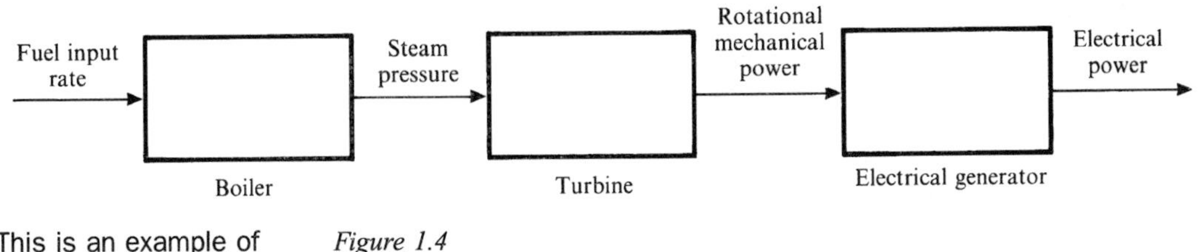

Figure 1.4
Block diagram for part of a power station

This is an example of where experience is important. There is a great deal of truth in the view that engineering is as much an art as a science.

Note that in Example 1.7 the choice of suitable signals was a little arbitrary. This is often the case when creating a block diagram. It is up to the engineer to decide which are the best signals to use. A lot depends on why the system is being analysed.

Example

1.8 Draw a block diagram for a central heating system typically found in a home.

Solution The block diagram is shown in Figure 1.5. The thermostat allows the desired temperature to be compared with the room temperature. If the room

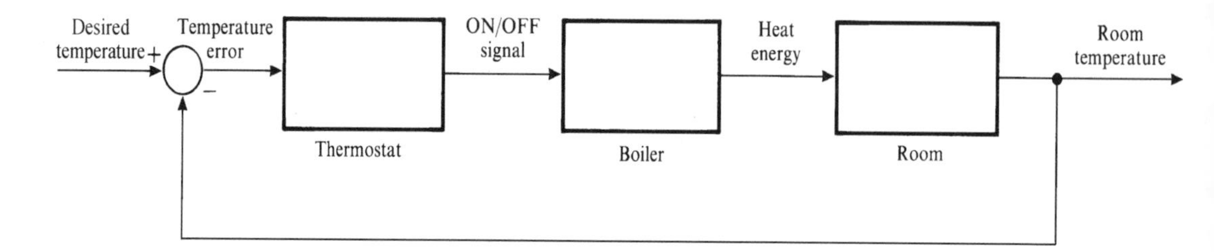

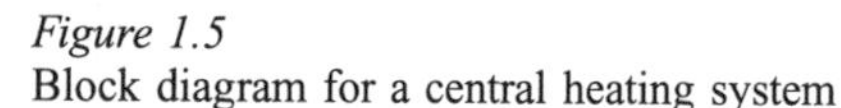

Figure 1.5
Block diagram for a central heating system

temperature is less than the desired temperature then the boiler is switched on. If the room temperature is greater than the desired temperature then the boiler is switched off.

Systems that we meet every day provide a good way of understanding engineering concepts.

Self-assessment questions 1.3

1. What are the main components of a block diagram?
2. Describe what the output from a summing junction consists of.
3. What effect does a take-off point have on a signal that is being tapped?

Exercises 1.3

1. Draw a block diagram for the bus of Exercises 1.2.1 and 1.2.6.
2. Draw a block diagram for the electronic calculator of Exercises 1.2.2 and 1.2.7.
3. Draw a block diagram for the windmill of Exercises 1.2.3 and 1.2.8.
4. Draw a block diagram for the control of the temperature of a laboratory furnace. Assume the furnace is heated and controlled by a thermostat.
5. Draw a block diagram for the control of the liquid level in a tank. Assume the tank has a float switch which shuts off the liquid flow when the correct level is reached.

1.4 Mathematical models

Computers may be used to solve mathematical models, thus extending their versatility.

Engineers often build models in order to understand how a system works. For example, physical models of cars are used in wind tunnel tests to determine the best aerodynamic shape for a car. The most versatile type of model used by an engineer is the **mathematical model**. This type of model is one in which a system is represented by a set of mathematical equations.

The great advantage of a mathematical model is that many different variations in system characteristics and inputs can be analysed without having to build anything. A mathematical model is therefore versatile and cheap.

KEY POINT

A mathematical model consists of a set of mathematical equations.

The strong division between the arts and the sciences is mainly a feature of the twentieth century. The Renaissance artist and inventor, Leonardo Da Vinci, saw no such division. His work demonstrates a continuity between many areas of learning.

Whenever a mathematical model of a system is built it is necessary to make certain assumptions and simplifications and so a mathematical model is rarely exact. A balance is chosen between the desire for simplicity and the need for accuracy. There are no exact guidelines for producing a good mathematical model of an engineering system. This is one of the reasons that engineering is as much an art as a science.

We shall leave the derivation of mathematical models for a range of engineering systems to future chapters but it is convenient to discuss a simple example at this point in order to describe some features of mathematical models.

Example

1.9 If a force is applied to a spring then the amount by which it stretches, known as its **extension**, is proportional to the size of this force, provided the force is not too large. Carry out the following:

(a) Derive a mathematical model for the extension of a spring when a force is applied.
(b) Draw a block diagram for the spring with force as the input and extension as the output.
(c) Determine the mathematical rule for the block which relates the input signal to the output signal. Draw a block diagram for the spring which includes this rule.

Solution (a) Let us denote the force applied to the spring by f and the extension of the spring by x. We are told that f is proportional to x, that is

$$f \propto x$$

So we can write

$$f = Kx$$

where K is a constant known as the **spring stiffness**. This is the mathematical model for the behaviour of the spring when a force is applied.

(b) The block diagram for the spring is shown in Figure 1.6.

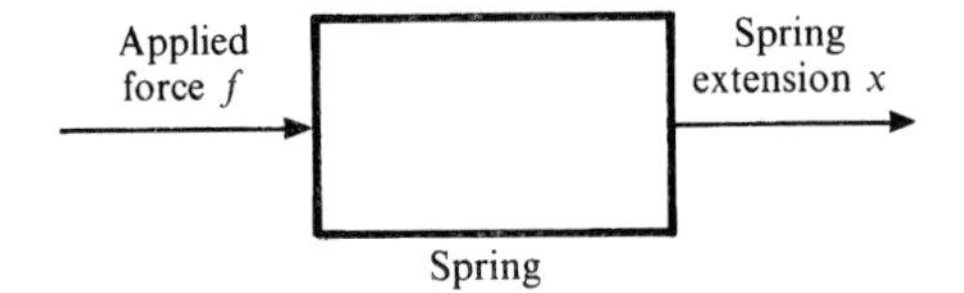

Figure 1.6
Block diagram for the spring

(c) The mathematical model for the spring is

$$f = Kx$$

Rearranging, we have

$$x = \frac{1}{K}f$$

We can now determine the mathematical rule for the block. We see that if we multiply the input f by $1/K$ then we obtain the output x. This is shown in Figure 1.7.

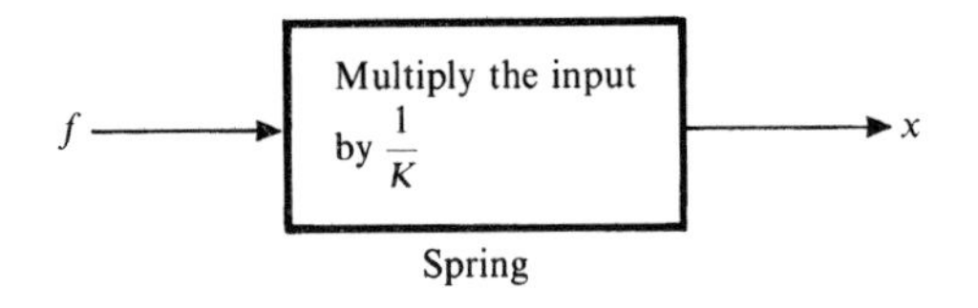

Figure 1.7
The rule for the extension of the spring is: multiply the input by $1/K$

It is instructive to think of a function as being rather like a machine. It takes an input, processes it and produces an output.

The previous example illustrates several points of interest. The mathematical rule derived in part (c) can be thought of as a mathematical function. It takes f as an input and produces x as an output. We see now why it is necessary to restrict block diagrams to one input signal and one output signal. This enables us to describe a block by means of a mathematical function. This is very useful to be able to do.

In order to form the mathematical model for the spring we assumed that force and extension were proportional no matter what force was applied. However, for a real spring, if too much force is applied then the spring will either snap or permanently deform. We see that the mathematical model is a simplified representation of the behaviour of a real spring. This is perfectly acceptable provided the model is not used to predict the behaviour of a spring when a large force is applied. The virtue of this model is that it is simple. We could have produced a more complicated mathematical model to take account of the regime in which a large force is applied but the simplicity of the first model would be lost and, unless it is necessary, the use of a complicated model should be avoided.

This model for the behaviour of the spring is known as a **linear model**. If we examine the graph of extension against force for the spring, shown in Figure 1.8, we see that it is a straight line; this is why it is called a linear

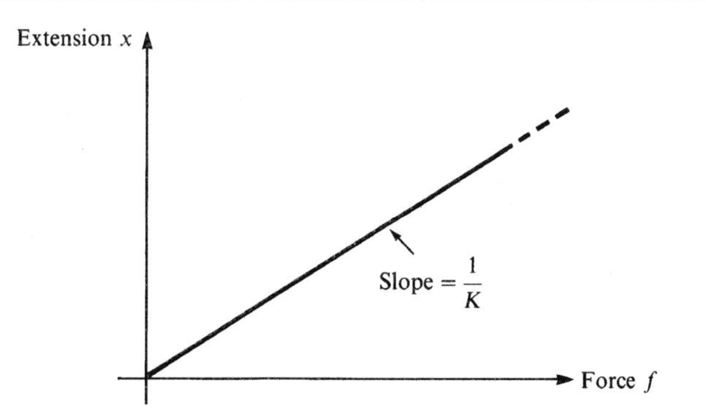

Figure 1.8
Graph of extension against force for a spring

model. It is common to call such a graph the **characteristic** of the spring. A characteristic is usually obtained by carrying out measurements on a system component.

It is difficult to convey how much easier it is to work with a linear model than a nonlinear model. It is only with experience that an engineer learns just how difficult it is to make real progress with a nonlinear model.

Linear models are particularly convenient for engineers to work with. So much so, that an engineer will strive to produce a linear model for a system whenever possible. If a mathematical model is not linear then it is known as a **nonlinear** model. Nonlinear models are much more difficult to deal with and so engineers try to avoid using them if possible.

To see why linear models are so convenient let us consider a further example.

Example

1.10 A spring has a spring stiffness constant K of 100 N m^{-1}. Calculate the extension of the spring when the following forces are applied:

(a) a force of 2 N
(b) a force of 3 N
(c) a force of 5 N

Solution (a) We have $f = Kx$ and $K = 100$ and so $f = 100x$. Therefore,

$$x = \frac{f}{100}$$

When a force of 2 N is applied, then

$$x = \frac{2}{100} = 0.02 \text{ m}$$

(b) We have a force of 3 N and so

$$x = \frac{3}{100} = 0.03 \text{ m}$$

(c) We have a force of 5 N and so

$$x = \frac{5}{100} = 0.05 \text{ m}$$

The previous example illustrates a key feature of a linear model. We see that we can calculate the extension due to a force of 5 N by adding together the extension due to a force of 2 N and the extension due to a force of 3 N. It is possible to generalise this principle. Before doing so, it is convenient to review the mathematical notation for a function.

KEY POINT

> If a mathematical function g, with input z, produces an output y, then we write
>
> $$y = g(z)$$
>
> In this case z is the **independent variable** and y is the **dependent variable**.

Suppose we have a mathematical model for a system defined by a function g such that for an input z we have an output y. This is illustrated in Figure 1.9.

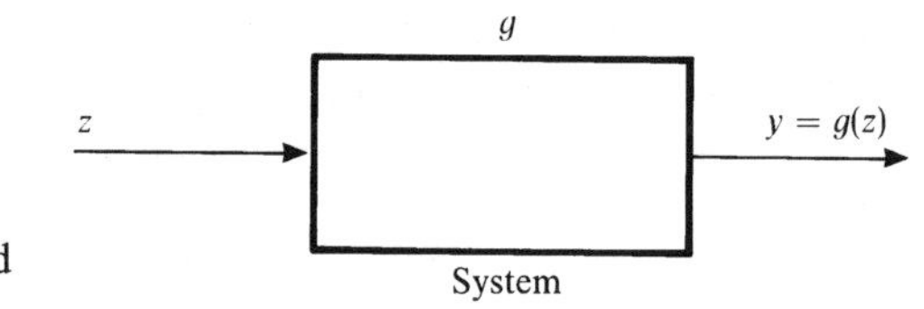

Figure 1.9
The system has a mathematical model defined by the function g

If an input z_1 gives an output y_1, that is, $y_1 = g(z_1)$, and an input z_2 gives an output y_2, that is, $y_2 = g(z_2)$, then the system is linear if the result of applying an input $az_1 + bz_2$, where a and b are constants, is to obtain an output $ay_1 + by_2$. Mathematically, we have

$$ay_1 + by_2 = g(az_1 + bz_2)$$

This is illustrated in Figure 1.10.

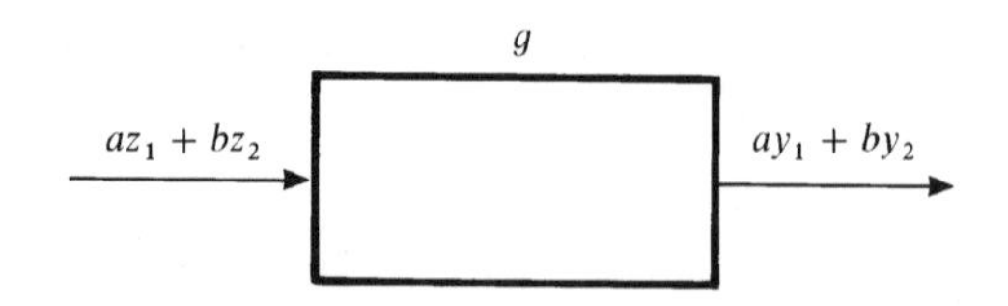

Figure 1.10
A linear model

Written in this general form the definition of a linear model appears to be complicated. In order to aid understanding, let us examine some simple cases to illustrate the definition.

If we put $a = b = 1$ then we have the situation shown in Figure 1.11. For this case, the output due to a sum of two individual inputs is simply the sum of

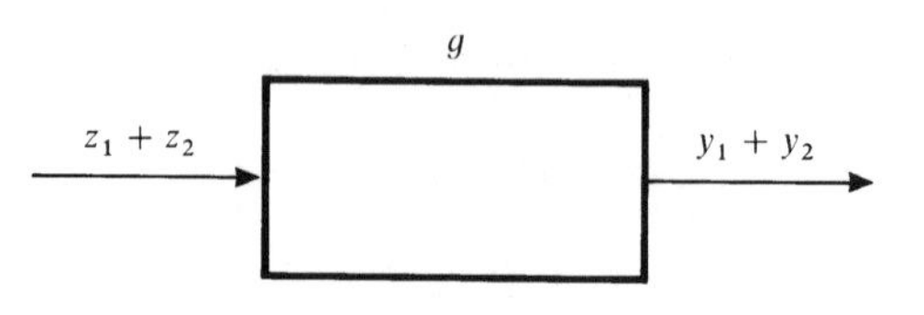

Figure 1.11
The output is obtained by adding together the two individual outputs

the individual outputs. Mathematically we have

$$y_1 + y_2 = g(z_1 + z_2)$$

We saw this in Example 1.10. There we had

$$\begin{aligned} 2\text{ N} &\rightarrow 0.02\text{ m} \\ 3\text{ N} &\rightarrow 0.03\text{ m} \\ (2+3)\text{ N} &\rightarrow (0.02+0.03)\text{ m} \\ 5\text{ N} &\rightarrow 0.05\text{ m} \end{aligned}$$

We have only considered the simple case of adding together numerical inputs. For many engineering systems the input may be a complicated signal that varies with time. For a linear model, it is possible to split this complicated signal into several simpler ones, calculate the outputs for each of these and then obtain the total output by simply adding them together. This is very convenient and makes the analysis of engineering systems much simpler. It is the main reason why linear models are so desirable.

Let us now consider the case when $b = 0$, that is, we apply a scaled version of an input to a linear model. Mathematically we have

$$ay_1 = g(az_1)$$

This is shown in Figure 1.12. We see that the output from the model is scaled by the same amount as the input. Simply put, doubling the input doubles the output, trebling the input trebles the output.

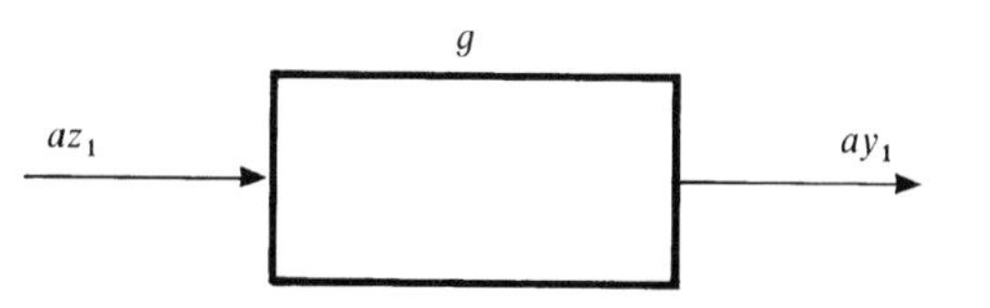

Figure 1.12
The output is scaled by the same amount as the input

The most general case is when we apply an input $az_1 + bz_2$. This is known as a **linear combination** of z_1 and z_2. We see that this is a mixture of addition and scaling of the inputs z_1 and z_2.

KEY POINT

For a linear model: if

$$y_1 = g(z_1)$$
$$y_2 = g(z_2)$$

then

$$ay_1 + by_2 = g(az_1 + bz_2) \qquad a, b \text{ constants}$$

In Example 1.10 it was easy to form a linear mathematical model for the spring because its characteristic was a straight line, as we saw in Figure 1.8. Sometimes the characteristic of a system may not be a straight line. It is still possible to form a linear model for such a system provided care is exercised.

The process of obtaining this linear model is known as **linearisation**. We shall illustrate this by means of an example.

Example

1.11 When a current i flows in a resistor of resistance R, then the power dissipated in the resistor is p. This is given by

$$p = i^2 R$$

Obtain a linear model relating p to i, centred on a current of 2 A, when $R = 5\ \Omega$.

Solution We are given $R = 5$. If we plot $p = 5i^2$ we see that the graph is curved. This is shown in Figure 1.13.

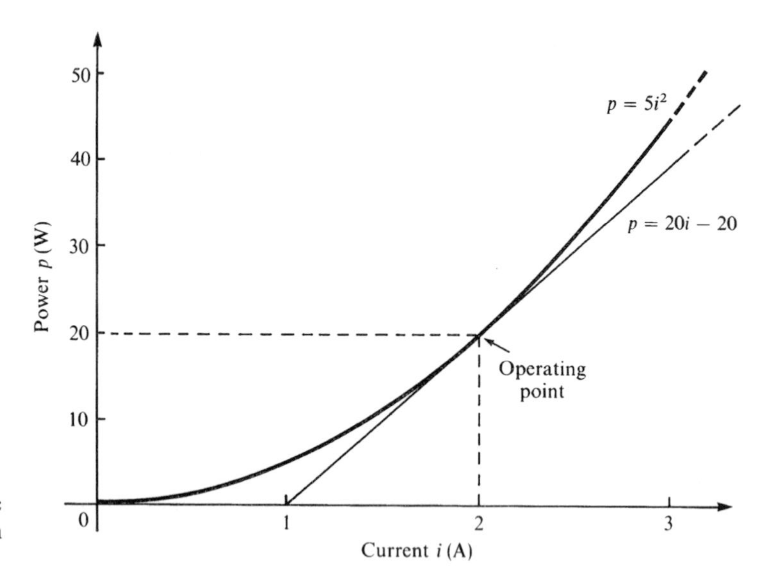

Figure 1.13
Power–current characteristic for the resistor together with a linear approximation

However, we require a linear model for the power dissipation in the resistor. The solution is to draw a straight line to the curve which touches the curve at the point where $i = 2$ and $p = 20$. Such a line is known as a **tangent** to the curve. The point at which the tangent touches the curve is known as the **operating point**. We see from Figure 1.13 that for values of i close to the operating point the function values on the curve are approximately equal to those on the straight line. We can now see that a suitable linear model for the resistor is the tangent to the curve at the operating point. Therefore if we obtain the equation for this tangent we then have a linear model for the resistor. In order to calculate the equation of this tangent we recall the equation for a straight line. In general this is

Try to imagine the tangent as being rather like a ruler that is brought towards the curve until it touches the curve.

$$y = mx + c$$

where x is the independent variable, y is the dependent variable, m is the slope of the line and c is the intercept with the y axis. For this example the independent variable is i and the dependent variable is p. So we have

$$p = mi + c$$

and we need to calculate m and c.

Recall from basic calculus that the slope of a tangent to a curve at a point is obtained by differentiating the function that describes the curve and substituting the value of the independent variable at the point. In this case we need to differentiate the function $p = 5i^2$. So we have

$$\frac{dp}{di} = \frac{d(5i^2)}{di} = 10i$$

At the operating point $i = 2$ and so

$$m = \left.\frac{dp}{di}\right|_{i=2} = 10 \times 2 = 20$$

Therefore, for the tangent we have

$$p = 20i + c$$

The tangent passes through the operating point. At the operating point, $i = 2$ and $p = 20$ so we can write

$$p = 20i + c$$
$$20 = 20 \times 2 + c$$
$$c = 20 - 40 = -20$$

So finally the linear model for the resistor is

$$p = 20i - 20$$

Examining Figure 1.13 we see that the linear model is quite a good approximation to the actual characteristic at points close to the operating point but becomes less accurate at points further away from the operating point. Let us examine this in more detail.

Example

1.12 In Example 1.11 we obtained a linear model for the power dissipation of a resistor centred at an operating point of $i = 2$ A. Calculate the accuracy of the linear model at the following current values:

(a) 2 A

(b) 2.01 A

(c) 2.1 A

(d) 3 A

Solution (a) Recall that the true characteristic is given by $p = 5i^2$ and the linear model is $p = 20i - 20$. At $i = 2$ the true value is $p = 5 \times 2^2 = 20$ W.

The linear approximation is $p = 20 \times 2 - 20 = 20$ W. We see that the linear model and the true characteristic agree at this point.

(b) At $i = 2.01$ the true value is $p = 5 \times 2.01^2 = 20.2005$ W. The linear approximation is $p = 20 \times 2.01 - 20 = 20.2$ W. We see that the linear model is very close to the true value. The percentage error is

$$\frac{20.2005 - 20.2}{20.2005} \times 100 = \frac{0.0005 \times 100}{20.2005} = 0.0025\%$$

(c) At $i = 2.1$ the true value is $p = 5 \times 2.1^2 = 22.05$ W. The linear approximation is $p = 20 \times 2.1 - 20 = 22$ W. Again the agreement is very close. The percentage error is

$$\frac{22.05 - 22.0}{22.05} \times 100 = \frac{0.05 \times 100}{22.05} = 0.23\%$$

(d) At $i = 3$ the true value is $p = 5 \times 3^2 = 45$ W. The linear approximation is $p = 20 \times 3 - 20 = 40$ W. We see that the linear approximation is now significantly different from the true value. The percentage error is

$$\frac{45 - 40}{45} \times 100 = \frac{5 \times 100}{45} = 11\%$$

Engineers can sometimes tolerate errors of a few per cent as many components are only manufactured to this accuracy.

The main thing to remember when obtaining a linear model of a system is that there is only a small region around the operating point for which the linear model is valid. The size of this region depends both on the shape of the characteristic of the system and the amount of error in the linear model that can be tolerated.

KEY POINT

> A linear model for a system with a curved characteristic can be obtained by calculating the equation of the tangent to the curve at the operating point.

Engineering systems that do not change with time are known as **static systems**.

Most of the engineering systems we shall be examining in this book are known as **dynamic systems**, so called because their behaviour changes with time and the word dynamic alludes to movement.

Dynamic systems can be mathematically modelled by means of one or more differential equations. A **differential equation** is a mathematical equation which contains derivatives of the engineering variables as well as the variables themselves. When only one differential equation is used to model a system then it is known as the **system differential equation**. The system differential equation describes the relationship between the input signal and the output signal for a system. The **order** of a system is defined as the order of the system differential equation that describes it. The order of a differential equation is equal to the order of its highest derivative.

Self-assessment questions 1.4

1. Explain what is meant by a mathematical model of a system.
2. Describe the advantages of a linear mathematical model.
3. Explain what is meant by the term linearisation.
4. What is a dynamic system?
5. What is the system differential equation?

Exercises 1.4

1. If a voltage is placed across a resistor then the amount of current that flows in the resistor is proportional to the applied voltage. Carry out the following:
   (a) Derive a mathematical model for the resistor.
   (b) Draw a block diagram for the resistor with voltage as the input and current as the output.
   (c) Determine the mathematical rule for the block which relates the input signal to the output signal. Draw a block diagram for the resistor which includes this rule.
2. When a voltage v is applied to a resistor of resistance R, then the power dissipated in the resistor is p. This is given by

$$p = \frac{v^2}{R}$$

   Carry out the following:
   (a) Obtain a linear model for the resistor relating p to v, centred on a voltage of 10 V, when $R = 20\ \Omega$.
   (b) Calculate the accuracy of the linear model at the following voltage values:
      (i) 10 V
      (ii) 10.01 V
      (iii) 10.1 V
      (iv) 20 V

1.5 Engineering units

Interestingly, North America has not universally adopted the SI system. They considered the changeover from the British system to be too costly. However, many American textbooks use it because of its importance.

Throughout this book there are many references to engineering units. A few remarks are therefore in order to clarify their usage. For several years it has been common practice in Europe and in many other countries to use the Système Internationale d'Unités, known as the SI system of units. We shall therefore use this system. Table 1.1 shows the **base units** of the SI system.

Base units are units obtained from a physical standard. For example, a kilogram is the mass of a platinum–iridium cylinder which is kept in France at Sèvres. Other units, known as **derived units**, can be obtained by combining together the base units. Therefore it is not necessary to have a physical standard for these units. For example, the unit of force, known as the newton, is kilogram.metre.second^{-2}. It is common practice to write kilogram.metre.second^{-2} rather than

$$\frac{\text{kilogram.metre}}{\text{second}^2}$$

Table 1.1
Base units of the SI system

Standard	Unit	Abbreviation
length	metre	m
mass	kilogram	kg
time	second	s
electric current	ampere	A
temperature	kelvin	K
luminous intensity	candela	cd
amount of substance	mole	mol

Table 1.2
Some derived units in the SI system

Quantity	Unit	Abbreviation
velocity	metre.second^{-1}	m s^{-1}
acceleration	metre.second^{-2}	m s^{-2}
force	kilogram.metre.second^{-2} (newton)	kg m s^{-2} (N)
energy	kilogram.metre^{2}.second^{-2} (joule)	kg m^{2} s^{-2} (J)
power	kilogram.metre^{2}.second^{-3} (watt)	kg m^{2} s^{-3} (W)
pressure	kilogram.metre^{-1}.second^{-2} (pascal)	kg m^{-1} s^{-2} (Pa)
frequency	second^{-1} (hertz)	s^{-1} (Hz)
charge	ampere.second (coulomb)	A s (C)
potential difference	watt.ampere^{-1} (volt)	W A^{-1} (V)
resistance	volt.ampere^{-1} (ohm)	V A^{-1} (Ω)
inductance	volt.second.ampere^{-1} (henry)	V s A^{-1} (H)
capacitance	ampere.second.volt^{-1} (farad)	A s V^{-1} (F)

It is important that physical standards are as accurate as possible because they form the basis of all measurement.

or kilogram.metre/second^2, although both alternatives are acceptable. If you are unsure about indices then refer to Appendix 1 for details. Table 1.2 shows some of the more common derived units. Often derived units are given names, for example the unit of force, which is the newton.

Finally, there are **supplementary units**. At present two are defined: the unit of plane angle, known as the **radian** (rad), and the unit of solid angle, known as the **steradian** (sr).

Units are written either in full or in abbreviated form. For example, we can write kilogram or kg to represent a unit of mass. It is common practice in engineering to use the abbreviated forms in order to save space as the units of some engineering quantities are quite long. We shall mainly use the abbreviated form for writing units unless inappropriate.

When two units are multiplied together this can be written as kg m or kg.m where the dot signifies multiplication. We shall use the dot when writing units in full but omit it when using the abbreviated form of units. We also adopt the convention of using a negative power to represent a unit that is to be divided, rather than using a slash symbol.

It is possible to obtain multiples and submultiples of units by using standard prefixes. The more common prefixes are shown in Table 1.3.

To avoid confusion, whenever a unit is given with a prefix we shall convert it into the main unit before use. For example, 3.5 mm is converted into 0.0035 m and then used. This reduces the chances of error when carrying out calculations. Note one confusion that cannot be avoided. The main unit of

Table 1.3
The main prefixes used in the SI system

Symbol	Prefix	Multiplication factor
G	giga	$1\ 000\ 000\ 000 = 10^9$
M	mega	$1\ 000\ 000 = 10^6$
k	kilo	$1\ 000 = 10^3$
c	centi	$0.01 = 10^{-2}$
m	milli	$0.001 = 10^{-3}$
μ	micro	$0.000\ 001 = 10^{-6}$
n	nano	$0.000\ 000\ 001 = 10^{-9}$
p	pico	$0.000\ 000\ 000\ 001 = 10^{-12}$

mass is the kilogram which also has a prefix in front of it. This does not cause a problem provided it is remembered that the kilogram is the main SI unit of mass.

Example

1.13 Convert the following subunits to their main units:

(a) 3 mm

(b) 8.1 cm

(c) 4.8 mm^2

(d) 3.2 mJ

(e) 6.3 g

(f) 10 kV

(g) 10.7 kΩ

(h) 2.5 km s^{-1}

Solution (a) Milli corresponds to 10^{-3} and so 3 mm $= 3 \times 10^{-3}$ m $= 0.003$ m.

(b) Centi corresponds to 10^{-2} and so 8.1 cm $= 8.1 \times 10^{-2}$ m $= 0.081$ m.

(c) Recall the laws of indices detailed in Appendix 1. In particular, $(a^m)^n = a^{mn}$. Here we have $(\text{milli})^2 = (\text{milli})(\text{milli})$ and so we have $(10^{-3})^2 = 10^{-3 \times 2} = 10^{-6}$ as the prefix. Therefore 4.8 mm^2 = 4.8×10^{-6} m^2.

(d) 3.2 mJ $= 3.2 \times 10^{-3}$ J $= 0.0032$ J.

(e) Recall that the main unit of mass is the kilogram. A kilogram is 1000 g and so

$$6.3 \text{ g} = \frac{6.3}{1000} \text{ kg} = 0.0063 \text{ kg}$$

(f) Kilo corresponds to 10^3 and so 10 kV $= 10 \times 10^3$ V $= 10\ 000$ V.

(g) 10.7 kΩ $= 10.7 \times 10^3$ Ω $= 10\ 700$ Ω.

(h) Kilo corresponds to 10^3 and so 2.5 km s^{-1} $= 2.5 \times 10^3$ m s^{-1}.

KEY POINT

This is an important point concerning physical equations. Many engineers forget this and thereby make more work for themselves.

> It is important to realise that an equation involving physical quantities must balance physically as well as numerically. In simple terms, three apples are not equal to three oranges even though $3 = 3$.

This can be useful when trying to work out the units a quantity has. For example, recall Example 1.9 which discusses the extension of a spring when a force is applied. The relationship is

$$f = Kx$$

Suppose we wish to work out the units associated with K. First we divide by x. This gives

$$K = \frac{f}{x}$$

Now we know that the unit of force is the newton and the unit of length is the metre. We therefore conclude that K has units newton/metre = newton.metre^{-1}. We see that this balances the units on both sides of the equation.

Example

1.14 Newton's second law states that the force f applied to an object is equal to its mass M multiplied by its acceleration a. In symbols we have

$$f = Ma$$

Mass has units of kilogram and acceleration has units of metre.second^{-2}. Determine the units of force from this information.

Solution The unit of force is immediately seen to be kilogram.metre.second^{-2}. Note that this is usually known as a newton as we have already seen.

The last example raises an interesting point. Often there are several different ways of expressing the units of a physical quantity because of the presence of abbreviated quantities. Care is therefore needed to check whether the units of an equation balance when several alternative representations are being used.

Self-assessment questions 1.5

1. What is the main system of units used in the world today?
2. Explain how a derived unit differs from a base unit.
3. Why is it desirable to convert subunits to their main units before carrying out a calculation?

Exercises 1.5

1. Convert the following subunits to their main units:
   (a) 3.5 cm
   (b) 5.6 mm^2
   (c) 6.23 kΩ
   (d) 25.6 g
   (e) 100.3 kV
   (f) 2.83 kJ
   (g) 6.32 cm^2
   (h) 8.76 mm^3
   (i) 8.25 cm s^{-1}
   (j) 9.853 mJ s^{-1}
   (k) 58.2 km s^{-2}

Test and assignment exercises 1

1. Suggest a suitable set of system components for a washing machine.
2. Suggest a suitable set of system components for an electric cooker.
3. Suggest a suitable set of system components for a space rocket.
4. Suggest possible system inputs and system outputs for a school.
5. Consider the washing machine of Test and assignment exercises 1.1. Suggest suitable signals and interconnections for the components of the washing machine.
6. Consider the electric cooker of Test and assignment exercises 1.2. Suggest suitable signals and interconnections for the components of the electric cooker.
7. Consider the space rocket of Test and assignment exercises 1.3. Suggest suitable signals and interconnections for the components of the space rocket.
8. Draw a block diagram for the washing machine of Test and assignment exercises 1.1 and 1.5.
9. Draw a block diagram for the electric cooker of Test and assignment exercises 1.2 and 1.6.
10. Draw a block diagram for the space rocket of Test and assignment exercises 1.3 and 1.7.
11. The flow of liquid through an orifice, q, is related to the pressure difference across the orifice, p, by the following:

$$q = 10p^{0.5}$$

    Obtain a linear model for the orifice centred on an operating point $p = 1$ N m^{-2}, $q = 10$ m^3 s^{-1}.
12. Convert the following subunits to their main units:
    (a) 10.6 mV
    (b) 12 kΩ
    (c) 35.2 g
    (d) 19.7 cm
    (e) 15 μF
    (f) 1.56 mH
    (g) 2.63 mm s^{-1}
    (h) 7.25 km s^{-2}
    (i) 0.35 kJ s^{-1}
    (j) 36.5 μW

2 Mechanical systems: translational

Objectives

This chapter:

- describes the model for the behaviour of a spring
- describes the model for the behaviour of a damper
- describes the model for the behaviour of a mass
- explains Newton's laws of motion
- explains the concept of an inertia force
- derives mathematical models for some translational mechanical systems

2.1 Introduction

Mechanical systems can be extremely complicated, as anyone who has tried to strip down a car engine will testify. Fortunately a large number of mechanical systems are much simpler and these are the ones we shall be dealing with. It is possible to break down many mechanical systems into just three types of components: the spring, the mass and the damper. Mechanical systems undergo a mixture of two types of motion. The first type is motion in a straight line, known as **translational motion**. The second type is motion about a fixed axis of rotation, known as **rotational motion**. For simplicity we only consider systems that involve one of these types of motion at any one time. This will allow many common systems to be analysed; systems where both types of motion occur at the same time require detailed study of mechanical engineering and so are beyond the scope of this book. This chapter will deal with translational systems and the following chapter will deal with rotational systems.

2.2 The spring

The important feature of a **spring** is that it stretches when a force is applied to it. The law governing the amount by which a spring stretches is known as Hooke's law. We briefly examined this law in Section 1.4.

KEY POINT

> The force f required to stretch a spring by an amount x is given by
>
> $$f = Kx \qquad \text{Eqn. [2.1]}$$
>
> where K is a constant known as the **spring stiffness**.

The SI unit of K is newton.metre^{-1}. Note that a more technical term for x is **displacement**, so called because it is the amount by which the spring is displaced from its original position, as shown in Figure 2.1.

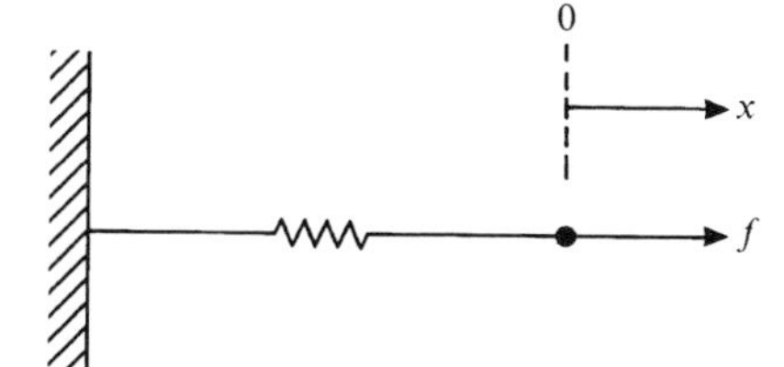

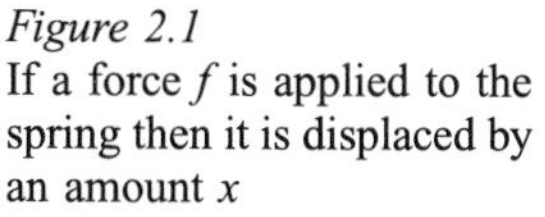

Figure 2.1
If a force f is applied to the spring then it is displaced by an amount x

There are very few conventions that command total agreement in engineering. It is important to be aware of this and look out for variations.

It is now convenient to discuss a few conventions when drawing figures for mechanical systems. The shaded part of Figure 2.1 represents a fixed anchor point for the spring and so the end of the spring attached to this point does not move. The other end of the spring has an arrow attached to it indicating that a force is being applied to the spring. The symbol for the spring is a zigzag line. The unstretched position of the spring is shown as a broken line with a zero above it although it is common practice to leave out the zero. This acts as the zero reference point for measuring the displacement of the spring. Also, the position of the broken line is not important as its role is purely symbolic and acts as a reminder that the spring has a position from which displacement measurements commence. The displacement of the spring is shown by means of an arrow. It is common practice to measure displacement from a position where the system is relaxed, termed the **equilibrium position**. This avoids the problem of zero offsets when constructing system equations. Let us now examine the behaviour of a spring in more detail.

Example

2.1 Plot a single graph of force against displacement for the following springs:

(a) stiffness $K = 20\ \text{N m}^{-1}$
(b) stiffness $K = 40\ \text{N m}^{-1}$
(c) stiffness $K = 50\ \text{N m}^{-1}$

Use a displacement range of 0 to 0.1 m.

Solution (a) Using Eqn. [2.1] we see that, when $x = 0$, $f = 20 \times 0 = 0$, and when $x = 0.1$, $f = 20 \times 0.1 = 2$. Therefore the range of force values is 0 to 2 N. We know that Eqn. [2.1] describes a straight line and so we can plot the graph which is shown in Figure 2.2. Note that the slope of this graph corresponds to the stiffness of the spring.

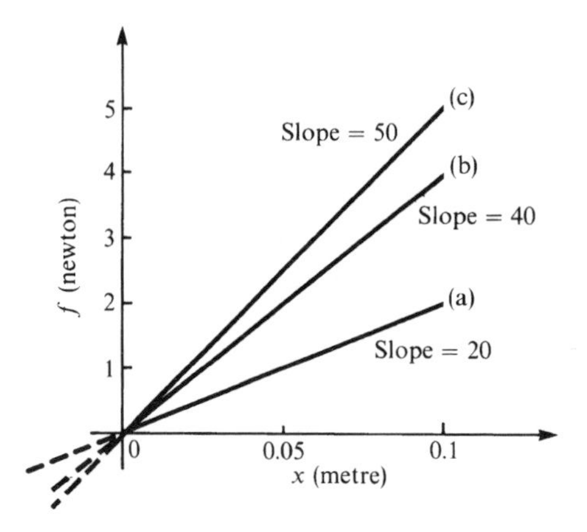

Figure 2.2
A force–displacement graph for three springs of different stiffnesses

(b) For this spring the range of force values is 0 to $40 \times 0.1 = 4$ N. This graph is shown in Figure 2.2.
(c) For this spring the range of force values is 0 to $50 \times 0.1 = 5$ N. The graph is shown in Figure 2.2.

From the previous example we see that for a stiffer spring a greater force is required to extend the spring by a given amount. When designing a spring this is one of the main characteristics that needs to be set. It is possible to buy springs with widely different values of K. For example, a car suspension spring has a much higher value of K than a spring used in kitchen weighing scales.

Some springs are designed to be compressed rather than extended and, indeed, some springs are used in both modes. The feature that distinguishes a spring that can be compressed is that its coils are open when no force is applied to it. For example, a car suspension spring has a shape rather like a giant corkscrew when unloaded and can be compressed as well as extended. It is still possible to use Hooke's law to model the behaviour of such a spring but care must be taken to use the correct directions for f and x. Recalling Example 2.1 we note that only positive values of force were applied to the springs and so the displacements were positive in each case. However, there was no reason why the direction of the force could not have been changed by 180°, corresponding to a negative force, and this would have led to a negative displacement for each of the springs. Examining Figure 2.2 we see that negative displacements for each of the springs have been marked on the graphs by broken lines. Note that negative displacements are the result of a negative force being applied to the springs in each case. Another possibility is to redefine f and x as shown in Figure 2.3.

For this case a compressed spring corresponds to a positive f and x and an expanded spring corresponds to a negative f and x. It does not matter which convention is used for defining f and x provided consistency is maintained after they have been defined.

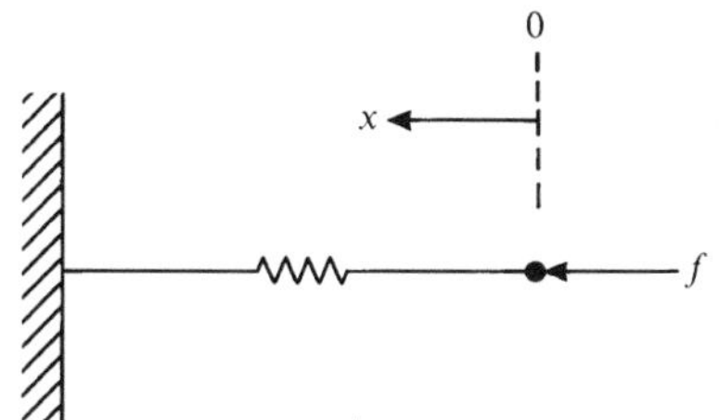

Figure 2.3
For this spring, a positive displacement of the spring corresponds to the spring being compressed

This is an important point to stress. Many errors occur because quantities have not been properly defined. A diagram is very useful in preventing this happening.

A number of assumptions lie behind this simple mathematical model for a spring that we have been using. We see that according to the model it is possible to obtain any displacement we need simply by applying a large enough force to the spring. In practice, experience shows that if too large a force is applied to the spring then it will break or permanently deform. We avoid this complication by only working with a range of forces on the spring for which Eqn. [2.1] is accurate. Within this range the mathematical model for the spring is linear, a concept we discussed in Section 1.4.

Most engineering components demonstrate more complicated behaviour than their mathematical models predict. However, these complexities can often be ignored, thus allowing simple models to be used.

Another assumption behind this simple model is that the spring has negligible mass. If the spring did have a mass then the relationship between f and x would be much more complicated. The spring is said to be an **ideal** spring. The reason we can make this assumption is because even though all real springs do have some mass in most cases its effect is negligible in comparison with that of its 'springiness' when determining the relationship between f and x. It is common practice in engineering to make use of these idealised components as it greatly simplifies the mathematical models needed to analyse engineering systems.

Self-assessment questions 2.2

1. Explain what is meant by the term 'spring stiffness'.
2. Why is it convenient to assume that a spring has negligible mass?

Exercises 2.2

1. Draw a force–displacement graph for a spring with a stiffness of $35\ \mathrm{N\ m^{-1}}$. Use a displacement range of -0.1 m to 0.2 m.

2.3 The damper

Before presenting the equation that models a damper it is worth examining its operation. Figure 2.4 shows a cross-section through a simple damper. It consists of a piston contained within a cylinder. The piston is free to travel in the cylinder but in order to do so oil needs to flow through the hole in the piston. This means that only a small force is required to expand the piston if it is moved slowly but a large force is required if the piston is to be expanded quickly.

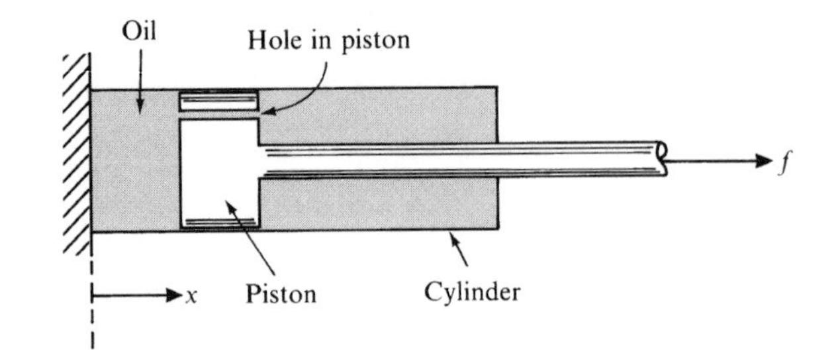

Figure 2.4
Cross-section through a damper

KEY POINT

> The force f required to expand a damper with a velocity v is given by
>
> $$f = Bv$$
>
> where B is a constant known as the **damping coefficient**.

The SI unit of B is newton.second.metre^{-1}. The same argument applies when compressing a damper but the directions of the quantities f and v are reversed. Recognising that velocity is the rate of change of displacement, that is, $v = \mathrm{d}x/\mathrm{d}t$, the relationship for the damper can be written as

$$f = B\frac{\mathrm{d}x}{\mathrm{d}t}$$

where x is the displacement of the piston. The symbol for a damper is shown in Figure 2.5.

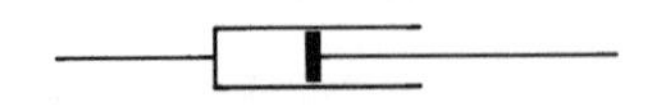

Figure 2.5
The symbol for a damper

In practice, most dampers do not have this simple linear relationship between force and velocity but it is usually a reasonable approximation to make for a simplified model. One assumption behind the linear model is that the piston is free to travel. Clearly this assumption breaks down when the piston hits the end of the cylinder. As with the spring, it is also convenient to assume that the damper has negligible mass, that is, it is an idealised component, in order to simplify the relationship between f and v. In practice, this is not a problem because the 'damping effect' usually dominates.

One of the most common examples of a practical damper is the shock absorber used in a car. In practice, there are all sorts of different designs for dampers. A 'homemade' version is a bicycle pump with the air hole partially covered by a finger. This provides an excellent way of experimenting with the characteristics of a damper, although it must be remembered that air is compressible which complicates the characteristics of the bicycle pump.

The common feature of all these phenomena is that force is proportional to velocity.

There are several other mechanical phenomena that can be modelled by means of a damper. One particularly important phenomenon is the friction that occurs between moving surfaces. This is known as **viscous friction**. This is a different frictional force to that between two surfaces that are in contact but not moving, known as **static friction**. A simple model for viscous friction is to assume that it is proportional to the relative velocity between the two surfaces. It can then be modelled by means of a damper. This effect is important when mechanical components slide relative to each other.

Self-assessment questions 2.3

1. Explain the difference between viscous friction and static friction.
2. Describe how a damper works.

Exercises 2.3

1. Plot a graph of force against velocity for a damper with a damping coefficient of $6\ \mathrm{N\ s\ m^{-1}}$. Use a velocity range of -0.5 to $1.0\ \mathrm{m\ s^{-1}}$. Note that a negative velocity corresponds to a direction change of 180°.

2.4 The mass

KEY POINT

If a force f is applied to a mass M then the motion of the mass is given by

$$f = Ma$$

where a is the acceleration of the mass.

Now acceleration is the rate of change of velocity, that is, $a = \mathrm{d}v/\mathrm{d}t$. So we can also write for the mass

$$f = M\frac{\mathrm{d}v}{\mathrm{d}t}$$

Recall from Section 2.3 that $v = \mathrm{d}x/\mathrm{d}t$. So another form of the equation for the motion of the mass is

$$f = M\frac{\mathrm{d}v}{\mathrm{d}t} = M\frac{\mathrm{d}}{\mathrm{d}t}\left(\frac{\mathrm{d}x}{\mathrm{d}t}\right) = M\frac{\mathrm{d}^2x}{\mathrm{d}t^2}$$

These three different forms for the motion of the mass are equivalent. Which one is used is a matter of convenience. Examining this law we note that, the greater the force applied to a mass, the faster it accelerates. Also, if a fixed force is applied to a mass then a small mass will accelerate faster than a large mass. This is illustrated in the following example.

KEY POINT

The following are equivalent:

$$\frac{\mathrm{d}^2x}{\mathrm{d}t^2} = \frac{\mathrm{d}v}{\mathrm{d}t} = a$$

where x is the displacement, v is the velocity and a is the acceleration.

Example

2.2 Calculate the acceleration of the following masses when a force of 5 N is applied to each of them.

(a) $M=0.5$ kg
(b) $M=2$ kg
(c) $M=10$ kg

Solution (a) We have $f=5$ and $M=0.5$. So,

$$f = Ma \qquad a = \frac{f}{M} = \frac{5}{0.5} = 10 \text{ m s}^{-2}$$

(b) We have $f=5$ and $M=2$. So,

$$a = \frac{f}{M} = \frac{5}{2} = 2.5 \text{ m s}^{-2}$$

(c) We have $f=5$ and $M=10$. So,

$$a = \frac{f}{M} = \frac{5}{10} = 0.5 \text{ m s}^{-2}$$

A mass is assumed to be **rigid** for simplicity. This means that it does not change its shape when a force is applied to it. Also, the mass is assumed to be concentrated at a single point. This simplifies analysis and means that the mass is an idealised component.

Self-assessment questions 2.4

1. Which will accelerate faster when a fixed force is applied to it: a small mass or a large mass?
2. It is required to accelerate a small mass and a large mass at a certain rate. Which mass will need the greater force to be applied?
3. State the relationship between displacement, velocity and acceleration.

Exercises 2.4

1. Calculate the acceleration of the following masses when a force of 4 N is applied to each of them:

   (a) $M = 23$ kg
   (b) $M = 6.3$ kg
   (c) $M = 2.73$ kg.

2.5 Newton's laws of motion

In 1686 Isaac Newton published the *Principia* which contained three laws of motion. These three laws form the basis for analysing mechanical systems. We shall examine each law in turn.

KEY POINT

Newton's first law of motion states:

Every body continues in a state of rest or constant velocity unless acted upon by a force.

This is the reason why a spaceship can travel in space at a constant velocity without the use of its rocket. There are no appreciable forces present to slow it down.

This law expresses the fact that an object will not start moving unless acted upon by a force. Also, if an object is already moving at a constant velocity then it will remain at the same velocity unless acted on by a force. Note that the terms body and object mean the same thing and are often used interchangeably by engineers.

This reluctance of an object to change its velocity is known as its **inertia**. A measure of the inertia of an object to translation is its **mass**. An object with a large mass is more reluctant to change velocity than an object with a small mass.

KEY POINT

Newton's second law of motion states:

If a force f is applied to a body of mass M then the motion of the body is given by

$$f = Ma$$

where a is the acceleration of the mass.

Einstein showed that the mass of an object changes as it approaches the velocity of light. The systems we analyse will be travelling at velocities where such an effect is negligible.

This statement is true if the mass of a body remains constant. This is the case for all of the mechanical systems we shall consider.

The law quantifies the force needed to accelerate an object. The larger the applied force the faster an object accelerates. Note that an object with a large mass requires a greater force to accelerate it at a certain rate than a small mass requires. Recall that we analysed this equation informally in Section 2.4.

Often there are several forces acting on an object. For this case it is necessary to work out the net force by adding the forces together using vectors. We say that the forces have been **combined** into a single force. We shall avoid the complications of vector addition by only considering systems in which the forces are acting in parallel. For this case the magnitude of the forces can be simply added or subtracted depending on which direction the forces are in.

KEY POINT

Newton's third law of motion states:

To every action there is always an equal and opposite reaction.

This law is useful because it states that the forces acting on a body must balance. So, for example, if a cup rests on a table then the force due to gravity is balanced by a reaction force from the desk. In this case the cup is stationary even though there are two forces acting on the cup. This is because the two forces cancel each other out.

It is still possible to think of forces acting on a body balancing each other out when an object is moving. Consider the example of a force f acting on an object of mass M, as shown in Figure 2.6(a). For convenience, assume the mass is located in deep space and so any gravitational forces are negligible. At

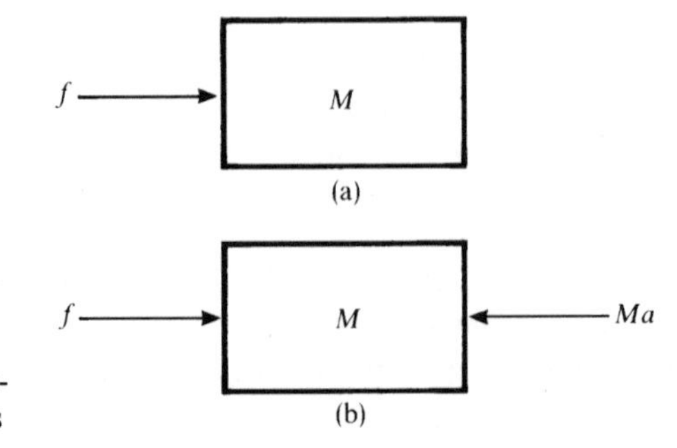

Figure 2.6
(a) A force f acting on an object of mass M; (b) a free-body diagram for this mass

first sight it appears that the force is unbalanced. However, the action of the force causes a reaction by the mass. This reaction is known as an **inertia force**. It represents the reluctance of a mass to change its velocity.

The magnitude of the inertia force is (mass) $\times$ (acceleration of mass), that is, Ma, and it acts in the opposite direction to f. An inertia force only occurs when there is an attempt to change the velocity of a mass. There are no inertia forces associated with springs and dampers because they are assumed to have negligible mass.

KEY POINT

> The only component we analyse that has an inertia force associated with it is the mass.

When there are several forces acting on an object then it is usual to draw a **free-body diagram**. This shows all the forces acting on an object, including any inertia forces. A free-body diagram provides a convenient way of isolating a component from its surroundings, thus making it easier to analyse. In effect, this means that any interactions between the component and other adjacent components have been taken care of in the free-body diagram. Therefore analysis of the free-body diagram, as the name implies, can take place as though the component was a free body, not dependent on other components.

We shall be constructing several free-body diagrams for components in the next section. A free-body diagram for the mass M is shown in Figure 2.6(b). Recall Newton's third law, which states that to every action there is an equal and opposite reaction. We can use Newton's third law to analyse the mass M. In this case the reaction to the applied force f is the inertia force Ma. Therefore we can equate these two forces, that is,

$$f = Ma$$

Alternatively we can write

$$f - Ma = 0$$

We see that the forces on the mass balance. This is true generally when there are several forces acting on a component. Provided we include any inertia forces, we can equate the forces acting on a component to zero.

KEY POINT

> The forces acting on a component can be equated to zero, that is, they balance, provided any inertia forces are included in the free-body diagram.

Self-assessment questions 2.5

1. What is meant by the inertia of an object?
2. Explain what a free-body diagram is.

3. State Newton's first law of motion.
4. State Newton's second law of motion.
5. State Newton's third law of motion.

2.6 Translational mechanical systems

Having examined the equations for the three basic elements of translational mechanical systems, as well as Newton's laws of motion, we are now in a position to analyse some mechanical systems. At this stage we are concerned with obtaining the system differential equation. This is the mathematical model for the system. We have already introduced the concept of the system differential equation in Section 1.4. In later chapters we shall see how to use the system differential equation to analyse the behaviour of a system. When attempting to obtain the system differential equation it is necessary to adopt a systematic procedure. We start with a simple example.

Example

2.3 This type of mechanical system occurs frequently in engineering. For example, a car suspension system is essentially a spring–mass–damper system.

Obtain a system differential equation for the system shown in Figure 2.7. The mass rests on a smooth frictionless surface. Assume that the input to the system is a force f applied to the mass and the output from the system is the displacement of the mass x.

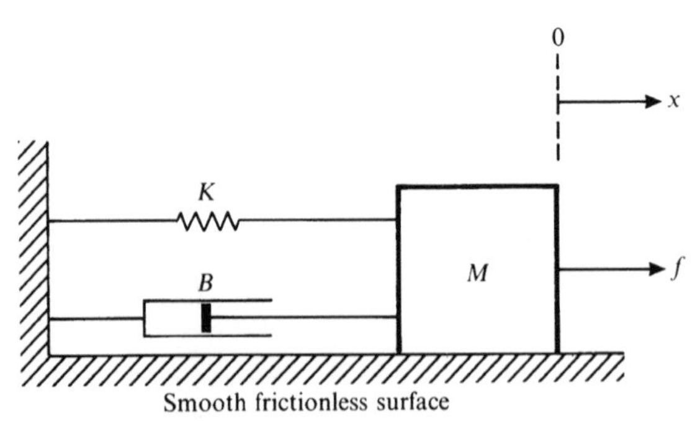

Figure 2.7
A spring–mass–damper system

Solution Let us assume that the mass has been displaced to the right and so x is positive and the spring is being stretched. When $x = 0$ the spring is neither stretched nor compressed. This corresponds to the equilibrium position. We now wish to know the forces acting on the mass. We therefore construct a free-body diagram for the mass. The free-body diagram for the mass is shown in Figure 2.8.

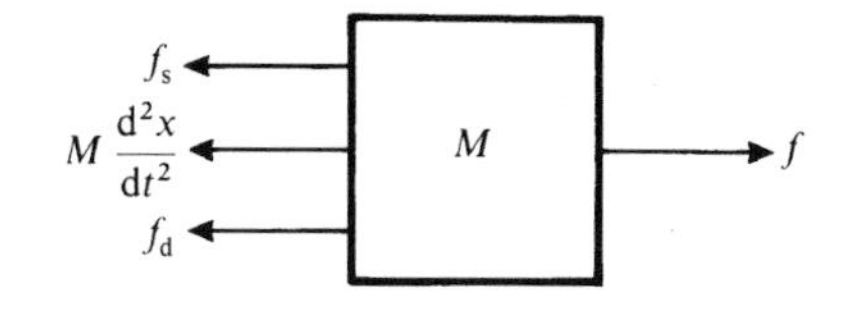

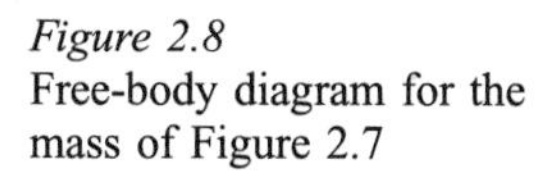
Figure 2.8
Free-body diagram for the mass of Figure 2.7

Recall that d^2x/dt^2 represents the acceleration of the mass.

The force due to the spring, f_s, opposes the applied force f because the spring is being stretched. Similarly, because the damper is being pulled apart, the force due to the damper, f_d, also opposes f. The free-body diagram also includes the inertia force Md^2x/dt^2 which opposes the applied force f and represents the reluctance of the mass to be accelerated.

As we have included the inertia force we can equate the forces acting on the mass to zero. So we have

$$f - f_s - f_d - M\frac{d^2x}{dt^2} = 0 \qquad \text{Eqn. [2.2]}$$

Note that forces in the positive x direction are positive because this is assumed to be the direction in which the acceleration is occurring. It is important to be consistent when setting up the system equations or otherwise mistakes will occur. In this case the positive x direction was defined in the diagram and so this is also the direction of positive velocity and acceleration.

We can write the equation governing the behaviour of the spring, which is

$$f_s = Kx \qquad \text{Eqn. [2.3]}$$

Note that this gives a positive value for f_s when x is positive. This is what we require as the fact that the spring is opposing the applied force was taken into account when setting up the free-body diagram. Finally for the damper we can write

$$f_d = B\frac{dx}{dt} \qquad \text{Eqn. [2.4]}$$

Combining Eqns [2.2], [2.3] and [2.4] to eliminate f_s and f_d we obtain

$$f - Kx - B\frac{dx}{dT} - M\frac{d^2x}{dt^2} = 0$$

$$f = M\frac{d^2x}{dt^2} + B\frac{dx}{dt} + Kx$$

This is the system differential equation with input f and output x. It is usual to place terms involving the input variable and its derivatives on one side of the equation and terms involving the output variable and its derivatives on the other side of the equation.

KEY POINT

> When writing a system differential equation it is conventional to place terms involving the input variable on one side of the equation and terms involving the output variable on the other side of the equation.

The previous example illustrates most of the points that need to be taken into account when developing a system differential equation for a simple mechanical system. Below is a summary of the three stages needed to create a system differential equation.

KEY POINT

Stages in obtaining a system differential equation for a mechanical system:

(i) Isolate the various components in the system and if necessary draw a free-body diagram for a component in order to decide what forces are acting on it.
(ii) Having identified the forces acting on a component, write the modelling equations for that component.
(iii) Combine the system equations together in order to obtain the system differential equation. The aim is to eliminate any intermediate variables and leave an equation just involving the input variable and its derivatives and the output variable and its derivatives.

Intermediate variables are so called because they lie in between the input and output variables.

It is conventional to draw a free-body diagram for each mass in the system as these tend to have more forces acting on them than other components owing to the presence of inertia forces. Sometimes, however, other components may need a free-body diagram in order to decide what forces are acting on them.

In the previous example, when all the equations had been written there were three equations and two intermediate variables, f_s and f_d, that needed to be eliminated. In order to eliminate all the intermediate variables there needs to be (1 + number of intermediate variables) equations.

KEY POINT

In order to eliminate all of the intermediate variables in a system and form the system differential equation there needs to be (1 + number of intermediate variables) equations.

It is good practice to identify the number of intermediate variables in a system before starting the process of elimination. This allows a check to be made that all the system equations have been stated and thus ensures that it is possible to obtain a system differential equation from the system equations. In Example 2.3 it was easy to eliminate the intermediate variables. Later we shall see examples where more skill and patience are needed.

Example

2.4 Obtain a system differential equation for the system shown in Figure 2.9. Assume the system input is the displacement x and the system output is the displacement y.

Solution From Figure 2.9 we see that a positive value of x corresponds to the damper being compressed. We therefore derive the system differential equation

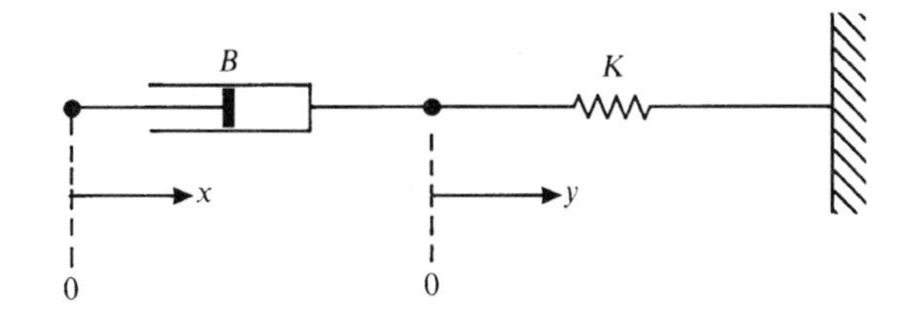

Figure 2.9
System for Example 2.4

assuming this is the case. Let us assume that a force f is being applied to the damper in order to make the damper close. This causes a reaction by the damper, f_d. As the damper has negligible mass there is no inertia force and so the only forces acting on the damper are f and f_d. By Newton's third law we can write

$$f - f_d = 0 \qquad \text{Eqn. [2.5]}$$

We now seek to obtain an expression for the magnitude of the damper force, f_d. A complication in this example is that both ends of the damper are free to move. Therefore it is the difference in velocity between the two ends of the damper that determines the amount of force needed to close the damper. The piston has velocity dx/dt and the cylinder has velocity dy/dt. So we can write

$$f_d = B\left(\frac{dx}{dt} - \frac{dy}{dt}\right) \qquad \text{Eqn. [2.6]}$$

Note that dx/dt is greater than dy/dt as the damper is closing owing to the applied force f.

Finally, the damper exerts a force f_d on the spring and so we can write for the spring

$$f_d = Ky \qquad \text{Eqn. [2.7]}$$

We now have three equations and two intermediate variables to eliminate, namely, f and f_d. Note that f_d occurs in all three equations and so we need to substitute Eqn. [2.5] into both Eqn. [2.6] and Eqn. [2.7] to eliminate f_d. Substituting Eqn. [2.5] into Eqn. [2.6] gives

$$f = B\left(\frac{dx}{dt} - \frac{dy}{dt}\right) \qquad \text{Eqn. [2.8]}$$

Substituting Eqn. [2.5] into Eqn. [2.7] gives

$$f = Ky \qquad \text{Eqn. [2.9]}$$

We now have two unused equations and one intermediate variable to eliminate. Combining Eqns [2.8] and [2.9] to eliminate f gives

$$Ky = B\left(\frac{dx}{dt} - \frac{dy}{dt}\right)$$

Removing the brackets gives

$$Ky = B\frac{dx}{dt} - B\frac{dy}{dt}$$

Finally we have

$$B\frac{dx}{dt} = B\frac{dy}{dt} + Ky$$

This is the system differential equation with input x and output y.

In Example 2.4 great care was taken to understand the forces acting in the system.

KEY POINT

> With practice, it is possible to write fewer system equations with fewer intermediate variables, thus making it easier to arrive at the system differential equation.

Example

2.5 Obtain a system differential equation for the system drawn in Figure 2.10. Assume the system input is the force f and the system output is the displacement y.

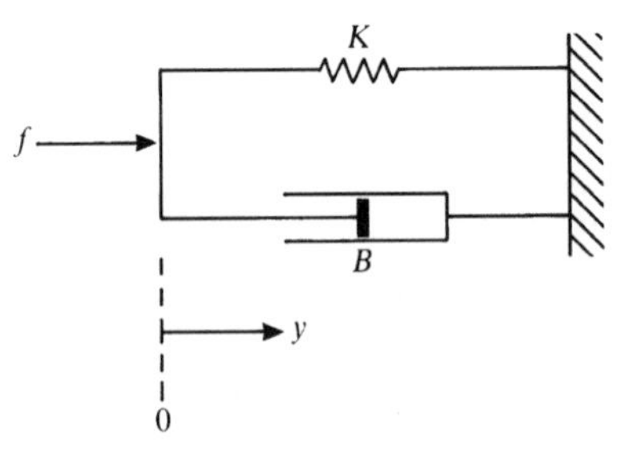

Figure 2.10
System for Example 2.5

Solution Because the force f is working to compress both the spring and the damper they each exert a force to oppose this compression. In order to avoid introducing too many intermediate variables we can note immediately that the reaction force due to the spring is Ky and the reaction force due to the damper is $B(dy/dt)$. There are no inertia forces present as the spring and damper are assumed to have negligible mass. So we can immediately write

$$f = B\frac{dy}{dt} + Ky$$

This is the system differential equation with input f and output y.

Let us now try to analyse a more complicated system containing two masses.

Example

2.6 Consider Figure 2.11 which shows two masses resting on a smooth frictionless surface, connected together by a spring of stiffness K_1. The second mass is connected to a fixed point by a spring of stiffness K_2. Obtain a system differential equation assuming the input is the force f and the output is the displacement of the second mass y. Assume that both the springs are relaxed when $x = 0$ and $y = 0$.

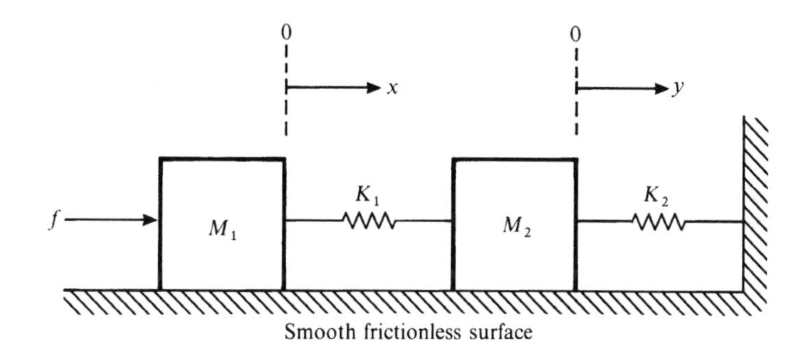

Figure 2.11
System for Example 2.6

Solution The input force f is being applied in a direction which suggests that both springs will be compressed. Therefore, for convenience, we assume that both x and y are positive and furthermore that x is greater than y. If at certain times y is greater than x then this does not invalidate the model. The main thing is to assume a convention and be consistent when setting up the equations. The equations will then correctly describe all the various possibilities. If later we wish to apply negative values for f, that is, change its direction by 180°, then this will not present a problem as the system equations will still be valid.

First we can construct a free-body diagram for the mass M_1. We define f_{s1} to be the force in spring 1 and f_{s2} to be the force in spring 2. Figure 2.12 shows the free-body diagram for M_1. Note that f_{s1} opposes f because when defining terms we have assumed spring 1 is being compressed.

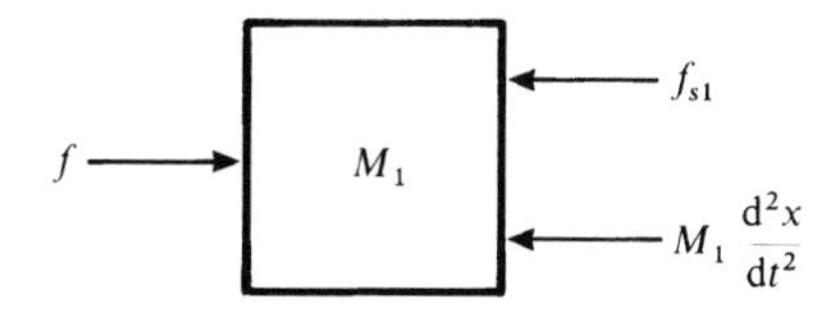

Figure 2.12
Free-body diagram for M_1

We have included the inertia force in the free-body diagram and so we can equate the forces on the mass to zero. Note that because spring 1 is being compressed it opposes the movement of the mass M_1 in the positive x direction. So we have

$$f - f_{s1} - M_1 \frac{d^2x}{dt^2} = 0 \qquad \text{Eqn. [2.10]}$$

Similarly, a free-body diagram can be constructed for the mass M_2. This is shown in Figure 2.13. We have included the inertia force in the free-body

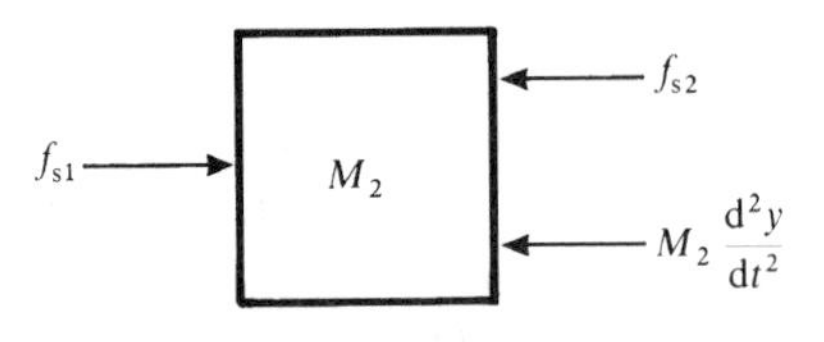

Figure 2.13
Free-body diagram for M_2

diagram and so we can equate the forces on the mass to zero. Note that the inertia force for this mass is $M_2(d^2y/dt^2)$ and so we have

$$f_{s1} - f_{s2} - M_2 \frac{d^2y}{dt^2} = 0 \qquad \text{Eqn. [2.11]}$$

Both ends of spring 1 are being displaced. However, we assumed earlier that the displacement x was greater than the displacement y. This is the reason why the spring is compressed. The amount by which it is being compressed is $x - y$ and so we can write

$$f_{s1} = K_1(x - y) \qquad \text{Eqn. [2.12]}$$

Only one end of spring 2 is being displaced; the other end is stationary. So we can simply write

$$f_{s2} = K_2 y \qquad \text{Eqn. [2.13]}$$

We now have four equations and three intermediate variables we wish to eliminate, namely, f_{s1}, f_{s2} and x. However, care is needed in order to carry this out correctly. First we can substitute Eqn. [2.12] into Eqn. [2.10] to eliminate f_{s1} from Eqn. [2.10]. This gives

$$f - K_1(x - y) - M_1 \frac{d^2x}{dt^2} = 0 \qquad \text{Eqn. [2.14]}$$

Unfortunately f_{s1} is also present in Eqn. [2.11] and so it is necessary to substitute Eqn. [2.12] into Eqn. [2.11]. This gives

$$K_1(x - y) - f_{s2} - M_2 \frac{d^2y}{dt^2} = 0 \qquad \text{Eqn. [2.15]}$$

We have now fully eliminated f_{s1} and have three unused equations, namely, Eqns [2.13], [2.14] and [2.15], and two intermediate variables to eliminate, that is, f_{s2} and x. It is extremely important not to reuse old equations or otherwise variables that have been eliminated will be reintroduced.

We can now substitute Eqn. [2.13] into Eqn. [2.15] to eliminate f_{s2}. This gives

$$K_1(x - y) - K_2 y - M_2 \frac{d^2y}{dt^2} = 0 \qquad \text{Eqn. [2.16]}$$

We now have two unused equations, namely, Eqn. [2.14] and Eqn. [2.16], and we have only one more intermediate variable to eliminate, namely, x. By examining Eqn. [2.14] and Eqn. [2.16] we see that it is easy to eliminate the term $x - y$ from the equations simply by adding them together. This gives

$$f - K_2 y - M_1 \frac{d^2x}{dt^2} - M_2 \frac{d^2y}{dt^2} = 0 \qquad \text{Eqn. [2.17]}$$

Unfortunately we are not quite there because we still have the d^2x/dt^2 term in Eqn. [2.17]. In order to eliminate this we have to use Eqn. [2.16] again but in a modified form. Rearranging Eqn. [2.16] we obtain

$$K_1x - K_1y - K_2y - M_2\frac{d^2y}{dt^2} = 0$$

$$K_1x = M_2\frac{d^2y}{dt^2} + (K_1 + K_2)y$$

Differentiating this equation twice we obtain

$$K_1\frac{d^2x}{dt^2} = M_2\frac{d^4y}{dt^4} + (K_1 + K_2)\frac{d^2y}{dt^2}$$

$$\frac{d^2x}{dt^2} = \frac{M_2}{K_1}\frac{d^4y}{dt^4} + \left(\frac{K_1 + K_2}{K_1}\right)\frac{d^2y}{dt^2} \qquad \text{Eqn. [2.16a]}$$

Note we have labelled this Eqn. [2.16a] because essentially it is the same equation as Eqn. [2.16]. Finally we can substitute Eqn. [2.16a] into Eqn. [2.17] to obtain the system differential equation

$$f - K_2y - M_1\left[\frac{M_2}{K_1}\frac{d^4y}{dt^4} + \left(\frac{K_1 + K_2}{K_1}\right)\frac{d^2y}{dt^2}\right] - M_2\frac{d^2y}{dt^2} = 0$$

Bringing the terms involving y over to the right-hand side we have

$$f = M_1\left[\frac{M_2}{K_1}\frac{d^4y}{dt^4} + \left(\frac{K_1 + K_2}{K_1}\right)\frac{d^2y}{dt^2}\right] + M_2\frac{d^2y}{dt^2} + K_2y$$

Multiplying through by K_1 we have

$$K_1f = M_1\left[M_2\frac{d^4y}{dt^4} + (K_1 + K_2)\frac{d^2y}{dt^2}\right] + K_1M_2\frac{d^2y}{dt^2} + K_1K_2y$$

Collecting terms gives the system differential equation

$$K_1f = M_1M_2\frac{d^4y}{dt^4} + (M_1K_1 + M_1K_2 + M_2K_1)\frac{d^2y}{dt^2} + K_1K_2y$$

The system input is f and the system output is y.

The last example may have seemed complicated but it is important to be able to handle this complexity because many practical engineering systems have high order system differential equations. The way to deal with the complexity is to be systematic in deriving the initial equations and then in eliminating the intermediate variables to obtain the system differential equation. Finally, consider an example with an engineering flavour.

Example

2.7 Hydraulic systems occur in many areas of engineering. Hydraulic valves are used to control the flow of fluid to the hydraulic cylinders. For example, earth-

Hydraulic systems tend to be used when large forces are required and yet precision is still important.

moving vehicles have hydraulic cylinders to position their shovels. It is common practice to use air to control the position of these valves. A system that uses air is known as a pneumatic system. Consider Figure 2.14 which shows a cross-section through a pneumatically controlled hydraulic spool valve.

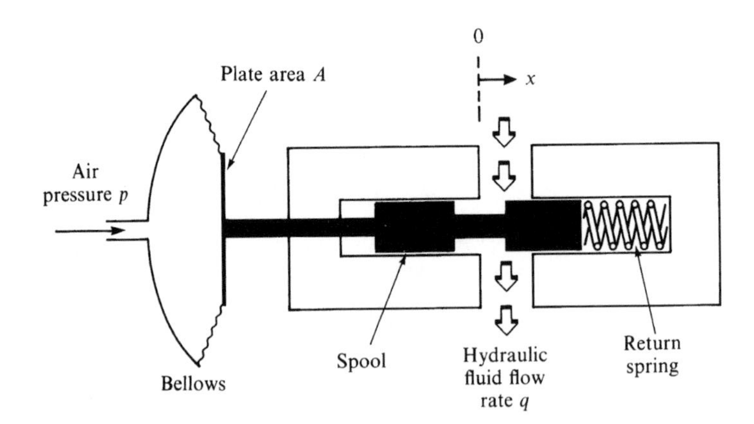

Figure 2.14
Pneumatically controlled hydraulic spool valve

When the air pressure is zero the spool is in a fully closed position and no hydraulic fluid can flow. As the air pressure is increased, the spool slides to the right allowing hydraulic fluid to flow. The further the spool moves to the right the higher the flow rate of hydraulic fluid through the valve. The relationship is given by

$$q = Gx$$

where q is the flow rate of hydraulic fluid, x is the displacement of the spool and G is a constant.

The position of the spool is adjusted by changing the air pressure p which is applied to the pneumatic bellows. The pressure plate of the bellows has area A. A return spring closes the spool as the air pressure is reduced. This spring has stiffness K and can be assumed to be relaxed when $x = 0$. There is viscous friction between the spool and the valve housing equivalent to a damping coefficient B. Static frictional effects can be ignored. The mass of the spool and connected movable parts is M. Derive a system differential equation with output the flow rate of hydraulic fluid, q, and input the control air pressure, p.

Solution The first stage is to consider the forces acting on the spool. For convenience we assume that $x > 0$, that is, the spool valve is partially open. There is a force f due to the bellows which is attempting to move the spool to the right. A force f_s due to the spring opposes this motion, as does a force f_d due to the viscous friction between the spool and the valve housing. Finally, there is an inertia force $M(d^2x/dt^2)$ which also opposes f. The free-body diagram for the spool and connected parts is shown in Figure 2.15.

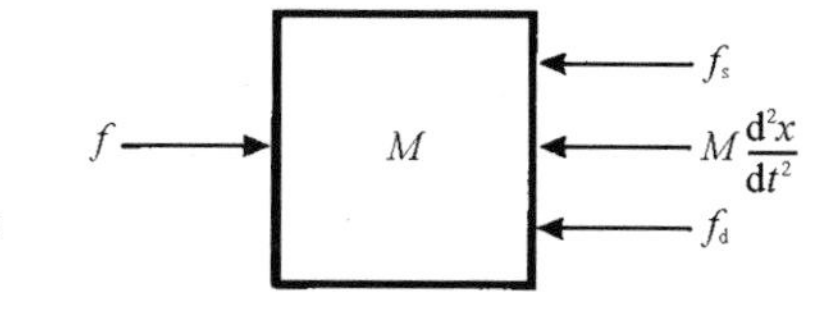

Figure 2.15
Free-body diagram for the spool and connected parts

Equating the forces on the mass to zero we can write

$$f - f_s - f_d - M\frac{d^2x}{dt^2} = 0$$

$$f = M\frac{d^2x}{dt^2} + f_d + f_s \qquad \text{Eqn. [2.18]}$$

For the spring we can write

$$f_s = Kx \qquad \text{Eqn. [2.19]}$$

For the viscous friction between the spool and the housing we can write

$$f_d = B\frac{dx}{dt} \qquad \text{Eqn. [2.20]}$$

Pressure has units of newton.metre^{-2} and area has units of metre^2.

The bellows have an area A.

Noting that, in general, (force) = (pressure) × (area) we can write

$$f = pA \qquad \text{Eqn. [2.21]}$$

Finally, we are given

$$q = Gx \qquad \text{Eqn. [2.22]}$$

We have five equations and four intermediate variables, namely, f, f_s, f_d and x, that we wish to eliminate and so this is possible. In this case the elimination is straightforward. We can immediately substitute Eqns [2.19], [2.20] and [2.21] into Eqn. [2.18] to eliminate f, f_s, and f_d. This gives

$$pA = M\frac{d^2x}{dt^2} + B\frac{dx}{dt} + Kx \qquad \text{Eqn. [2.23]}$$

We now only need to eliminate x and we have two unused equations, namely, Eqns [2.22] and [2.23]. Unfortunately x appears on its own and in differentiated form in Eqn. [2.23]. Therefore before we can proceed we need to differentiate Eqn. [2.22] twice. This gives

$$\frac{dq}{dt} = G\frac{dx}{dt} \qquad \text{Eqn. [2.22a]}$$

and

$$\frac{d^2q}{dt^2} = G\frac{d^2x}{dt^2} \qquad \text{Eqn. [2.22b]}$$

Note that these equations have been given the same number as Eqn. [2.22] as they are essentially the same. Finally we can substitute Eqns [2.22], [2.22a] and [2.22b] into Eqn. [2.23] to eliminate x. This gives

$$pA = \frac{M}{G}\frac{d^2q}{dt^2} + \frac{B}{G}\frac{dq}{dt} + K\frac{q}{G}$$

Multiplying through by G yields

$$GAp = M\frac{d^2q}{dt^2} + B\frac{dq}{dt} + Kq$$

This is the system differential equation with input p and output q.

Self-assessment questions 2.5

1. What is an intermediate variable?
2. How many equations are needed to eliminate n intermediate variables?
3. What is the procedure for obtaining the system differential equation of a translational mechanical system?

Exercises 2.5

1. Derive the system differential equation for the system shown in Figure 2.16. Assume the input is the displacement x and the output is the displacement y.

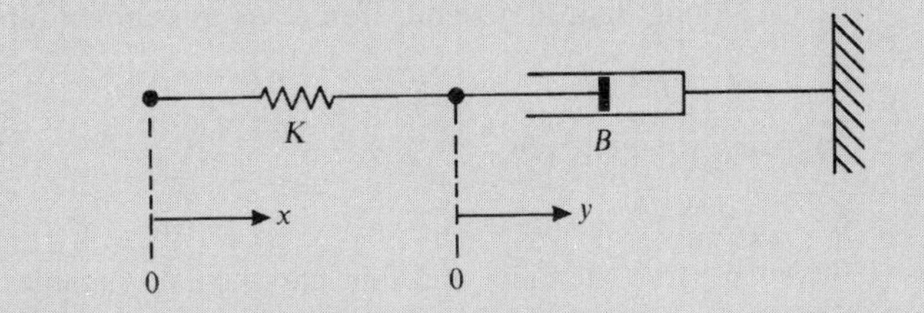

Figure 2.16 System for Exercise 2.5.1

2. Derive the system differential equation for the system shown in Figure 2.17. Assume the input is the displacement x and the output is the displacement y.

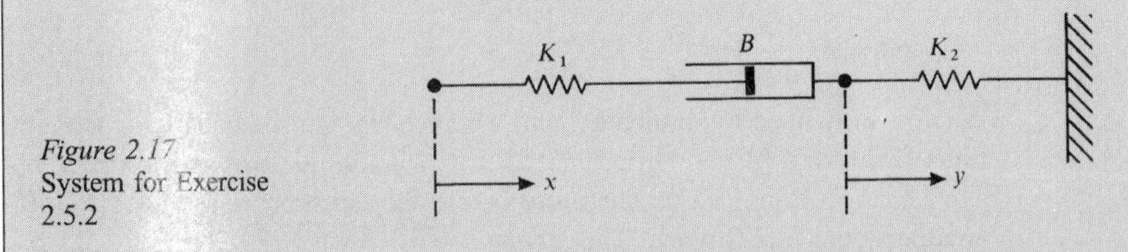

Figure 2.17 System for Exercise 2.5.2

3. Derive the system differential equation for the system shown in Figure 2.18. Assume the input is the displacement x and output is the displacement y.

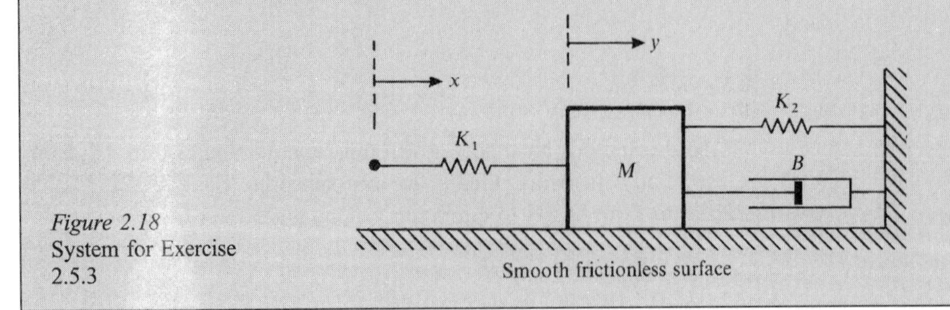

Figure 2.18 System for Exercise 2.5.3

4. Derive the system differential equation for the system shown in Figure 2.19. Assume the input is the force f and the output is the displacement y.

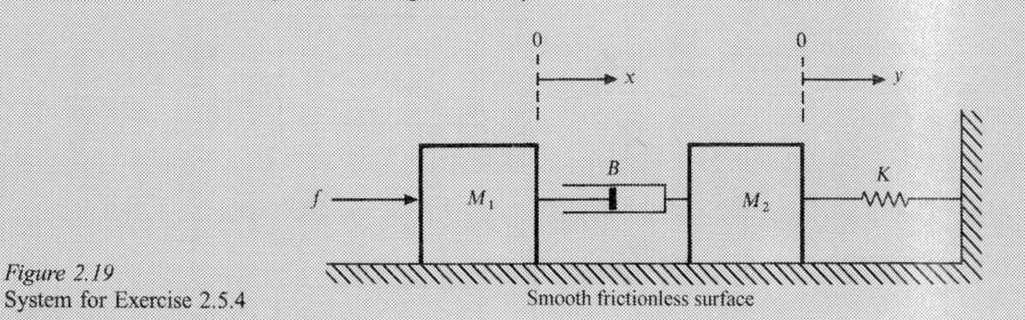

Figure 2.19
System for Exercise 2.5.4

Test and assignment exercises 2

1. Plot a graph of force against displacement for a spring with a stiffness of 15 N m^{-1}. Use a displacement range of −0.2 to 0.2 m.
2. Plot a graph of force against velocity for a damper with a damping coefficient of 13 N s m^{-1}. Use a velocity range of −1.5 to 2.8 m s^{-1}.
3. A mass is required to accelerate at a rate of 20 m s^{-2} when a force of 5 N is applied. Calculate the magnitude of the mass needed.
4. Figure 2.20(a) shows a wheel of a railway carriage together with the part of the carriage it can be assumed to support. Figure 2.20(b) shows a simple model of this system. Derive a system differential equation for this system with input the displacement of the railtrack, x, and output the displacement of the carriage, y. Note that x departs from zero owing to irregularities in the rails and because of the gaps between the rails. For convenience, ignore the force due to gravity acting on the mass. This does not introduce any error as its only effect is to change the equilibrium position of the spring.

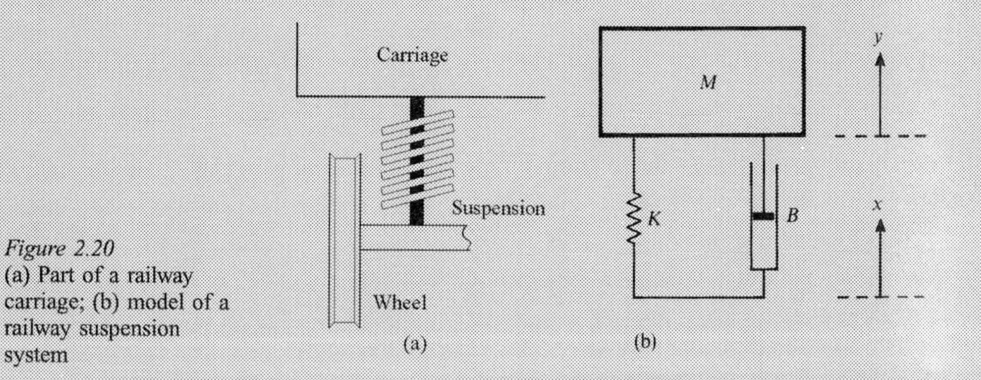

Figure 2.20
(a) Part of a railway carriage; (b) model of a railway suspension system

5. Figure 2.21(a) shows a wheel of a car together with the part of the car it can be assumed to support. Figure 2.21(b) shows a simple model of this system.

   The wheel hub has an equivalent mass M_1 and the car tyre can be modelled by a spring of stiffness K_1. Derive a system differential equation for this system with input the displacement of the road, x, and output the displacement of the car body, y. For convenience, ignore the force due to gravity acting on the masses.

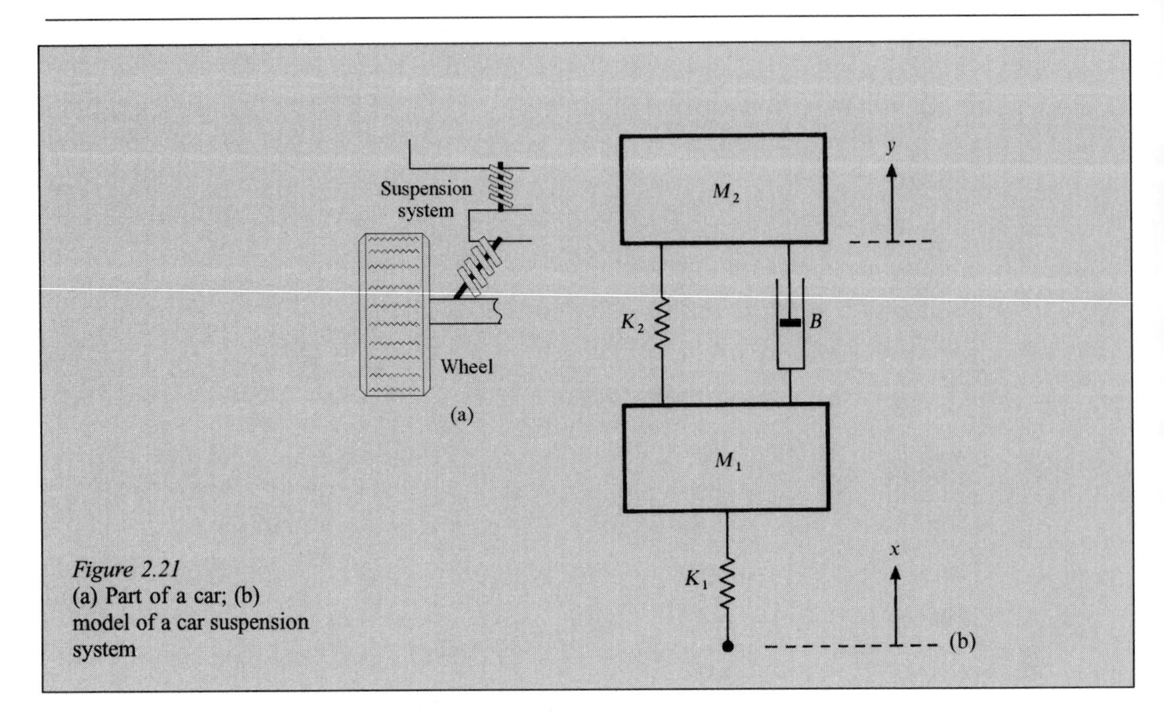

Figure 2.21
(a) Part of a car; (b) model of a car suspension system

3 Mechanical systems: rotational

Objectives

This chapter:

- describes the model for the behaviour of a torsional spring
- describes the model for the behaviour of a rotational damper
- describes the model for the behaviour of a rotating mass
- explains the concept of moment of inertia
- describes the rotational form of Newton's laws of motion
- explains the concept of an inertia torque
- derives mathematical models for some rotational mechanical systems

3.1 Introduction

The concepts associated with rotational mechanical systems are directly equivalent to those of translational mechanical systems. Again the three main components are the spring, the mass and the damper. Therefore the material in this chapter forms a natural development from the material of Chapter 2.

3.2 Basic concepts

Before analysing the main components of rotational mechanical systems it is worth reviewing some basic concepts. We saw that the concept of force was

This is another example of where everyday objects can be utilised to understand engineering concepts.

central to the analysis of translational systems. The equivalent quantity for rotational systems is **torque**. A torque has the ability to rotate objects about an axis. Consider Figure 3.1 which shows a cross-section through a door. Assume the door is of the type that has a unit connected at the top to close the door automatically and thus provides some resistance to the door being opened. If a force f_1 is applied to the door then it causes the door to rotate about the hinge.

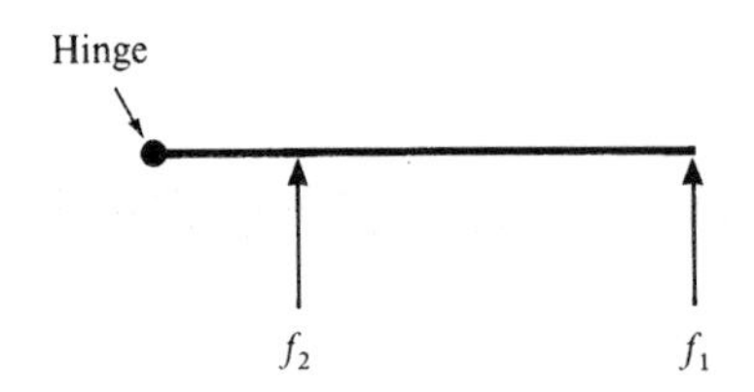

Figure 3.1
Cross-section through a door

Note, however, that applying the force f_2 instead would still cause the door to rotate but f_2 would need to be greater than f_1 because it has been applied nearer the hinge. Try this for yourself. In fact it is very difficult to open a door if it is pushed close to the hinge. The important point to note is that, when applying a force to cause an object to rotate, the position of the force relative to the point of rotation is as important as the magnitude of the force. Some objects are not fixed along an axis so they can translate as well as rotate. For such objects, two forces are required, of equal magnitude and opposite direction, if the object is to rotate and not translate. Consider Figure 3.2 which illustrates this case for a simple rectangular object.

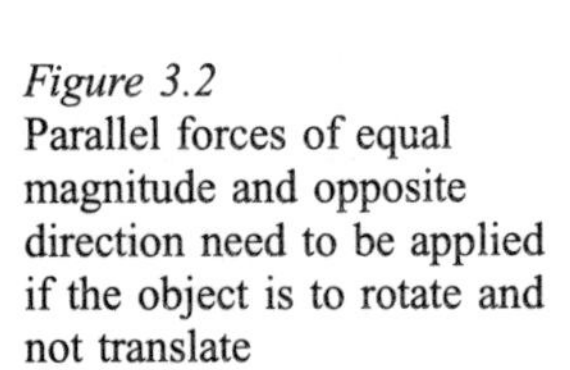

Figure 3.2
Parallel forces of equal magnitude and opposite direction need to be applied if the object is to rotate and not translate

In order for this object merely to rotate, the direction of the applied forces must change as the object rotates so that the forces remain perpendicular to the object. This figure illustrates the essence of a torque. It is an arrangement of two parallel forces which merely causes an object to rotate. Both the magnitude of the forces and the distance between the points of application of the forces is important in determining the size of the torque. A common way of viewing a torque is that it is a twisting force. The SI unit of torque is the newton.metre.

3.3 The torsional spring

Figure 3.3 shows an example of a torsional spring. It consists of a cylindrical piece of material clamped at one end and free to rotate at the other. A torque T

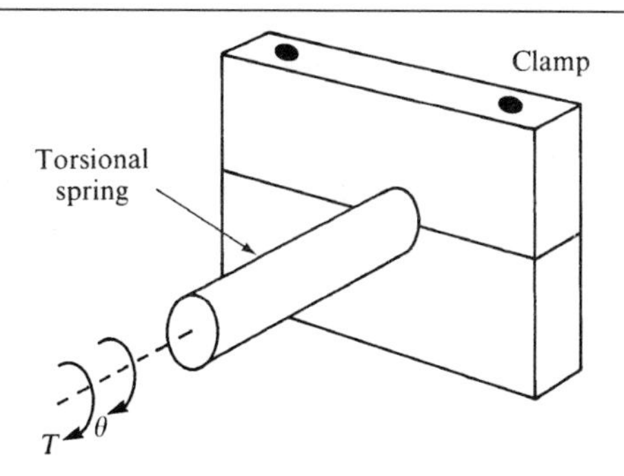

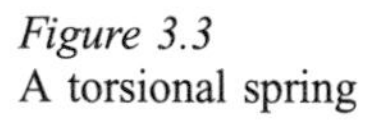

Figure 3.3
A torsional spring

An example of a torsional spring is the torsion bar in a car used to reduce rolling.

is applied to the free end of the spring. This causes the spring to twist. The amount of twist of the free end, θ, is known as the **angular displacement**. A negative value for θ corresponds to rotation in the opposite direction. It is usual to measure θ in units of radians so that calculus can be used without any complications. The amount by which the shaft twists is proportional to the applied torque, provided the torque is not too large.

KEY POINT

The torque T required to twist a spring by an angular displacement θ is given by

$$T = K\theta \qquad \text{Eqn. [3.1]}$$

where K is a constant known as the **spring stiffness**.

We have to use upper case T to represent torque because lower case t is universally reserved by engineers to represent time.

Note the similarity between this equation and that for a translational spring discussed in Section 2.2. The SI unit for K is newton.metre.radian^{-1}.

By examining Eqn. [3.1] we can see that if K is large then a large torque is required to rotate the spring by a certain amount. Conversely, if K is small then the spring will twist by the same amount when a much smaller torque is applied. Choosing a value of K depends on the application for which the spring is intended. There are many different designs of torsional spring with widely varying values of K.

As was the case with the translational spring, the torsional spring is assumed to have negligible mass and so is an idealised component. This simplifies the analysis of rotational systems. The symbol for the torsional spring of Figure 3.3 is shown in Figure 3.4. Note the use of shading to denote

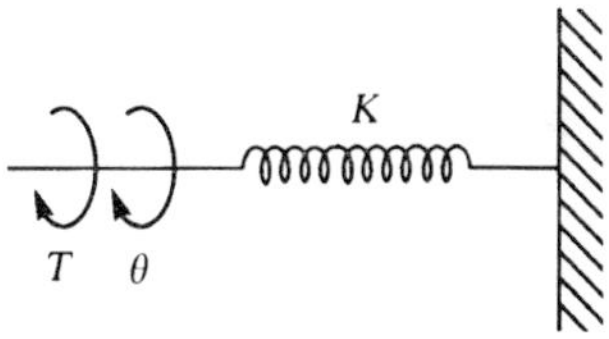

Figure 3.4
The symbol for a torsional spring is a coiled line

This is a common convention which simplifies calculations. In effect, it ensures that the graph of torque against angular displacement for the spring passes through the origin and so there is no offset.

a fixed anchor point that does not move. The spring itself is shown as a coiled line suggesting rotation. Compare this with the zigzag line used to denote a translational spring, discussed in Section 2.2. The curved lines used to show the directions of T and θ indicate a turning motion. They also allow the direction of motion to be indicated, that is, clockwise or anticlockwise. Usually θ is taken to be zero when the spring is relaxed and has no torque applied to it.

Self-assessment questions 3.3

1. Explain what is meant by the spring stiffness of a torsional spring.
2. Write a relationship for the behaviour of a torsional spring when a torque T is applied to the spring for the case when both ends are free to move and have angular displacements θ_1 and θ_2. Assume the spring has stiffness K.

Exercises 3.3

1. Draw a graph of torque against angular displacement for the following torsional springs:

   (a) $K = 20$ N m rad^{-1}
   (b) $K = 30$ N m rad^{-1}
   (c) $K = 50$ N m rad^{-1}

   Plot all the curves on the same graph. Use an angular displacement range of -2 rad to 3 rad.

3.4 The rotational damper

The angular velocity of an object, ω, is given by

$$\omega = \frac{d\theta}{dt}$$

where θ is the angular displacement of the object.

Figure 3.5 shows an example of a rotational damper. It consists of a casing, attached to a shaft, containing a fluid such as oil. Another shaft enters the casing and has paddles attached to it. It is fairly easy to rotate the two shafts relative to each other if this is done slowly, that is, the difference in angular

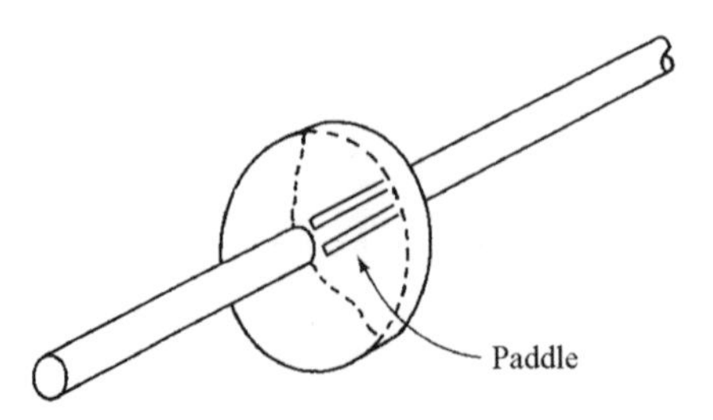

Figure 3.5
A rotational damper

velocities of the two shafts is small. However, trying to rotate the two shafts quickly relative to each other is much more difficult. If we assume that one of the shafts is fixed then the torque T required to turn the other shaft is proportional to the angular velocity ω at which the shaft is rotated.

KEY POINT

> The torque T required to rotate a damper with an angular velocity ω is given by
>
> $$T = B\omega$$
>
> where B is a constant known as the **damping coefficient**.

Note the similarity between this equation and that for a translational damper, discussed in Section 2.3. The SI unit for B is newton.metre.second.radian^{-1}.

We can also write this equation as

$$T = B\frac{d\theta}{dt}$$

because

$$\omega = \frac{d\theta}{dt}$$

Many different types of rotational damper exist. For example, a shaft rotating in a bearing housing can be modelled as a rotational damper. This is because, the faster the shaft rotates relative to the bearing housing, the greater the damping torque. Therefore it is the moving frictional force between the shaft and the housing that provides the damping torque.

The rotational damper is assumed to have negligible mass and so is an idealised component. This is usually acceptable because for most real dampers the damping properties dominate the mass properties and so little error is incurred in assuming the damper has negligible mass.

The symbol for the rotational damper of Figure 3.5 is shown in Figure 3.6. For convenience the shaft connecting the casing is assumed to be stationary and anchored to a fixed point. Note the difference between this symbol and that for a translational damper.

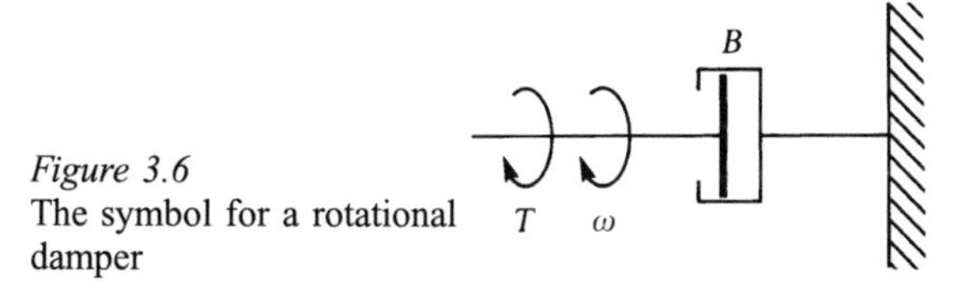

Figure 3.6
The symbol for a rotational damper

Self-assessment questions 3.4

1. Explain the action of a rotational damper.
2. Explain what is meant by the term 'damping coefficient'.
3. Derive an expression for the behaviour of a rotational damper when a torque T is applied to it and both ends are free to rotate and have angular velocities ω_1 and ω_2. The damping coefficient of the damper is B.

Exercises 3.4

1. Draw a torque–angular velocity graph for the following rotational dampers:

   (a) $B = 10$ N m s rad^{-1}
   (b) $B = 20$ N m s rad^{-1}
   (c) $B = 50$ N m s rad^{-1}

   Draw each of the curves on the same graph and use an angular velocity range of -15 rad s^{-1} to 25 rad s^{-1}.

3.5 The mass

When rotating a mass the amount of torque required to achieve a given acceleration depends not only on the mass but on its distribution about the centre of rotation. Consider Figure 3.7 which shows cross-sections through two cylindrical objects of equal masses but different orientations.

Object A requires a larger torque to accelerate it at a given rate than object B because the mass distribution of object A is further away from the axis of rotation than that of object B. A measure of the inertia of an object to rotation about a particular axis is the **moment of inertia** which has symbol J. This is a measure of the resistance of the mass to rotation about an axis. Note that the moment of inertia of an object changes if the axis of rotation is changed.

We shall not get involved in calculating the moment of inertia of particular objects as we shall not need them. Their values are readily available as reference in most textbooks on mechanics. Also, we assume a particular

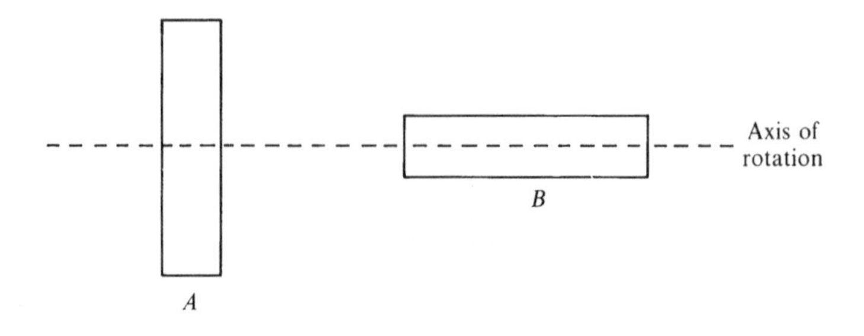

Figure 3.7
Cross-section through two cylindrical objects of equal masses but different orientations

value of J in our calculations or assume it remains to be specified. For example, motor manufacturers supply a value of moment of inertia for the motor rotors in their catalogue and so it is not necessary to calculate them.

KEY POINT

If a torque T is applied to a mass with moment of inertia J then the motion of the mass is given by

$$T = J\alpha$$

where α is the angular acceleration of the mass.

This equation can also be written as

$$T = J\frac{d\omega}{dt}$$

$$T = J\frac{d^2\theta}{dt^2}$$

where ω is the angular velocity of the mass and θ is the angular displacement of the mass.

KEY POINT

The following are equivalent:

$$\frac{d^2\theta}{dt^2} = \frac{d\omega}{dt} = \alpha$$

where θ is the angular displacement, ω is the angular velocity and α is the angular acceleration.

We note that the larger the value of J the more torque is required to accelerate the mass at a particular rate. The SI unit for J is kilogram.metre^2.

Example

3.1 Calculate the torque required to rotate an object of moment of inertia 13 kg m^2 with an angular acceleration of 2.5 rad s^{-2}.

Solution Substituting values into the equation $T = J\alpha$ we have

$$T = 13 \times 2.5 = 32.5 \text{ N m}$$

An interesting point can be made about the previous example. Recall from Chapter 1 that equations involving physical quantities need to balance physically as well as numerically. If we examine the product $J\alpha$ it has units

$$\text{right-hand side units} = (\text{kilogram.metre}^2)(\text{radian.second}^{-2})$$

$$= \text{kilogram.radian.metre}^2.\text{second}^{-2}$$

However, T has units

$$\text{left-hand side units} = \text{newton.metre}$$

Recall from Section 1.5 that the newton can also be written as kilogram.metre.second^{-2}. So we can write

$$\text{left-hand side units} = \text{kilogram.metre.second}^{-2}\text{.metre}$$
$$= \text{kilogram.metre}^2\text{.second}^{-2}$$

$\theta = l/r$ radians where l is the arc length subtended by the angle θ and r is the radius of the circle.

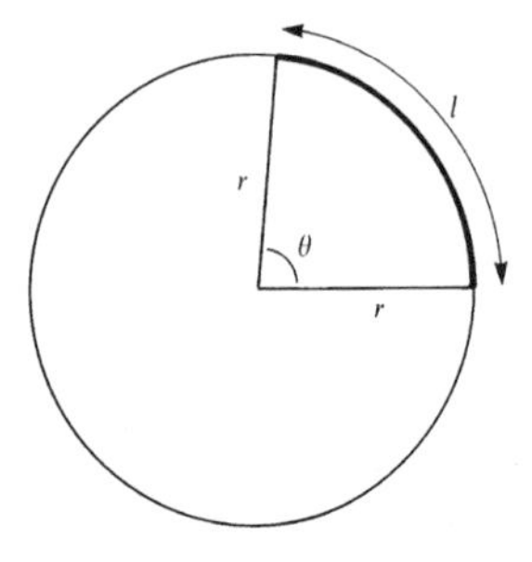

At first sight it looks as though the units do not balance on both sides of the equation. However, the reason for this is that the radian is what is known as a **dimensionless quantity**. Recall that a radian is defined as the ratio of two lengths. As the radian does not have any physical dimension it is not necessary for it to balance in an equation. In order to understand fully the physical balancing of equations it is necessary to study the topic of **dimensional analysis**. Because of lack of space it is not possible to do so here but there are several good textbooks dealing with this topic for the interested reader.

Self-assessment questions 3.5

1. Two objects have the same mass. Explain why one may require a larger torque to rotate it at a given acceleration than the other.
2. State the relationship between angular displacement, angular velocity and angular acceleration.
3. Define the radian.

Exercises 3.5

1. Calculate the angular acceleration of the following objects when a torque of 7 N m is applied to each of them:

   (a) $J = 18$ kg m^2
   (b) $J = 9.32$ kg m^2
   (c) $J = 1.62$ kg m^2

3.6 Newton's laws of motion

We examined the translational form of Newton's laws of motion in Section 2.5. The rotational form of Newton's laws of motion are directly analogous and so can be summarised briefly.

KEY POINT

Newton's first law of motion states:

Every body continues in a state of rest or constant angular velocity unless acted upon by a torque.

This law expresses the fact that an object will not start rotating unless acted upon by a torque. Also, if an object is already moving at a constant angular velocity then it will remain at the same angular velocity unless acted upon by a torque.

The reluctance of an object to change its angular velocity is known as its **inertia**. A measure of the inertia of an object in rotation is the **moment of inertia**.

KEY POINT

> Newton's second law of motion states:
>
> If a torque T is applied to a body with moment of inertia J then the motion of the body is given by
>
> $$T = J\alpha$$
>
> where α is the angular acceleration of the mass.

For all the mechanical systems we consider, the mass of the object remains constant and so the moment of inertia is constant provided the axis of rotation is not changed. This law quantifies the amount of torque needed to accelerate an object rotationally. When several torques are acting on an object then it is the net torque that determines the angular acceleration. In order to keep things simple we only consider torques acting around a single axis of rotation. Determining the net torque is then a simple matter of deciding whether a torque is acting in a clockwise or anticlockwise direction and either adding or subtracting the torque depending on its direction.

KEY POINT

> Newton's third law of motion states:
>
> To every action there is always an equal and opposite reaction.

This law is useful because it states that the torques acting on a body must balance. In order to make this law generally applicable it is important to include **inertia torque** which represents the reluctance of a body to change its angular velocity. The inertia torque has magnitude $J\alpha$ and acts in the opposite direction to the applied torque. Using this concept it is then possible to draw a free-body diagram for a rotating body and equate the torques to zero by Newton's third law of motion. The procedure is directly analogous to that discussed in Section 2.5 in connection with translational mechanical systems.

KEY POINT

> The torques acting on a rotating component can be equated to zero, that is, they must balance, provided any inertia torques are included in the free-body diagram.

There are no inertia torques associated with springs and dampers because they are assumed to have negligible mass.

KEY POINT

The only component we will analyse that has an inertia torque associated with it is the mass.

Self-assessment questions 3.6

1. Describe the rotational form of Newton's second law of motion.
2. Explain what is meant by Newton's third law of motion.
3. Explain what is meant by an inertia torque.

3.7 Rotational mechanical systems

We are now in a position to examine some rotational mechanical systems.

Example

3.2 Consider Figure 3.8 which shows a torsional pendulum. A mass is contained in a housing but can rotate about a vertical axis. It is suspended from a fixed point by means of a torsional spring of stiffness K. There is damping between the mass and the housing and the damping coefficient is B. The mass has a moment of inertia J and a torque is being applied to it of magnitude T. The angular position of the mass is θ. Derive a system differential equation assuming the system input is T and the system output is θ.

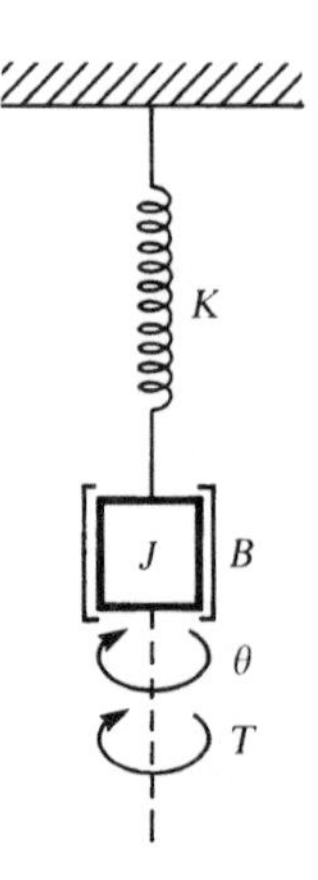

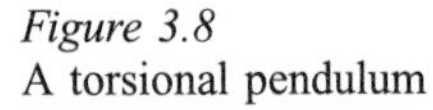
Figure 3.8
A torsional pendulum

Solution The first stage in solving this problem is to draw a free-body diagram for the mass. As this is a rotational system we need to examine the torques acting on the mass. For convenience we assume the mass is being rotated in the positive θ direction. Provided we are consistent in observing this convention then negative values of θ will automatically be taken care of by the equations. Because the torsional spring is being twisted it will exert a torque T_s opposing this twist. If you are not sure why this is the correct direction then consider the alternative that the spring helps the twist. This is clearly ludicrous. However, this approach of considering the alternative can be useful when trying to decide the direction of particular torques for more complicated systems. We also note that the damper produces a damping torque T_d which opposes the applied torque because it tries to slow down the movement of the mass – the alternative that it speeds up the movement of the mass is silly. Finally, there is an inertia torque, $J(d^2\theta/dt^2)$ which opposes the applied torque. So now we can construct the free-body diagram to show the torques acting on the mass. This is shown in Figure 3.9. Note that the axis of rotation needs to be shown on the free-body diagram in order to be able to show the direction of the torques.

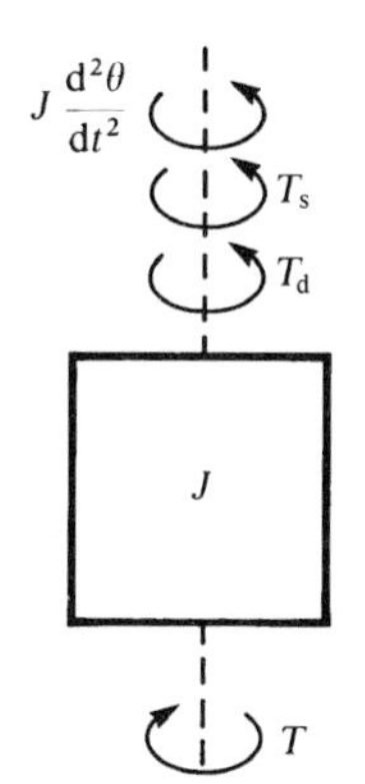

Figure 3.9
A free-body diagram for the mass

We can now apply Newton's third law to the mass and equate the torques to zero. So we have

$$T - T_s - T_d - J\frac{d^2\theta}{dt^2} = 0 \qquad \text{Eqn. [3.2]}$$

We can also write for the torsional spring

$$T_s = K\theta \qquad \text{Eqn. [3.3]}$$

For the damper we have

$$T_d = B\frac{d\theta}{dt} \qquad \text{Eqn. [3.4]}$$

We now have three equations for the system and we need to eliminate two intermediate variables, namely, T_s and T_d, in order to obtain the system differential equation. Recall that we need one more equation than the number of intermediate variables to proceed successfully. We see that this is so. In fact, in this case, elimination of T_s and T_d is trivial and so we can immediately write

$$T - K\theta - B\frac{d\theta}{dt} - J\frac{d^2\theta}{dt^2} = 0$$

$$T = J\frac{d^2\theta}{dt^2} + B\frac{d\theta}{dt} + K\theta$$

This is the system differential equation with input T and output θ.

We now consider another example.

Example

3.3 Consider the system shown in Figure 3.10. Derive a system differential equation for the system with input the angular displacement θ_i and output the angular displacement θ_o.

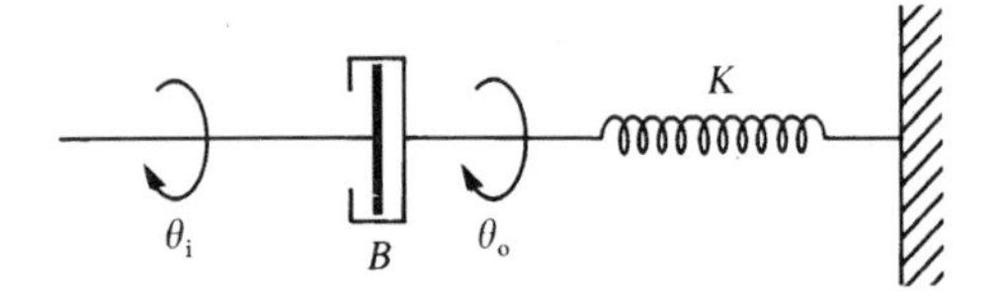

Figure 3.10
System for Example 3.3

Solution Rotating the input shaft will require a torque, which we denote by T. So for the rotational damper we can write

$$T = B\left(\frac{d\theta_i}{dt} - \frac{d\theta_o}{dt}\right)$$

This equation emerges because both of the damper shafts are moving and so it is the difference between the two shaft angular velocities that is important.

The damper has negligible mass and so there is no inertia torque associated with it. So by Newton's third law of motion, the torque T is transmitted directly onto the torsional spring. For the spring we can write

$$T = K\theta_o$$

Combining these two equations to eliminate T gives

$$B\left(\frac{d\theta_i}{dt} - \frac{d\theta_o}{dt}\right) = K\theta_o$$

$$B\frac{d\theta_i}{dt} = B\frac{d\theta_o}{dt} + K\theta_o$$

This is the system differential equation with input θ_i and output θ_o.

In the previous example we tried to keep the number of intermediate variables to a minimum. This required insight into how the torques were distributed in the system. Compare this approach with that of Example 2.4 which dealt with an analogous translational mechanical system. There the approach was more rigorous but the result was the introduction of two intermediate variables, rather than one. Now let us examine a more complicated rotational system.

Example

3.4 Consider the system shown in Figure 3.11. Derive a system differential equation for the system relating the input angular displacement θ_i to the output angular displacement θ_o.

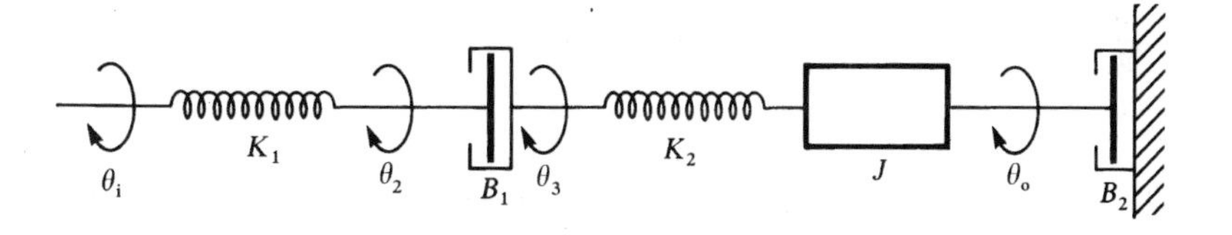

Figure 3.11
System for Example 3.4

Solution A torque is required to move the input shaft. Let us denote this torque by T. So for the first spring we can write

$$T = K_1(\theta_i - \theta_2) \qquad \text{Eqn. [3.5]}$$

The spring has negligible mass and so there is no inertia torque. So by Newton's third law the torque is transmitted to the first damper.

The damper has negligible mass and so there is no inertia torque to oppose T. So we can write

$$T = B_1\left(\frac{d\theta_2}{dt} - \frac{d\theta_3}{dt}\right) \qquad \text{Eqn. [3.6]}$$

The torque T is transmitted on unchanged to the second spring by Newton's third law. So we can write

$$T = K_2(\theta_3 - \theta_o) \qquad \text{Eqn. [3.7]}$$

Note that the displacement of the part of the second spring connected to the mass is θ_o as the mass is assumed to be a rigid body which does not deform.

Next we need to examine the torques acting on the mass. There is the driving torque T. The second damper exerts a torque T_d which opposes the driving torque. There is also an inertia torque $J(d^2\theta_o/dt^2)$ which opposes the driving torque. The free-body diagram for the mass is shown in Figure 3.12.

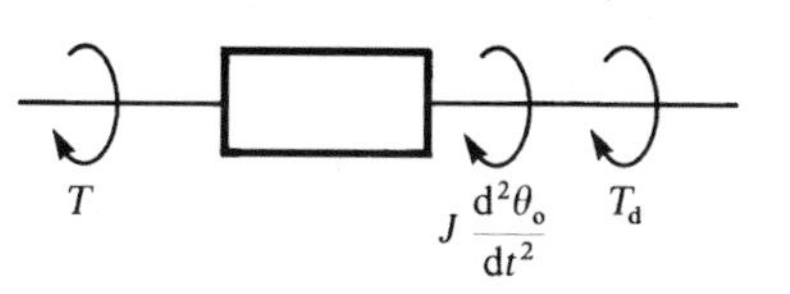

Figure 3.12
Free-body diagram for Example 3.4

Using Newton's third law of motion we can write

$$T - T_{\mathrm{d}} - J\frac{\mathrm{d}^2\theta_{\mathrm{o}}}{\mathrm{d}t^2} = 0$$

$$T = J\frac{\mathrm{d}^2\theta_{\mathrm{o}}}{\mathrm{d}t^2} + T_{\mathrm{d}} \qquad \text{Eqn. [3.8]}$$

Finally, for the second damper we can write

$$T_{\mathrm{d}} = B_2\frac{\mathrm{d}\theta_{\mathrm{o}}}{\mathrm{d}t} \qquad \text{Eqn. [3.9]}$$

We now have five equations and four intermediate variables to eliminate, namely, T, T_{d}, θ_2, θ_3, and so this is possible.

We first eliminate T_{d} by combining Eqns. [3.8] and [3.9]. This gives

$$T = J\frac{\mathrm{d}^2\theta_{\mathrm{o}}}{\mathrm{d}t^2} + B_2\frac{\mathrm{d}\theta_{\mathrm{o}}}{\mathrm{d}t} \qquad \text{Eqn. [3.10]}$$

We now have four unused equations, namely, Eqns [3.5], [3.6], [3.7] and [3.10], and three intermediate variables to eliminate. Next we eliminate T by substituting Eqn. [3.5] into [3.6], [3.7] and [3.10]. Substituting Eqn. [3.5] into Eqn. [3.6] gives

$$K_1(\theta_{\mathrm{i}} - \theta_2) = B_1\left(\frac{\mathrm{d}\theta_2}{\mathrm{d}t} - \frac{\mathrm{d}\theta_3}{\mathrm{d}t}\right) \qquad \text{Eqn. [3.11]}$$

Substituting Eqn. [3.5] into Eqn. [3.7] gives

$$K_1(\theta_{\mathrm{i}} - \theta_2) = K_2(\theta_3 - \theta_{\mathrm{o}}) \qquad \text{Eqn. [3.12]}$$

Substituting Eqn. [3.5] into Eqn. [3.10] gives

$$K_1(\theta_{\mathrm{i}} - \theta_2) = J\frac{\mathrm{d}^2\theta_{\mathrm{o}}}{\mathrm{d}t^2} + B_2\frac{\mathrm{d}\theta_{\mathrm{o}}}{\mathrm{d}t} \qquad \text{Eqn. [3.13]}$$

Now we have three equations, namely, Eqns [3.11], [3.12] and [3.13] and two intermediate variables to eliminate. Next we eliminate θ_3. This only appears in Eqns [3.11] and [3.12]. Unfortunately it appears in Eqn. [3.11] in differentiated form. So we first need to differentiate Eqn. [3.12]. This gives

$$K_1\left(\frac{\mathrm{d}\theta_{\mathrm{i}}}{\mathrm{d}t} - \frac{\mathrm{d}\theta_2}{\mathrm{d}t}\right) = K_2\left(\frac{\mathrm{d}\theta_3}{\mathrm{d}t} - \frac{\mathrm{d}\theta_{\mathrm{o}}}{\mathrm{d}t}\right)$$

$$\frac{\mathrm{d}\theta_3}{\mathrm{d}t} = \frac{K_1}{K_2}\left(\frac{\mathrm{d}\theta_{\mathrm{i}}}{\mathrm{d}t} - \frac{\mathrm{d}\theta_2}{\mathrm{d}t}\right) + \frac{\mathrm{d}\theta_{\mathrm{o}}}{\mathrm{d}t} \qquad \text{Eqn. [3.12a]}$$

This is labelled Eqn. [3.12a] as it is essentially the same as Eqn. [3.12]. Substituting Eqn. [3.12a] into Eqn. [3.11] gives

$$K_1(\theta_{\mathrm{i}} - \theta_2) = B_1\left\{\frac{\mathrm{d}\theta_2}{\mathrm{d}t} - \left[\frac{K_1}{K_2}\left(\frac{\mathrm{d}\theta_{\mathrm{i}}}{\mathrm{d}t} - \frac{\mathrm{d}\theta_2}{\mathrm{d}t}\right) + \frac{\mathrm{d}\theta_{\mathrm{o}}}{\mathrm{d}t}\right]\right\}$$

Removing brackets gives

$$K_1\theta_i - K_1\theta_2 = B_1\frac{d\theta_2}{dt} - \frac{B_1K_1}{K_2}\frac{d\theta_i}{dt} + \frac{B_1K_1}{K_2}\frac{d\theta_2}{dt} - B_1\frac{d\theta_o}{dt}$$

Finally,

$$K_1\theta_i + \frac{B_1K_1}{K_2}\frac{d\theta_i}{dt} = K_1\theta_2 + \left(B_1 + \frac{B_1K_1}{K_2}\right)\frac{d\theta_2}{dt} - B_1\frac{d\theta_o}{dt} \qquad \text{Eqn. [3.14]}$$

Now we have two equations, namely, Eqns [3.13] and [3.14], and one intermediate variable, θ_2, to eliminate. Unfortunately, θ_2 occurs as itself and in differentiated form in Eqn. [3.14]. So we need to differentiate Eqn. [3.13]. This gives

$$K_1\theta_i - K_1\theta_2 = J\frac{d^2\theta_o}{dt^2} + B_2\frac{d\theta_o}{dt}$$

$$K_1\theta_2 = K_1\theta_i - J\frac{d^2\theta_o}{dt^2} - B_2\frac{d\theta_o}{dt}$$

$$K_1\frac{d\theta_2}{dt} = K_1\frac{d\theta_i}{dt} - J\frac{d^3\theta_o}{dt^3} - B_2\frac{d^2\theta_o}{dt^2}$$

$$\frac{d\theta_2}{dt} = \frac{d\theta_i}{dt} - \frac{J}{K_1}\frac{d^3\theta_o}{dt^3} - \frac{B_2}{K_1}\frac{d^2\theta_o}{dt^2} \qquad \text{Eqn. [3.13a]}$$

Finally, we substitute Eqns [3.13] and [3.13a] into Eqn. [3.14] to eliminate θ_2. This gives

$$K_1\theta_i + \frac{B_1K_1}{K_2}\frac{d\theta_i}{dt} = K_1\theta_i - J\frac{d^2\theta_o}{dt^2} - B_2\frac{d\theta_o}{dt} + \left(B_1 + \frac{B_1K_1}{K_2}\right)\left(\frac{d\theta_i}{dt} - \frac{J}{K_1}\frac{d^3\theta_o}{dt^3} - \frac{B_2}{K_1}\frac{d^2\theta_o}{dt^2}\right) - B_1\frac{d\theta_o}{dt}$$

Cancelling the $K_1\theta_i$ terms and putting the first bracketed term over a common denominator gives

$$\frac{B_1K_1}{K_2}\frac{d\theta_i}{dt} = -J\frac{d^2\theta_o}{dt^2} - B_2\frac{d\theta_o}{dt} + \frac{B_1}{K_2}(K_2 + K_1) \times \left(\frac{d\theta_i}{dt} - \frac{J}{K_1}\frac{d^3\theta_o}{dt^3} - \frac{B_2}{K_1}\frac{d^2\theta_o}{dt^2}\right) - B_1\frac{d\theta_o}{dt}$$

Note that cancelling of the $K_1\theta_i$ terms means there are no terms that are not differentiated. So we can integrate once to simplify things. This gives

$$\frac{B_1K_1}{K_2}\theta_i = -J\frac{d\theta_o}{dt} - B_2\theta_o + \frac{B_1}{K_2}(K_1 + K_2)\left(\theta_i - \frac{J}{K_1}\frac{d^2\theta_o}{dt^2} - \frac{B_2}{K_1}\frac{d\theta_o}{dt}\right) - B_1\theta_o$$

Removing the brackets gives

$$\frac{B_1K_1}{K_2}\theta_i = -J\frac{d\theta_o}{dt} - B_2\theta_o + \frac{B_1}{K_2}(K_1+K_2)\theta_i$$
$$-\frac{JB_1}{K_1K_2}(K_1+K_2)\frac{d^2\theta_o}{dt^2} - \frac{B_1B_2}{K_1K_2}(K_1+K_2)\frac{d\theta_o}{dt} - B_1\theta_o$$

Gathering terms gives

$$\left[\frac{B_1}{K_2}(K_1+K_2) - \frac{B_1K_1}{K_2}\right]\theta_i = \frac{JB_1}{K_1K_2}(K_1+K_2)\frac{d^2\theta_o}{dt^2}$$
$$+\left[\frac{B_1B_2}{K_1K_2}(K_1+K_2)+J\right]\frac{d\theta_o}{dt}$$
$$+(B_1+B_2)\theta_o$$

$$B_1\theta_i = \frac{JB_1}{K_1K_2}(K_1+K_2)\frac{d^2\theta_o}{dt^2}$$
$$+\left[\frac{B_1B_2}{K_1K_2}(K_1+K_2)+J\right]\frac{d\theta_o}{dt}$$
$$+(B_1+B_2)\theta_o$$

Multiplying through by K_1K_2 and simplifying gives

$$B_1K_1K_2\theta_i = JB_1(K_1+K_2)\frac{d^2\theta_o}{dt^2} + [B_1B_2(K_1+K_2)+JK_1K_2]\frac{d\theta_o}{dt}$$
$$+K_1K_2(B_1+B_2)\theta_o$$

This is the system differential equation with input θ_i and output θ_o.

The last example was complicated. It was included to show that by being systematic it is possible to obtain the system differential equation of quite complicated systems. The key is to keep track of what intermediate variables have been eliminated and to take care not accidentally to reuse equations that have been used before. In Chapter 8 we introduce the Laplace transform. Using this it is possible to simplify the work needed to obtain a system differential equation of a complicated system.

Self-assessment questions 3.7

1. Write an equation for the relationship between applied torque and angular velocity for a damper in which both shafts are free to rotate.
2. Describe the procedure for obtaining a system differential equation of a rotational mechanical system.

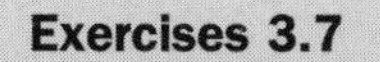

Exercises 3.7

1. Obtain a system differential equation for the system shown in Figure 3.13. Use the angular displacement θ_i as the input variable and the angular displacement, θ_o as the output variable.

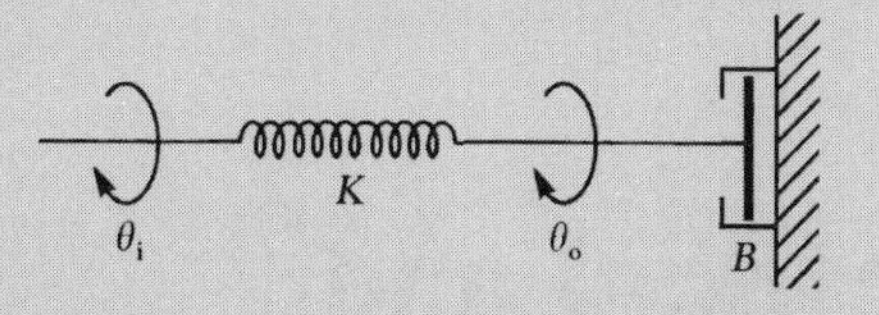

Figure 3.13
System for Exercise 3.7.1

2. Obtain a system differential equation for the system shown in Figure 3.14. Use the driving torque T as the input variable and the angular displacement θ as the output variable.

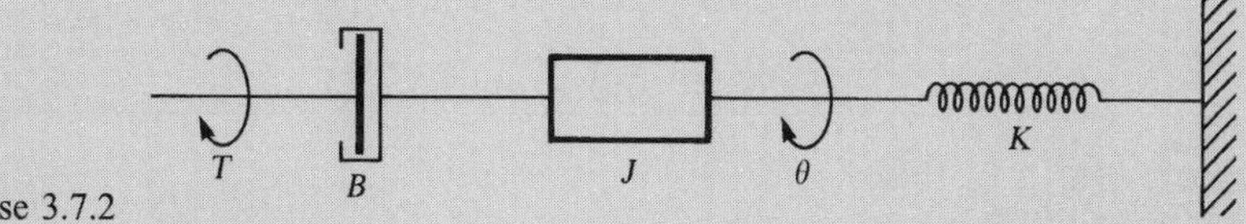

Figure 3.14
System for Exercise 3.7.2

Test and assignment exercises 3

1. Draw a graph of torque against angular displacement for a torsional spring with a stiffness $K = 40$ N m rad^{-1}. Use an angular displacement range of -2 to 4 rad.
2. Draw a graph of torque against angular velocity for a rotational damper with damping coefficient $B = 20$ N m s rad^{-1}. Use an angular velocity range of -5 to 10 rad s^{-1}.
3. Obtain a system differential equation for the system shown in Figure 3.15. Use the angular displacement θ_i as the input variable and the angular displacement θ_o as the output variable.

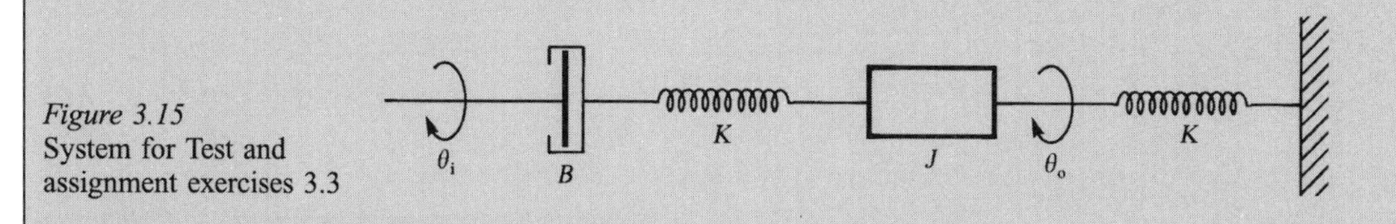

Figure 3.15
System for Test and assignment exercises 3.3

4. Obtain a system differential equation for the system shown in Figure 3.16. Use the angular displacement θ_i as the input variable and the angular displacement θ_o as the output variable.

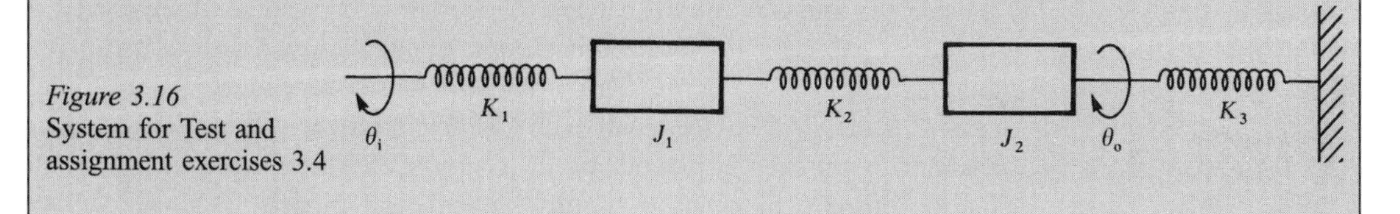

Figure 3.16
System for Test and assignment exercises 3.4

4 Electrical and electromechanical systems

Objectives

This chapter:

- describes the model for the behaviour of a resistor
- describes the model for the behaviour of a capacitor
- describes the model for the behaviour of an inductor
- explains Kirchhoff's circuit laws
- derives mathematical models of some electrical systems
- derives mathematical models of some electromechanical systems

4.1 Introduction

The electrical systems that we examine contain three types of components: the resistor, the inductor and the capacitor, together with a source of electricity. A variety of sources exist in electrical engineering. We use a source that maintains a fixed voltage across its output terminals independent of how much current is drawn. Engineers call such a source an **ideal voltage source**. Such sources do not exist in practice but it is very useful to imagine such a source for the purpose of creating simple mathematical models. Practical voltage sources that come close to ideal voltage sources can be created, if necessary, by engineers. The term **load** is used to describe any component that draws current from a source of electricity. A load may consist of a resistor, capacitor or inductor, or any combinations of these.

4.2 The resistor

A **resistor** is a device consisting of two terminals separated by a material that resists the flow of current. The main law governing the behaviour of a resistor is Ohm's law. This relates the amount of current flowing through a resistor to the magnitude of the voltage applied across the resistor.

KEY POINT

For a resistor, the voltage v across the resistor is given by

$$v = iR$$

where i is the current through the resistor and R is the **resistance** of the resistor and is a constant.

In general, for a voltage source, the current flows in the same direction as the arrow. For a load, the current flows in the opposite direction to the arrow.

The SI unit of resistance is the ohm. Ohm's law is based on the assumption that the resistor is able to withstand the current flowing through it no matter how large it is. In practice, beyond a certain current level the resistor will break down. This illustrates that Ohm's law is merely a mathematical model of the resistor which is applicable over a certain operating regime. It is important not to confuse the mathematical model with the resistor itself. Figure 4.1 shows the circuit for a resistor connected to an ideal voltage source. Note the convention that the tip of the voltage arrow represents a higher voltage value. Note, also, the symbols for a resistor and an ideal voltage source. The voltage source is drawn as a circle with + and − used to show the polarity of the source.

Example

4.1 (a) Calculate the voltage across a 3 Ω resistor if a current of 5 A flows through the resistor.

(b) Calculate the current flowing through a 147 Ω resistor if the voltage across the resistor is 15 V.

Solution (a) We have $i = 5$ and $R = 3$ and so

$$v = iR = 5 \times 3 = 15 \text{ V}$$

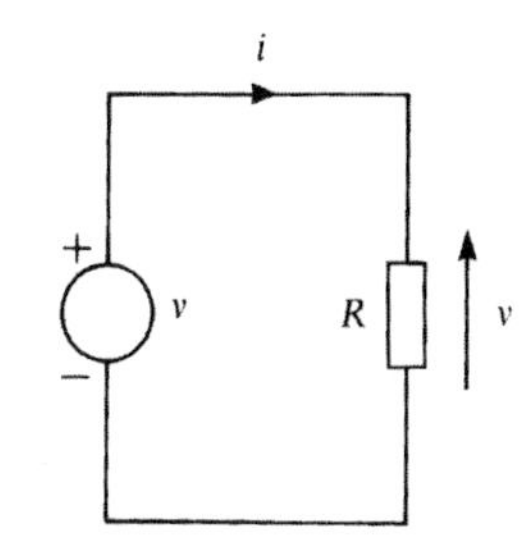

Figure 4.1
A resistor connected to an ideal voltage source

(b) We have $v = 15$ and $R = 147$ and so

$$i = \frac{v}{R} = \frac{15}{147} = 0.102 \text{ A}$$

Self-assessment questions 4.2

1. Describe the circumstances under which Ohm's law breaks down for a resistor.
2. What does a voltage arrow on a circuit show?
3. Describe the characteristics of an ideal voltage source.

Exercises 4.2

1. Calculate the voltage across a 28 Ω resistor when a current of 0.564 A flows through it.
2. Calculate the current flowing through a 45 Ω resistor if there is a voltage of 23 V across it.
3. Draw a voltage–current graph for a resistor with a resistance of 50 Ω. Use a current range of -0.5 A to 1.5 A.

4.3 The capacitor

A **capacitor** consists of two parallel metallic plates separated by an insulating material.

KEY POINT

For a capacitor, the current through a capacitor, i, is given by

$$i = C\frac{\mathrm{d}v}{\mathrm{d}t}$$

where v is the voltage across the capacitor and C is the **capacitance** of the capacitor.

In practice, all capacitors have some leakage current but this can be made very small.

Note that $\mathrm{d}v/\mathrm{d}t$ is the rate of change of voltage with time. The SI unit of capacitance is the farad. Figure 4.2 shows the symbol for a capacitor.

In order to create a simple model the capacitor is assumed to have material between the plates that has an infinite resistance. Therefore there is no **leakage current** through the capacitor when a voltage is applied. Current flow

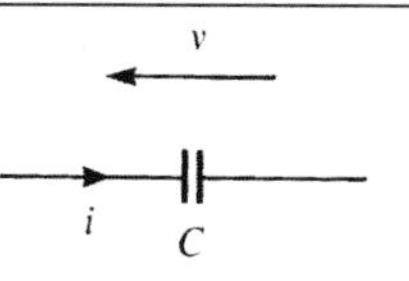

Figure 4.2
The symbol for a capacitor

in the capacitor merely results in a charge buildup on the plates, thus increasing the voltage across the capacitor. Current flow only occurs if the applied voltage is not equal to the capacitor voltage. So the capacitor we use is an ideal component, a concept we examined in our discussion of mechanical components in Chapters 2 and 3.

Example

4.2 (a) Calculate the current flowing through a 0.2 F capacitor when the voltage across the capacitor is changing at the rate of 2 V s^{-1}.

(b) Calculate the rate of change of voltage across a 0.3 F capacitor when a current of 2 A flows through it.

Solution (a) We have $\mathrm{d}v/\mathrm{d}t = 2$ and $C = 0.2$ and so

$$i = C\frac{\mathrm{d}v}{\mathrm{d}t} = 0.2 \times 2 = 0.4 \text{ A}$$

(b) We have $i = 2$ and $C = 0.3$ and so

$$\frac{\mathrm{d}v}{\mathrm{d}t} = \frac{i}{C} = \frac{2}{0.3} = 6.67 \text{ V s}^{-1}$$

Self-assessment questions 4.3

1. Explain the mechanism by which current flows through a capacitor.
2. Explain the term 'leakage current'.

Exercises 4.3

1. Calculate the current flowing through a 0.023 F capacitor if the rate of change of voltage across the capacitor is 0.12 V s^{-1}.
2. Calculate the rate of change of voltage across a 0.001 F capacitor if a current of 0.3 A flows through it.

4.4 The inductor

An **inductor** consists of a coil of wire. Any wire through which current flows has a magnetic field surrounding it but with an inductor the density of the magnetic field is especially strong owing to the concentrating effect of coiling the wire. It can be shown experimentally that any attempt to change the density of this magnetic field leads to a voltage being **induced** to oppose this change.

KEY POINT

For the inductor, the relationship between the induced voltage v and the current flow i through the inductor is given by

$$v = L\frac{\mathrm{d}i}{\mathrm{d}t}$$

where L is the **inductance** of the inductor.

The SI unit of inductance is the henry. Figure 4.3 shows the symbol for an inductor. Note that in order to produce a simple model for the inductor it is assumed to have negligible resistance and so it is an ideal component.

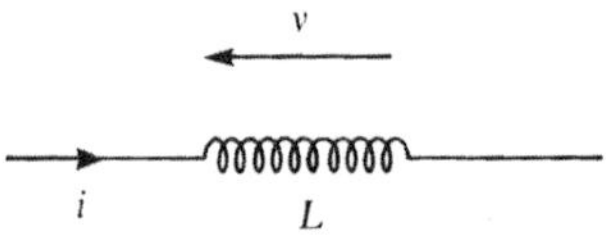

Figure 4.3
The symbol for an inductor

Example 4.3

4.3 (a) Calculate the voltage across a 0.134 H inductor when the current in the inductor is changing at the rate of 0.5 A s^{-1}.

(b) Calculate the rate of change of current through a 0.25 H inductor when it has a voltage of 2 V across it.

Solution (a) We have $L=0.134$ and $\mathrm{d}i/\mathrm{d}t=0.5$ and so

$$v = L\frac{\mathrm{d}i}{\mathrm{d}t} = 0.134 \times 0.5 = 0.067 \text{ V}$$

(b) We have $v=2$ and $L=0.25$ and so

$$\frac{\mathrm{d}i}{\mathrm{d}t} = \frac{v}{L} = \frac{2}{0.25} = 8 \text{ A s}^{-1}$$

Self-assessment questions 4.4

1. What effect does changing the current through an inductor have?
2. What effect does coiling a wire have on the magnetic field that surrounds the wire?

Exercises 4.4

1. Calculate the voltage across an inductor of 0.5 H when the rate of change of current through the inductor is 0.1 A s^{-1}.
2. Calculate the rate of change of current through a 0.0125 H inductor when it has a voltage of 6 V across it.

4.5 Electrical systems

Before analysing some electrical systems it is necessary to introduce Kirchhoff's circuit laws.

KEY POINT

The first of these laws is **Kirchhoff's current law**. This states that:
The sum of the currents flowing into a junction equals the sum of the currents flowing out of a junction.

This law arises because electric charge cannot be created or destroyed and so any flow of charge into a junction must equal the flow of charge out of the junction.

KEY POINT

The second of the circuit laws is **Kirchhoff's voltage law**. This states that:
Around any closed loop in an electrical circuit the sum of the voltage rises must equal the sum of the voltage falls.

This law arises out of the need for there to be an energy balance in a closed loop of a circuit. A crude analogy that sometimes helps is to think of an electrical circuit as being like a roller coaster with rises in height and falls in height. For the roller coaster to work the sum of the height rises must equal the sum of the height falls or there will be a break in the level of the roller coaster at a certain point, with disastrous results.

Let us now analyse some electrical systems and obtain their system differential equations. Before doing so it is worth mentioning one convention that is commonly used by engineers. The input to a system may be a voltage signal. When this is the case an arrow is often drawn on a sketch of the circuit to indicate this input signal, rather than drawing a voltage source of the type shown in Figure 4.1. One reason for this is that it is common to cascade several of these systems to form a larger system. As such, the output from one **stage** may form the input to the next stage and so the voltage source for a particular stage may arise from an earlier stage, rather than from a simple voltage source. We shall use this convention.

Example

4.4 Obtain a system differential equation for the circuit shown in Figure 4.4. The system input is the voltage v_i and the system output is the voltage v_o.

Solution When analysing this circuit we make the assumption that no current is being drawn by an external load and so the current flowing through the resistor is

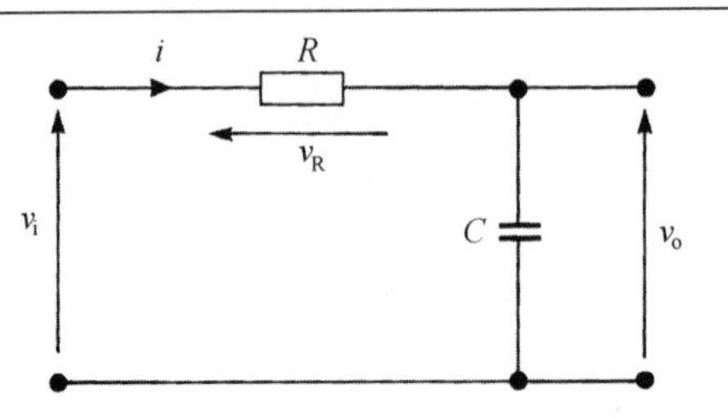

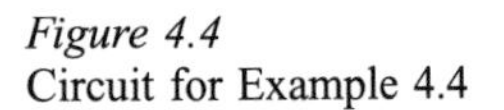
Figure 4.4
Circuit for Example 4.4

equal to the current flowing through the capacitor. Using Kirchhoff's voltage law we have

$$v_i = v_R + v_o \qquad \text{Eqn. [4.1]}$$

Using Ohm's law for the resistor gives

$$v_R = iR \qquad \text{Eqn. [4.2]}$$

Finally, for the capacitor we have

$$i = C\frac{dv_o}{dt} \qquad \text{Eqn. [4.3]}$$

We now have three equations and two intermediate variables, i and v_R, that we wish to eliminate, and so this is possible. Combining Eqn. [4.1] and Eqn. [4.2] to eliminate v_R we obtain

$$v_i = iR + v_o \qquad \text{Eqn. [4.4]}$$

We now have two unused equations, namely, Eqns [4.3] and [4.4] and one intermediate variable to eliminate. Combining Eqn. [4.3] and Eqn. [4.4] to eliminate i we obtain

$$v_i = RC\frac{dv_o}{dt} + v_o$$

This is the system differential equation with input v_i and output v_o.

Example

4.5

Obtain a system differential equation for the circuit shown in Figure 4.5. The system input is the voltage v_i and the system output is the voltage v_o.

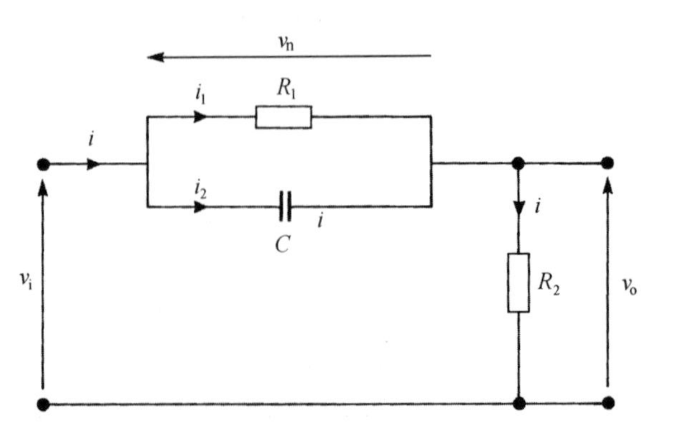

Figure 4.5
Circuit for Example 4.5

Solution The appropriate currents and voltages have been labelled in Figure 4.5. v_n is the voltage across the RC network. Using Kirchhoff's current law we have

$$i = i_1 + i_2 \qquad \text{Eqn. [4.5]}$$

Using Ohm's law for R_1 we have

$$v_n = i_1 R_1 \qquad \text{Eqn. [4.6]}$$

Using Ohm's law for R_2 we have

$$v_o = iR_2 \qquad \text{Eqn. [4.7]}$$

For the capacitor we have

$$i_2 = C\frac{dv_n}{dt} \qquad \text{Eqn. [4.8]}$$

Finally, using Kirchhoff's voltage law we have

$$v_i = v_n + v_o \qquad \text{Eqn. [4.9]}$$

We now have five equations and four intermediate variables, i, i_1, i_2 and v_n, to eliminate and so this is possible. We can eliminate i_1 by combining Eqn. [4.5] and Eqn. [4.6]. This gives

$$i = \frac{v_n}{R_1} + i_2 \qquad \text{Eqn. [4.10]}$$

We can eliminate i_2 by combining Eqn. [4.8] and Eqn. [4.10]. This gives

$$i = \frac{v_n}{R_1} + C\frac{dv_n}{dt} \qquad \text{Eqn. [4.11]}$$

We now have three unused equations, namely, Eqns. [4.7], [4.9] and [4.11], and two intermediate variables left to eliminate. We can eliminate i by combining Eqn. [4.7] and Eqn. [4.11]. This gives

$$\frac{v_o}{R_2} = \frac{v_n}{R_1} + C\frac{dv_n}{dt} \qquad \text{Eqn. [4.12]}$$

Finally, in order to eliminate v_n we need to combine Eqn. [4.9] and Eqn. [4.12]. However, we need Eqn. [4.9] in a differentiated form to do this. Therefore differentiating Eqn. [4.9] we have

$$\frac{dv_i}{dt} = \frac{dv_n}{dt} + \frac{dv_o}{dt} \qquad \text{Eqn. [4.9a]}$$

Finally combining Eqns [4.9], [4.9a] and [4.12] to eliminate v_n we have

$$\frac{v_o}{R_2} = \frac{v_i - v_o}{R_1} + C\left(\frac{dv_i}{dt} - \frac{dv_o}{dt}\right)$$

Multiplying by R_1R_2 gives

$$R_1 v_o = R_2(v_i - v_o) + R_1R_2C\left(\frac{dv_i}{dt} - \frac{dv_o}{dt}\right)$$

$$R_1R_2C\frac{dv_i}{dt} + R_2 v_i = R_1R_2C\frac{dv_o}{dt} + (R_1 + R_2)v_o$$

This is the system differential equation with input v_i and output v_o.

Example

4.6 Obtain a system differential equation for the circuit shown in Figure 4.6. The system input is the voltage v_i and the system output is the voltage v_o.

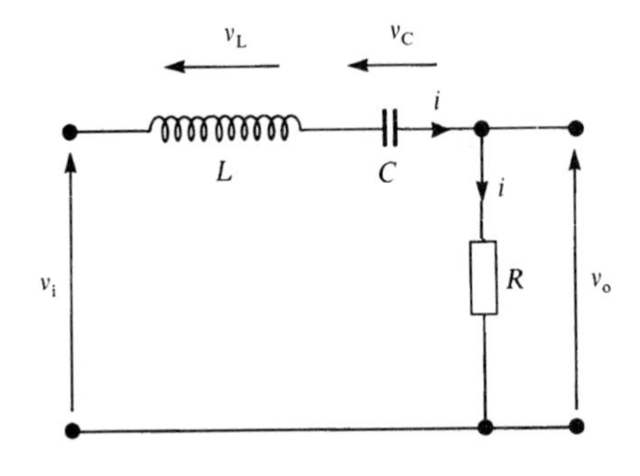

Figure 4.6
Circuit for Example 4.6

Solution Again we assume no load current is being drawn. Using Kirchhoff's voltage law we have

$$v_i = v_L + v_C + v_o \qquad \text{Eqn. [4.13]}$$

Using Ohm's law for the resistor gives

$$v_o = iR \qquad \text{Eqn. [4.14]}$$

For the inductor we have

$$v_L = L\frac{di}{dt} \qquad \text{Eqn. [4.15]}$$

For the capacitor we have

$$i = C\frac{dv_C}{dt} \qquad \text{Eqn. [4.16]}$$

We have four equations and three intermediate variables, v_L, v_C and i, that we wish to eliminate and so this is possible. Combining Eqn. [4.13] and Eqn. [4.15] to eliminate v_L we obtain

$$v_i = L\frac{di}{dt} + v_C + v_o \qquad \text{Eqn. [4.17]}$$

By examining Eqn. [4.16] we observe that v_C has been differentiated and so if we wish to eliminate it we must first differentiate Eqn. [4.17]. We then have

$$\frac{dv_i}{dt} = L\frac{d^2i}{dt^2} + \frac{dv_C}{dt} + \frac{dv_o}{dt} \qquad \text{Eqn. [4.17a]}$$

This has been labelled Eqn. [4.17a] because it is essentially the same as Eqn. [4.17]. Combining Eqn. [4.16] and Eqn. [4.17a] to eliminate v_C we obtain

$$\frac{dv_i}{dt} = L\frac{d^2i}{dt^2} + \frac{i}{C} + \frac{dv_o}{dt} \qquad \text{Eqn. [4.18]}$$

Finally, to eliminate i we need to combine Eqn. [4.14] and Eqn. [4.18]. Before doing so we note that Eqn. [4.18] has i differentiated twice and so we need to differentiate Eqn. [4.14] twice. This gives

$$\frac{d^2v_o}{dt^2} = R\frac{d^2i}{dt^2} \qquad \text{Eqn. [4.14a]}$$

Finally, combining Eqns [4.14a] and [4.18] to eliminate i gives

$$\frac{dv_i}{dt} = \frac{L}{R}\frac{d^2v_o}{dt^2} + \frac{v_o}{RC} + \frac{dv_o}{dt}$$

$$RC\frac{dv_i}{dt} = LC\frac{d^2v_o}{dt^2} + RC\frac{dv_o}{dt} + v_o$$

This is the system differential equation.

Self-assessment questions 4.5

1. State Kirchhoff's current law.
2. Give a simple analogy that helps to explain Kirchhoff's voltage law.

Exercises 4.5

1. Obtain a differential equation relating the input voltage v_i and the output voltage v_o for the system shown in Figure 4.7. Assume that the load current is zero.
2. Obtain a differential equation relating the input voltage v_i and the output voltage v_o for the system shown in Figure 4.8. Assume that the load current is zero.

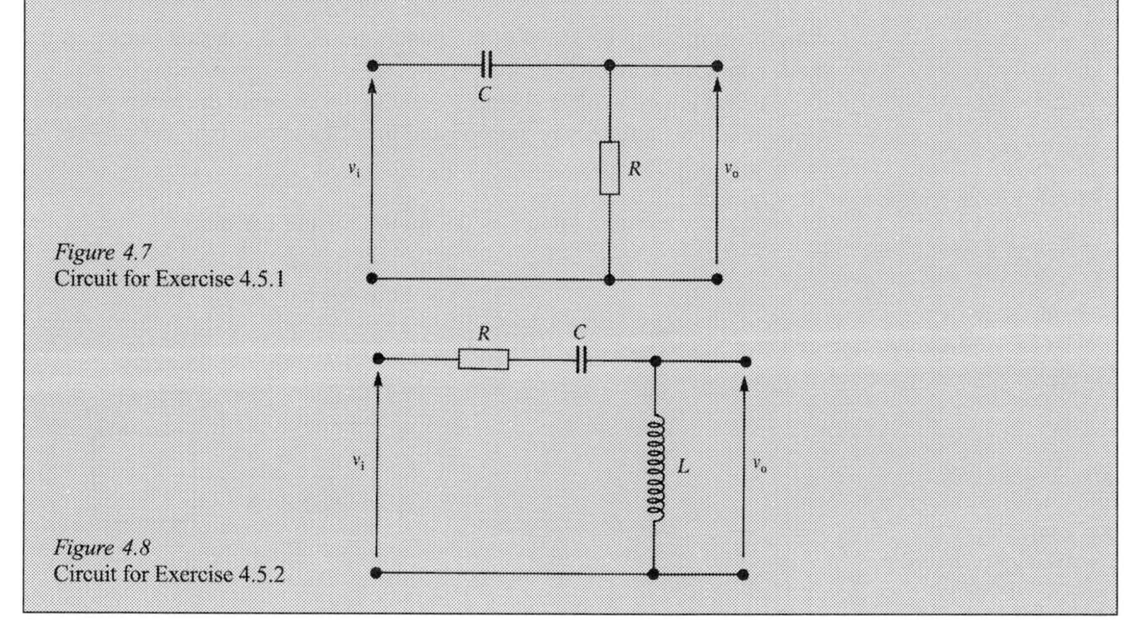

Figure 4.7
Circuit for Exercise 4.5.1

Figure 4.8
Circuit for Exercise 4.5.2

4.6 Electromechanical systems

Let us now examine electromechanical systems. There are many different types of electromechanical systems. These include systems containing motors, systems containing generators as well as systems containing a mixture of the two. In addition, the machines may be alternating current or direct current. Space only permits a discussion of systems containing a d.c. motor, but the principles are the same for other types of electromechanical systems. A d.c. motor is so called because it is connected to a direct current power supply. It consists of two main components. There is a rotating component, known as the **armature**, which consists of a coil of wire mounted on a mechanical construction. The other main component is the **stator**, which contains a coil known as the **field winding**. The field winding surrounds the armature and it is the interaction between the fields of the armature and the stator that causes the motor to turn. There are several ways of connecting a d.c. motor. We shall examine the type where the speed of the motor is varied by varying the voltage on the armature circuit. This is known as **armature control**. The voltage supply to the field winding is kept constant and this is known as a **fixed field** configuration.

It is possible to change the speed of a d.c. motor by varying the field current. This is known as **field control**.

Figure 4.9 shows a schematic diagram of an armature-controlled d.c. motor connected to mechanical load. A voltage v_a is applied to the armature and this gives rise to an armature current i_a. The armature has a resistance R_a and an inductance L_a. When the motor rotates a voltage is generated which opposes the applied voltage and is known as the **back e.m.f.**, e_b. This is marked in Figure 4.9 with an arrow showing the correct polarity of this voltage. The back e.m.f. is proportional to the speed with which the motor rotates and is given by

$$e_b = K_e\omega$$

where ω is the angular speed of the motor and K_e is a constant known as the **back e.m.f. constant**.

As a result of a voltage being applied to the armature the motor produces a torque T. This torque is proportional to the armature current and is given by

$$T = K_T i_a$$

where K_T is a constant known as the **motor torque constant**.

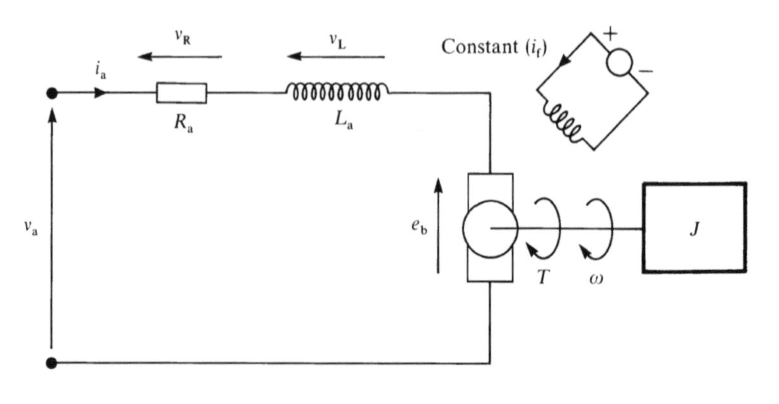

Figure 4.9
An armature-controlled d.c. motor connected to mechanical load

KEY POINT

For an armature-controlled d.c. motor:

$$e_b = K_e\omega$$
$$T = K_T i_a$$

where K_e and K_T are constants for a fixed field current.

Example

Calculate a system differential equation for the system shown in Figure 4.9. This consists of a d.c. armature-controlled motor connected to a load and the combined moment of inertia of the motor shaft and load is J. The system input is v_a and the system output is ω. Assume that the armature inductance is negligible.

Solution For convenience the appropriate system variables have been marked on the figure. The armature inductance is negligible and so we can assume $v_L = 0$. So, applying Kirchhoff's voltage law to the armature circuit gives

$$v_a = v_R + e_b \qquad \text{Eqn. [4.19]}$$

For the armature resistor we have

$$v_R = i_a R_a \qquad \text{Eqn. [4.20]}$$

We know that for an armature-controlled d.c. motor,

$$e_b = K_e\omega \qquad \text{Eqn. [4.21]}$$

and

$$T = K_T i_a \qquad \text{Eqn. [4.22]}$$

Finally, examining the torques on the mass we note that the driving torque T is opposed by the inertia torque $J(\mathrm{d}\omega/\mathrm{d}t)$, and so we can write

$$T - J\frac{\mathrm{d}\omega}{\mathrm{d}t} = 0$$

and hence

$$T = J\frac{\mathrm{d}\omega}{\mathrm{d}t} \qquad \text{Eqn. [4.23]}$$

We now have five equations and four intermediate variables, v_R, e_b, i_a and T, to eliminate and so this is possible. First of all we can eliminate v_R by combining Eqn. [4.19] and Eqn. [4.20]. This gives

$$v_a = i_a R_a + e_b \qquad \text{Eqn. [4.24]}$$

Combining Eqn. [4.21] and Eqn. [4.24] to eliminate e_b gives

$$v_a = i_a R_a + K_e\omega \qquad \text{Eqn. [4.25]}$$

Combining Eqn. [4.22] and Eqn. [4.23] to eliminate T gives

$$K_T i_a = J\frac{d\omega}{dt} \qquad \text{Eqn. [4.26]}$$

Finally, combining Eqn. [4.25] and Eqn. [4.26] to eliminate i_a gives

$$v_a = \frac{JR_a}{K_T}\frac{d\omega}{dt} + K_e\omega$$

$$K_T v_a = JR_a\frac{d\omega}{dt} + K_e K_T \omega$$

This is the system differential equation.

Let us now examine a more complicated electromechanical system.

Example

4.8 Figure 4.10 shows an armature-controlled d.c. motor connected to a load that has rotational damping. Control of the position of the motor is required and so the system input is v_a and the system output is the angular position of the motor, θ. The motor has an armature inductance L_a and the load is seated in bearings with a damping coefficient B. Derive a system differential equation.

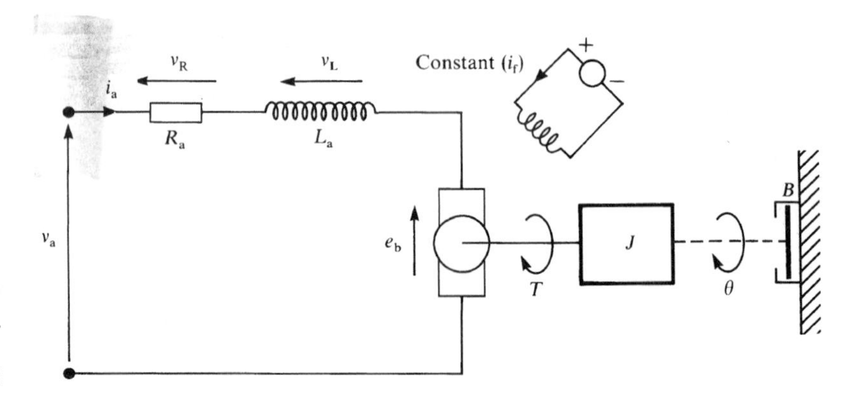

Figure 4.10
An armature-controlled d.c. motor connected to a load that has rotational damping

Solution Applying Kirchhoff's voltage law to the armature circuit gives

$$v_a = v_R + v_L + e_b \qquad \text{Eqn. [4.27]}$$

For the armature resistor we have

$$v_R = i_a R_a \qquad \text{Eqn. [4.28]}$$

For the armature inductor we have

$$v_L = L_a\frac{di_a}{dt} \qquad \text{Eqn. [4.29]}$$

For an armature-controlled d.c. motor we have

$$e_b = K_e \omega$$

That is,

$$e_b = K_e \frac{d\theta}{dt} \qquad \text{Eqn. [4.30]}$$

and

$$T = K_T i_a \qquad \text{Eqn. [4.31]}$$

Finally, examining the torques on the mass we note that the driving torque T is opposed by the damping torque $B(d\theta/dt)$ and the inertia torque $J(d^2\theta/dt^2)$, and so we can write

$$T - B\frac{d\theta}{dt} - J\frac{d^2\theta}{dt^2} = 0$$

and so

$$T = B\frac{d\theta}{dt} + J\frac{d^2\theta}{dt^2} \qquad \text{Eqn. [4.32]}$$

We have six equations and five intermediate variables to eliminate, namely, v_R, v_L, e_b, i_a, T, and so this is possible. Substituting Eqn. [4.28] and Eqn. [4.29] into Eqn. [4.27] to eliminate v_R and v_L we obtain

$$v_a = i_a R_a + L_a \frac{di_a}{dt} + e_b \qquad \text{Eqn. [4.33]}$$

Combining Eqn. [4.30] and Eqn. [4.33] to eliminate e_b we have

$$v_a = i_a R_a + L_a \frac{di_a}{dt} + K_e \frac{d\theta}{dt} \qquad \text{Eqn. [4.34]}$$

Combining Eqn. [4.31] with Eqn. [4.32] to eliminate T we obtain

$$K_T i_a = B\frac{d\theta}{dt} + J\frac{d^2\theta}{dt^2}$$

$$i_a = \frac{B}{K_T}\frac{d\theta}{dt} + \frac{J}{K_T}\frac{d^2\theta}{dt^2} \qquad \text{Eqn. [4.35]}$$

Finally, we need to eliminate i_a. We note that it appears in both its native and differentiated form in Eqn. [4.34]. Therefore we need to differentiate Eqn. [4.35]. This gives

$$\frac{di_a}{dt} = \frac{B}{K_T}\frac{d^2\theta}{dt^2} + \frac{J}{K_T}\frac{d^3\theta}{dt^3} \qquad \text{Eqn. [4.35a]}$$

Finally, substituting Eqn. [4.35] and Eqn. [4.35a] into Eqn. [4.34] we obtain

$$v_a = \left(\frac{B}{K_T}\frac{d\theta}{dt} + \frac{J}{K_T}\frac{d^2\theta}{dt^2}\right)R_a + L_a\left(\frac{B}{K_T}\frac{d^2\theta}{dt^2} + \frac{J}{K_T}\frac{d^3\theta}{dt^3}\right) + K_e\frac{d\theta}{dt}$$

$$K_T v_a = L_a J\frac{d^3\theta}{dt^3} + (L_a B + JR_a)\frac{d^2\theta}{dt^2} + (BR_a + K_e K_T)\frac{d\theta}{dt}$$

This is the system differential equation. We note that this is a third order system.

Self-assessment questions 4.5

1. Explain what is meant by the back e.m.f. of a d.c. motor.
2. State the equations that model the behaviour of an armature-controlled d.c. motor.

Exercises 4.6

1. An armature-controlled d.c. motor, with negligible armature inductance, is connected to a rotational mechanical system. The arrangement is shown in Figure 4.11. Derive a system differential equation relating the armature voltage v_a to the angular speed of the load, ω.

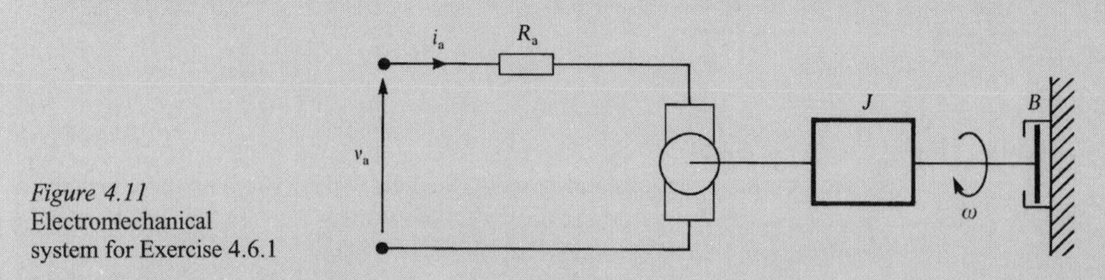

Figure 4.11
Electromechanical system for Exercise 4.6.1

2. An armature-controlled d.c. motor is connected to a rotational mechanical system as shown in Figure 4.12. Derive a system differential equation relating the armature voltage v_a to the angular speed of the load, ω_0. The angular speed of this motor is ω.

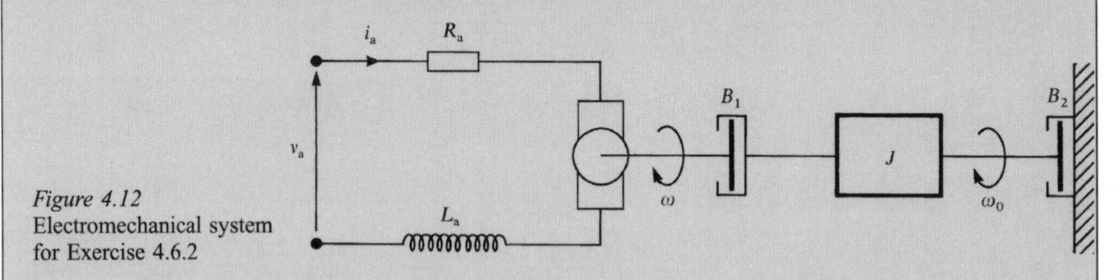

Figure 4.12
Electromechanical system for Exercise 4.6.2

Test and assignment exercises 4

1. Derive a system differential equation for the electrical system shown in Figure 4.13. The system input is v_i and the system output is v_o.

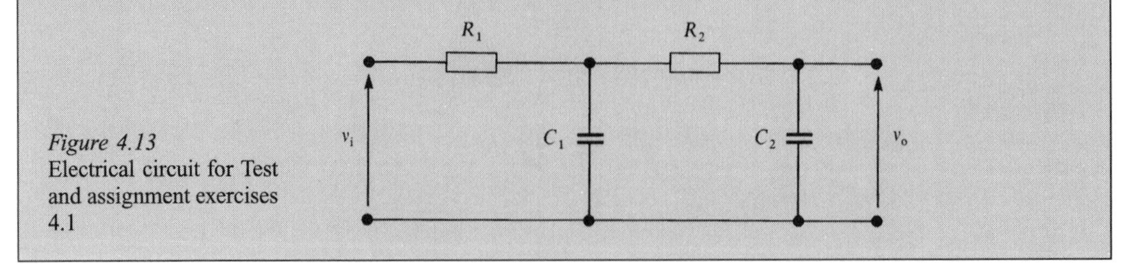

Figure 4.13
Electrical circuit for Test and assignment exercises 4.1

2. Derive a system differential equation for the electrical system shown in Figure 4.14. The system input is v_i and the system output is v_o.

3. Derive a system differential equation for the electromechanical system shown in Figure 4.15. The system input is the armature voltage v_a and the system output is the angular position of the load, θ_0. The angular position of the motor is θ.

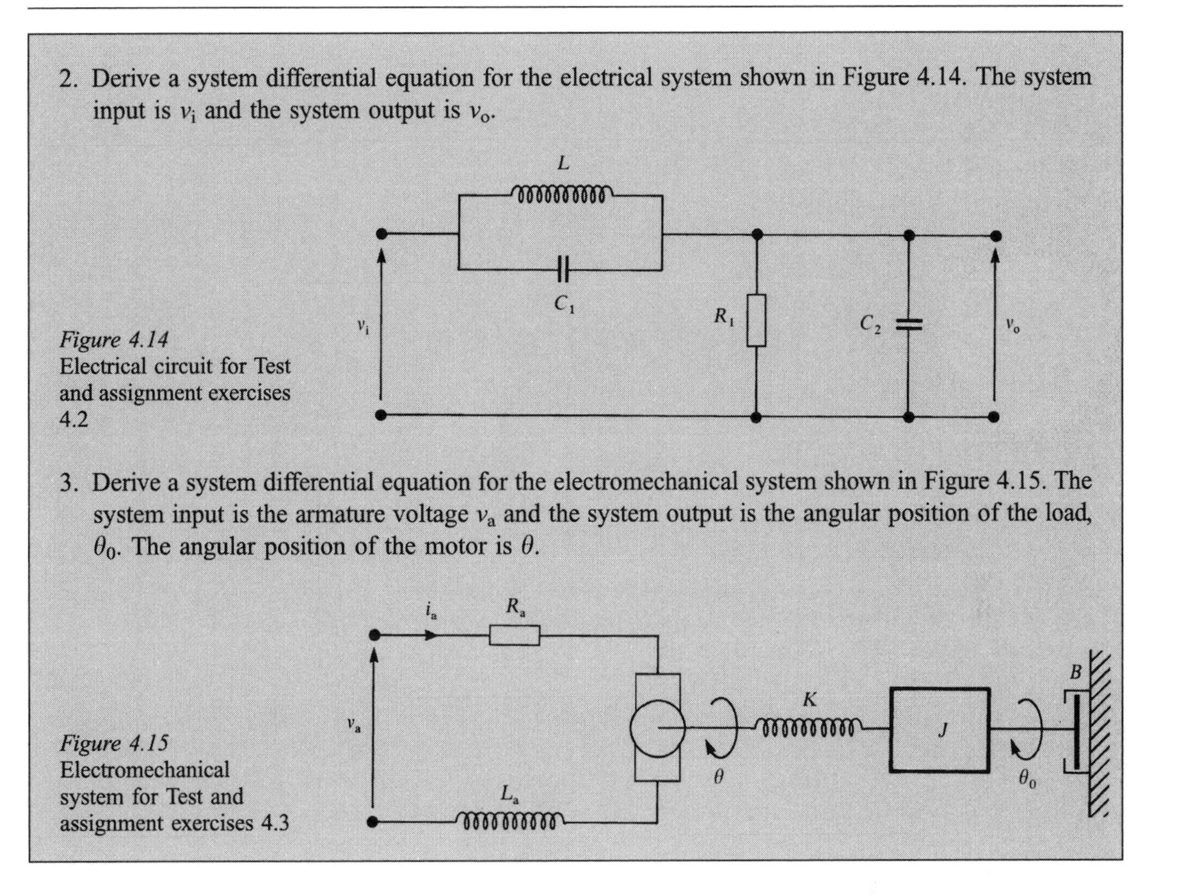

Figure 4.14
Electrical circuit for Test and assignment exercises 4.2

Figure 4.15
Electromechanical system for Test and assignment exercises 4.3

5 Liquid systems

Objectives

This chapter:

- describes the model for the behaviour of a storage tank
- describes the model for the behaviour of a flow resistor
- derives mathematical models for some liquid systems
- explains the limitations of the linear models used

5.1 Introduction

Liquid systems form a small part of the more general topic of fluid systems. The term **fluid** is used to describe materials that can significantly change their shape. There are two types of fluids, namely, gases and liquids. A **gas** is a compressible fluid. A **liquid** is an incompressible fluid which is not significantly affected by temperature. We are confining our attention to liquid systems for reasons of space and complexity. Liquid systems are commonly encountered in many industrial settings. For example, the chemical industry makes extensive use of storage tanks to store chemicals which are then mixed together to allow chemical reactions to occur. A major problem with liquid systems is that many of the components, for example pipes and valves, are nonlinear and so obtaining a linear model can be difficult, if not impossible, in many cases. We shall only examine two system components, namely, the storage tank and the flow resistor. Other components do exist but modelling them requires a treatment beyond the scope of this book.

5.2 The storage tank

Consider Figure 5.1 which shows a **storage tank** capable of holding a liquid. The tank has vertical sides and the cross-sectional area of the tank is A. The

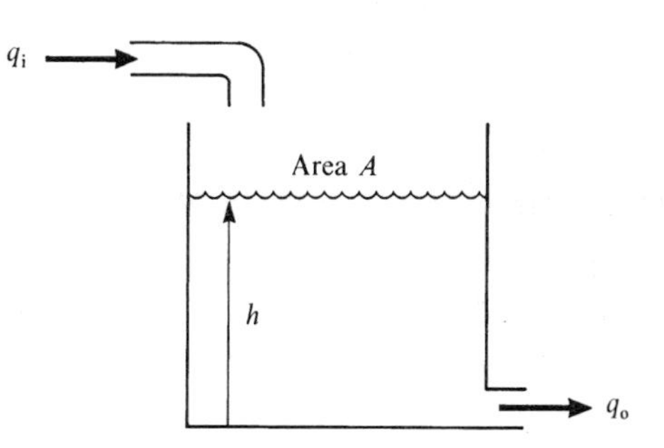

Figure 5.1
A liquid storage tank

height of the liquid in the tank is h. The flow rate of liquid into the tank is q_i and the flow rate of liquid out of the tank is q_o. The total volume of liquid in the tank is V. The SI unit for flow rate is metre^3.second^{-1}.

If we examine the behaviour of the tank then we see that if liquid is flowing into and out of the tank then the difference between these two flow rates gives the rate at which liquid is being added to the volume already in the tank. In other words, it gives the rate of increase of the volume of liquid in the tank.

KEY POINT

For a storage tank we can write

$$q_i - q_o = \frac{dV}{dt} \qquad \text{Eqn. [5.1]}$$

where q_i is the flow rate into the tank, q_o is the flow rate out of the tank and V is the volume of liquid in the tank.

This law is known as the **conservation of mass**. Note that the tank has vertical sides and so the volume of liquid in the tank is equal to the height of liquid in the tank multiplied by its cross-sectional area. So we have

$$V = Ah \qquad \text{Eqn. [5.2]}$$

Furthermore the cross-sectional area is a constant and independent of the height of the liquid. Combining Eqn. [5.1] and Eqn. [5.2] we obtain

$$q_i - q_o = \frac{dV}{dt} = \frac{d(Ah)}{dt} = A\frac{dh}{dt}$$

because A is a constant.

KEY POINT

For a storage tank with vertical sides:

$$q_i - q_o = A\frac{dh}{dt}$$

where A is the cross-sectional area of the tank and h is the height of liquid in the tank.

Self-assessment questions 5.2

1. State the law of conservation of mass for a liquid storage tank.
2. What simplification can be made to the law of conservation of mass when a storage tank has vertical sides?

5.3 The flow resistor

The pressure drop is due to a loss of energy in the liquid as a result of frictional losses.

Whenever a liquid flows through a component such as a pipe, valve or orifice there is some resistance to the flow of the liquid, which leads to a pressure drop in the liquid across the component. Such a component is known as a **flow resistor**. Consider Figure 5.2 which shows liquid flowing through a pipe that contains a restriction. The flow rate of the liquid is q. The pressure upstream of the restriction is p_1 and the pressure downstream of the restriction is p_2.

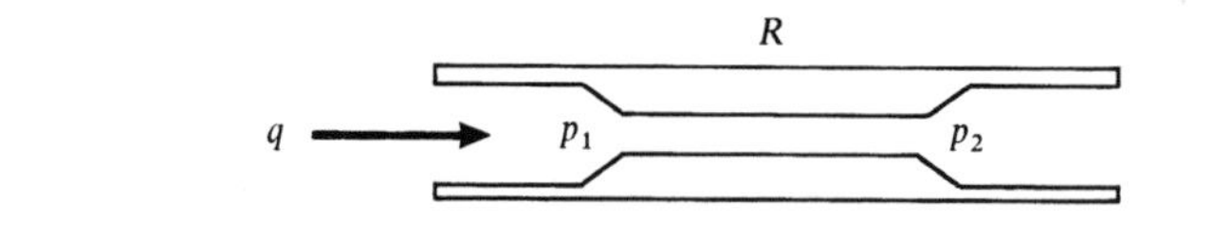

Figure 5.2
A flow resistor

KEY POINT

For the flow resistor with upstream pressure p_1 and downstream pressure p_2 we can write

$$p_1 - p_2 = qR$$

where R is the **resistance** of the flow resistor and q is the flow rate of liquid through the flow resistor.

The SI unit of resistance is newton.second.metre^{-5}. For simple models R is taken to be a constant. When R is constant, then the flow resistor is a **linear component**. We shall adopt this approach. In practice the value of R often varies with q for many flow resistors, that is, the resistor is a **nonlinear component**. Analysis of systems containing nonlinear components is beyond the scope of this book.

This law is the liquid equivalent of Ohm's law. We see that the liquid pressure drop corresponds to a voltage drop, the liquid flow rate corresponds to a current and the fluid resistance is equivalent to an electrical resistance.

Self-assessment questions 5.3

1. State the law relating pressure drop and flow rate for a linear flow resistor.
2. Explain the analogy between Ohm's law and the flow of liquid through a resistor.

5.4 Liquid systems

We are now in a position to examine some liquid systems and to obtain their system differential equations. Before doing so it is worth stating the formula for the pressure due to a column of liquid as we shall be making frequent use of it.

KEY POINT

For a column of liquid of density ρ and height h the pressure p at the bottom of the column of liquid is given by

$$p = \rho g h$$

where g is the acceleration due to gravity

Example

5.1 Consider the system shown in Figure 5.3. The flow out of the tank is restricted by a valve of resistance R which can be assumed to be a constant. Note the symbol for a valve. The liquid flow rate into the tank is q_i and the liquid flow rate out of the tank is q_o. The height of the liquid in the tank is h and the tank

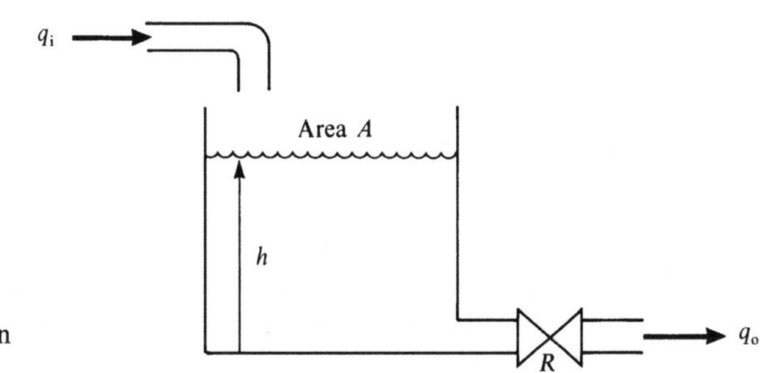

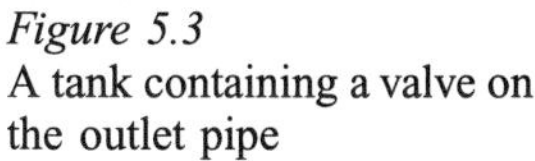
Figure 5.3
A tank containing a valve on the outlet pipe

has vertical sides and a cross-sectional area A. Obtain a system differential equation for the system with q_i as the input and q_o as the output.

Solution Using the law of conservation of mass for the tank we can write

$$q_i - q_o = A\frac{dh}{dt} \qquad \text{Eqn. [5.3]}$$

If we examine the valve then we see that the pressure on the valve upstream of the flow is atmospheric pressure plus the pressure due to the liquid. Downstream of the valve the only pressure is that due to the atmosphere. Therefore the pressure difference across the valve is ρgh where ρ is the density of the liquid, g is the acceleration due to gravity and h is the height of the liquid. So for the valve we can write

$$(\rho gh + \text{atmospheric pressure} - \text{atmospheric pressure}) = Rq_o$$

and so

$$\rho gh = Rq_o \qquad \text{Eqn. [5.4]}$$

It is important to be aware of the effect of atmospheric pressure but usually, as in this case, it cancels out. From now on we shall assume this to be the case unless specifically stated otherwise. In order to obtain a system differential equation we need to combine Eqn. [5.3] and Eqn. [5.4] to eliminate h. However, we see that in Eqn. [5.3] the variable h is in derivative form. We therefore need to differentiate Eqn. [5.4] with respect to time to obtain dh/dt. This gives

$$\rho g\frac{dh}{dt} = R\frac{dq_o}{dt}$$

Rearranging gives

$$\frac{dh}{dt} = \frac{R}{\rho g}\frac{dq_o}{dt} \qquad \text{Eqn. [5.4a]}$$

We have labelled this Eqn. [5.4a] because it is essentially the same as Eqn. [5.4]. Finally, combining Eqn. [5.3] and Eqn. [5.4a] we obtain

$$q_i - q_o = \frac{AR}{\rho g}\frac{dq_o}{dt}$$

and so

$$q_i = \frac{AR}{\rho g}\frac{dq_o}{dt} + q_o$$

The word capacitance is used because a tank stores liquid in the same way that an electrical capacitor stores charge.

It is convenient to define a quantity $C = A/\rho g$, known as the **tank capacitance**. The SI unit of tank capacitance is metre^5.newton^{-1}. We then have

$$q_i = CR\frac{dq_o}{dt} + q_o$$

This is the system differential equation. Note the similarity between this equation and that of the RC circuit analysed in Example 4.4.

KEY POINT

The capacitance C of a storage tank with vertical sides is given by

$$C = \frac{A}{\rho g}$$

where A is the cross-sectional area of the tank, ρ is the density of the liquid and g is the acceleration due to gravity.

We now examine a more complicated system.

Example

5.2 **A coupled tank system**

This arrangement is like a small scale version of a chemical process line.

Consider the system shown in Figure 5.4. Two tanks are coupled together. They have cross-sectional areas A_1 and A_2. The height of liquid in the tanks is h_1 and h_2. They have outlet valves with resistances, R_1 and R_2. Liquid flows into tank 1 with a flow rate q_i. Liquid flows out of tank 2 with a flow rate q_o. The flow rate of liquid between tank 1 and tank 2 is q. Obtain a system differential equation relating the input flow q_i to the output flow q_o.

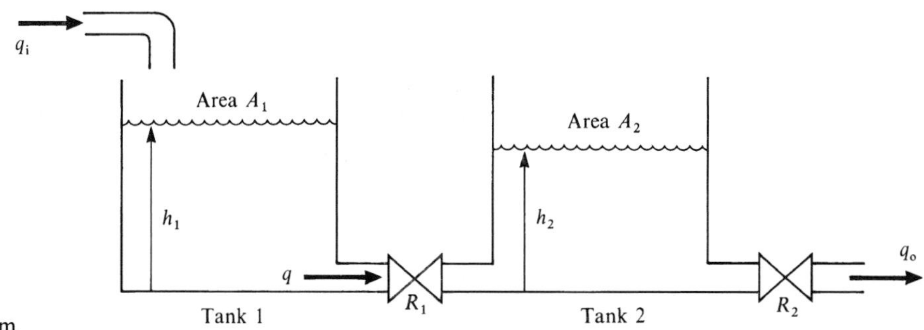

Figure 5.4
A coupled tank system

Solution Here we have a reasonably complicated system, so care must be taken to be systematic in the setting up of the system equations and the elimination of intermediate variables. For tank 1, using the conservation of mass, we have

$$q_i - q = A_1 \frac{dh_1}{dt} \qquad \text{Eqn. [5.5]}$$

For the outlet valve from tank 1 we have a pressure difference of $\rho g h_1 - \rho g h_2$ and so we can write

$$\rho g h_1 - \rho g h_2 = R_1 q$$

and so

$$\rho g(h_1 - h_2) = R_1 q \qquad \text{Eqn. [5.6]}$$

For tank 2, using the conservation of mass, we have

$$q - q_o = A_2 \frac{dh_2}{dt} \qquad \text{Eqn. [5.7]}$$

For the outlet valve from tank 2 we have

$$\rho g h_2 = R_2 q_o \qquad \text{Eqn. [5.8]}$$

There are therefore four system equations and we need to eliminate three intermediate variables, namely, h_1, h_2 and q, and so this is possible. Some variables occur in native and in derivative form. We leave the elimination of these until later as they tend to be more difficult to deal with. Therefore we first eliminate q. We see this occurs in Eqn. [5.5], Eqn. [5.6] and Eqn. [5.7]. For convenience we arrange Eqn. [5.6] to give

$$q = \frac{\rho g}{R_1}(h_1 - h_2) \qquad \text{Eqn. [5.6a]}$$

We have labelled this Eqn. [5.6a] as it is essentially the same equation as Eqn. [5.6]. We now substitute this in turn into Eqn. [5.5] and Eqn. [5.7]. This gives

$$q_i - \frac{\rho g}{R_1}(h_1 - h_2) = A_1 \frac{dh_1}{dt} \qquad \text{Eqn. [5.9]}$$

and

$$\frac{\rho g}{R_1}(h_1 - h_2) - q_o = A_2 \frac{dh_2}{dt} \qquad \text{Eqn. [5.10]}$$

We have now eliminated q and used Equations [5.5], [5.6] and [5.7]. We have Eqn. [5.8] left and have created two new equations, namely, Eqns [5.9] and [5.10]. It is important not to reuse old equations as this will reintroduce variables that have already been eliminated. So now we have three equations left and two intermediate variables to eliminate, namely, h_1 and h_2. If we rearrange Eqn. [5.8] then we have

$$h_2 = \frac{R_2}{\rho g} q_o \qquad \text{Eqn. [5.8a]}$$

and we can substitute this into Eqn. [5.9] and Eqn. [5.10] to eliminate h_2. Before doing so we note that h_2 occurs on its own and in derivative form. We therefore also need to use the differentiated form of Eqn. [5.8a], that is,

$$\frac{dh_2}{dt} = \frac{R_2}{\rho g} \frac{dq_o}{dt} \qquad \text{Eqn. [5.8b]}$$

So substituting Eqn. [5.8a] and Eqn. [5.8b] into Eqn. [5.9] and Eqn. [5.10] in turn we obtain

$$q_i - \frac{\rho g}{R_1}\left(h_1 - \frac{R_2}{\rho g} q_o\right) = A_1 \frac{dh_1}{dt} \qquad \text{Eqn. [5.11]}$$

and also

$$\frac{\rho g}{R_1}\left(h_1 - \frac{R_2}{\rho g} q_o\right) - q_o = \frac{A_2 R_2}{\rho g} \frac{dq_o}{dt} \qquad \text{Eqn. [5.12]}$$

We now have two equations left and one intermediate variable to eliminate, that is, h_1. We need to rearrange Eqn. [5.12] to make the h_1 term the subject of the equation. First, removing brackets,

$$\frac{\rho g}{R_1}h_1 - \frac{\rho g R_2}{\rho g R_1}q_o - q_o = \frac{A_2 R_2}{\rho g}\frac{dq_o}{dt}$$

Then collecting terms involving q_o

$$\frac{\rho g}{R_1}h_1 = q_o\left(1 + \frac{R_2}{R_1}\right) + \frac{A_2 R_2}{\rho g}\frac{dq_o}{dt}$$

Rearranging gives

$$\frac{\rho g}{R_1}h_1 = q_o\left(\frac{R_1 + R_2}{R_1}\right) + \frac{A_2 R_2}{\rho g}\frac{dq_o}{dt}$$

Multiplying by $R_1/\rho g$ we get

$$h_1 = \frac{R_1 + R_2}{\rho g}q_o + \frac{R_1 A_2 R_2}{(\rho g)^2}\frac{dq_o}{dt} \qquad \text{Eqn. [5.12a]}$$

We also need a derivative form of this equation to eliminate dh_1/dt. So differentiating Eqn. [5.12a] gives

$$\frac{dh_1}{dt} = \frac{R_1 + R_2}{\rho g}\frac{dq_o}{dt} + \frac{R_1 A_2 R_2}{(\rho g)^2}\frac{d^2 q_o}{dt^2} \qquad \text{Eqn. [5.12b]}$$

Finally we substitute Eqn. [5.12a] and Eqn. [5.12b] into Eqn. [5.11] to eliminate h_1 and obtain

$$q_i - \frac{\rho g}{R_1}\left[\frac{R_1 + R_2}{\rho g}q_o + \frac{R_1 A_2 R_2}{(\rho g)^2}\frac{dq_o}{dt} - \frac{R_2}{\rho g}q_o\right] = A_1\left[\frac{R_1 + R_2}{\rho g}\frac{dq_o}{dt} + \frac{R_1 A_2 R_2}{(\rho g)^2}\frac{d^2 q_o}{dt^2}\right]$$

Removing the square brackets gives

$$q_i - \frac{R_1 + R_2}{R_1}q_o - \frac{A_2 R_2}{\rho g}\frac{dq_o}{dt} + \frac{R_2}{R_1}q_o = \frac{A_1(R_1 + R_2)}{\rho g}\frac{dq_o}{dt} + \frac{A_1 R_1 A_2 R_2}{(\rho g)^2}\frac{d^2 q_o}{dt^2}$$

Making q_i the subject of the equation gives

$$q_i = \frac{A_1 R_1 A_2 R_2}{(\rho g)^2}\frac{d^2 q_o}{dt^2} + \frac{A_2 R_2 + A_1 R_1 + A_1 R_2}{\rho g}\frac{dq_o}{dt} + \frac{R_1 + R_2 - R_2}{R_1}q_o$$

Finally, multiplying by $(\rho g)^2$ gives

$$(\rho g)^2 q_i = A_1 R_1 A_2 R_2 \frac{d^2 q_o}{dt^2} + \rho g(A_1 R_1 + A_1 R_2 + A_2 R_2)\frac{dq_o}{dt} + (\rho g)^2 q_o$$

This is the system differential equation.

Self-assessment questions 5.4

1. Explain why it is possible to ignore atmospheric pressure when calculating the pressure difference across a tank outlet valve.
2. State the formula for the capacitance of a tank.

Exercises 5.4

1. Obtain a system differential equation for the system shown in Figure 5.5. Assume the input to the system is the liquid flow rate q_i and the output from the system is the liquid flow rate q_o.

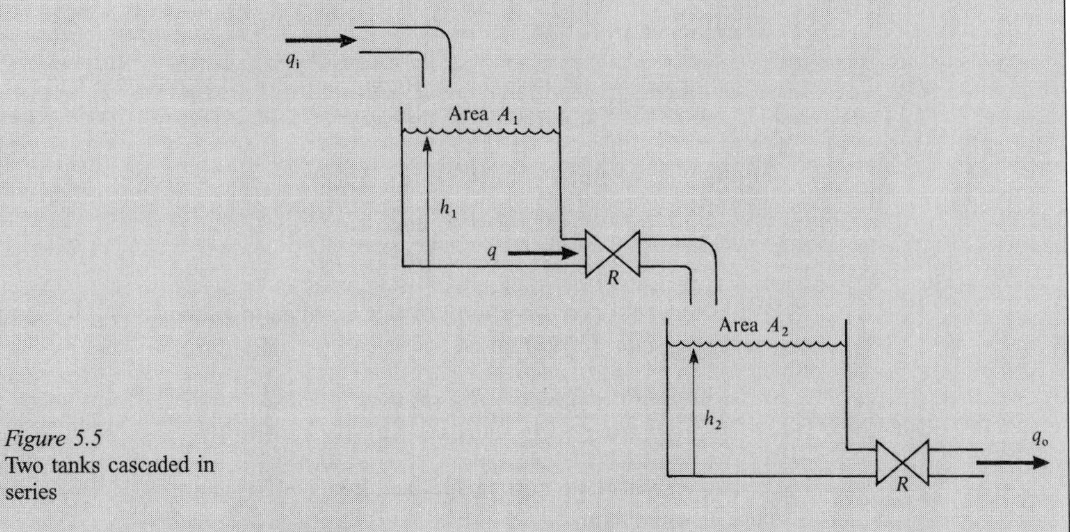

Figure 5.5
Two tanks cascaded in series

Test and assignment exercises 5

1. Obtain a system differential equation for the system shown in Figure 5.6. Assume the input to the system is the liquid flow rate q_i and the output from the system is the liquid flow rate q_o.
2. Obtain a system differential equation for the system shown in Figure 5.7. Assume the input to the system is the liquid flow rate q_i and the output from the system is the liquid flow rate q_o.

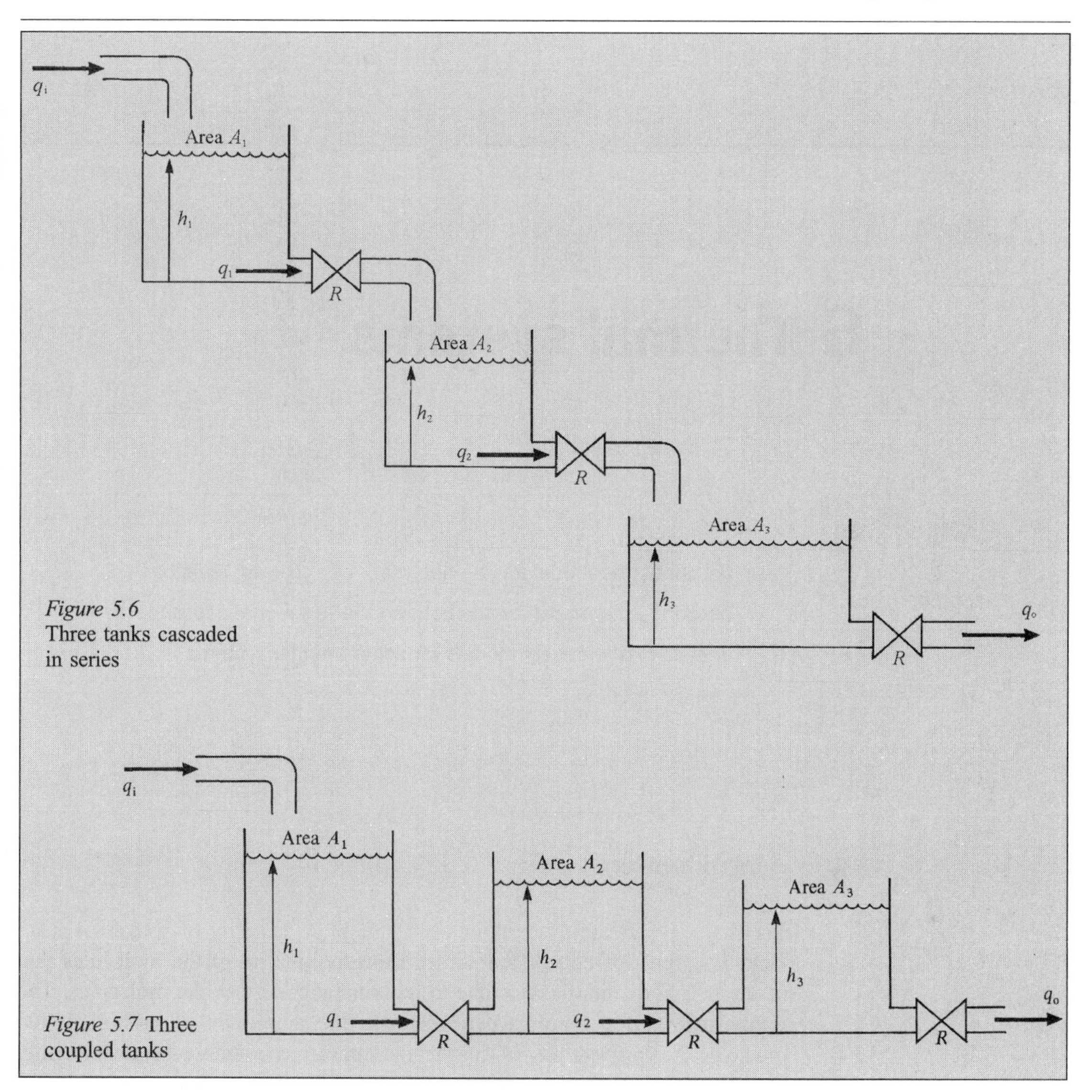

Figure 5.6
Three tanks cascaded in series

Figure 5.7 Three coupled tanks

6 Thermal systems

Objectives

This chapter:

- describes the model for the behaviour of a thermal resistor
- describes the model for the behaviour of a thermal capacitor
- derives mathematical models for some thermal systems

6.1 Introduction

Thermal systems are systems in which the main quantity of interest is heat. **Heat** is a form of energy that arises from the motion of the molecules that make up a body, that is, it is due to the kinetic energy of the molecules. The **temperature** of a body is dependent on the amplitude of motion of the molecules. A large variety of thermal systems are encountered in engineering. One example is a building heated by central heating. Calculating the boiler size required for the building involves a consideration of heat losses through the walls, ceilings and floor as well as the amount of heat needed to warm up items in the building such as furniture. Such an analysis can be complicated but it is possible to develop simple models which are suitable for many requirements.

This is often the case in engineering. It is surprising how useful a very simple model can be in many situations.

6.2 The thermal resistor

Figure 6.1 shows a section of material of area A and thickness h. The temperature of one side is T_1; the temperature of the other side is T_2 and we assume $T_1 > T_2$. The rate of heat flow through this material is q and its direction is

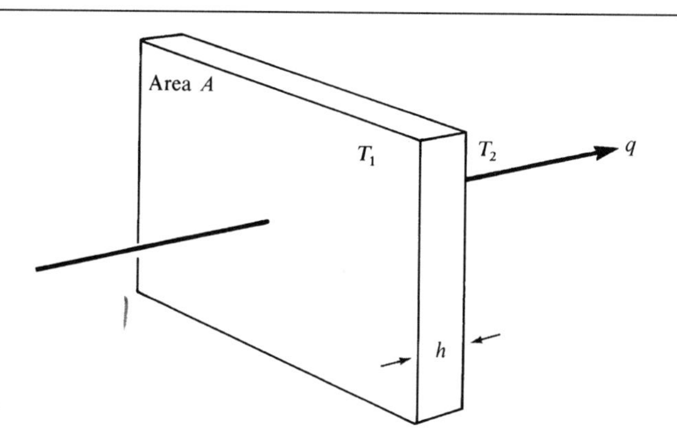

Figure 6.1
A section of material with different temperatures either side

Materials exhibit many interesting properties close to absolute zero. For example, some materials become **superconductors** and have no electrical resistance. Cables can transport electricity without any energy losses if they are made of superconducting materials.

from the high temperature side to the low temperature side. Heat is a form of energy and so the phrase 'heat flow' and 'energy flow' can be used interchangeably. They both have the SI unit of joule.second^{-1} or watt. Note the SI unit of temperature is the kelvin. This unit was chosen to measure the temperature of an object with a scale starting at **absolute zero**. Absolute zero is the lowest temperature that an object can have. It corresponds to the temperature at which the kinetic energy of a molecule of a body is zero. On this scale the melting point of ice corresponds to a temperature of 273.15 kelvin.

Many engineers use the alternative temperature measurement scale of degree Celsius (°C) . This has the same spacing as the kelvin scale but starts at 273.15 kelvin. So to convert from kelvin to degrees Celsius it is necessary to subtract the number 273.15. We shall use the SI unit of temperature measurement to maintain consistency with the other SI units used throughout the book.

The section of material shown in Figure 6.1 resists the flow of heat and so is known as a **thermal resistor**.

KEY POINT

For the thermal resistor we can state

$$T_1 - T_2 = \frac{hq}{kA} \qquad \text{Eqn. [6.1]}$$

where k is a constant known as the **thermal conductivity** of the material.

The SI unit for k is watt.metre^{-1}.kelvin^{-1}.

KEY POINT

We define the **thermal resistance** R of the section of material as

$$R = \frac{h}{kA} \qquad \text{Eqn. [6.2]}$$

The SI unit for R is kelvin.watt^{-1}. Eqn. [6.1] then becomes

KEY POINT

$$T_1 - T_2 = qR \qquad \text{Eqn. [6.3]}$$

House designers now take more care to minimise heat losses as this is a way to reduce fuel bills. As the price of fuel has increased, this has become a strong selling point in favour of a well-designed new house. A modern house is usually much more energy efficient than a house built as little as 20 years ago. One thing that has now become very popular is floor insulation. Here a layer of insulating material is put down before the floor boards are fitted.

This can be thought of as a thermal equivalent of Ohm's law. Instead of a voltage difference we have a temperature difference. This gives rise to a heat flow instead of a current flow. The amount of heat flow for a given temperature difference depends on the thermal resistance of the material.

Examining Eqn. [6.2] we see that increasing the thickness of the material increases its thermal resistance. This makes sense as we know that the loft insulation in a house is better if it is thicker and putting more blankets on a bed increases the insulating properties of the bedding. Increasing the area of the material reduces the thermal resistance. Again this is common sense because there is now a larger area over which heat can flow and so for a given temperature difference more heat will flow. This is the reason why rooms with bigger wall areas tend to require more heat to remain at a particular temperature; the heat losses are greater from a large room than from a small room. Finally, the value of thermal conductivity k affects the thermal resistance. If a material with a low value of k is used then this corresponds to a higher thermal resistance. The reason loft insulation consists of materials such as fibre glass and polystyrene is that they have a low thermal conductivity.

Example

6.1 Calculate the thermal resistance of a rectangular sheet of polystyrene of length 2.2 m, width 1.7 m and thickness 0.015 m. For polystyrene, $k = 0.08$ W m^{-1} K^{-1}.

Solution The area A of the sheet of polystyrene is

$$A = 2.2 \times 1.7 = 3.74 \text{ m}^2$$

So writing Eqn. [6.2] we have

$$R = \frac{h}{kA} = \frac{0.015}{0.08 \times 3.74} = 0.05013 = 5.013 \times 10^{-2} \text{ K m}^{-1}$$

Example

6.2 Calculate the heat flow rate through a rectangular sheet of perspex of length 2.7 m, width 1.45 m and thickness 0.06 m when the temperature difference across the sheet is 60 K. For perspex, $k = 0.2$ W m^{-1} K^{-1}.

Solution We have

$$T_1 - T_2 = 60 \text{ K}$$

The thermal resistance of the sheet of perspex is

$$R = \frac{h}{kA} = \frac{0.06}{0.2 \times 2.7 \times 1.45} = 0.07663 = 7.663 \times 10^{-2}\ \mathrm{K\ W^{-1}}$$

Recalling Eqn. [6.3] we have

$$T_1 - T_2 = qR$$
$$q = \frac{T_1 - T_2}{R}$$

So

$$q = \frac{60}{7.663 \times 10^{-2}} = 783.0\ \mathrm{W}$$

Self-assessment questions 6.2

1. Explain the analogy between Ohm's law and the law for heat flow through a material due to a temperature difference across the material.
2. Describe the ways in which the thermal resistance of a section of material can be increased.

Exercises 6.2

1. Calculate the thermal resistance of the following sections of material:

   (a) a rectangular sheet of aluminium of length 3 m, width 2 m and thickness 3×10^{-3} m (for aluminium, $k = 2.01 \times 10^{-4}\ \mathrm{W\ m^{-1}\ K^{-1}}$);
   (b) a rectangular sheet of polystyrene of length 3 m, width 2 m and thickness 3×10^{-3} m (for polystyrene, $k = 0.08\ \mathrm{W\ m^{-1}\ K^{-1}}$).
2. Calculate the heat flow rate through the following pieces of material when a temperature difference of 80 K is applied across them:

   (a) a circular sheet of rubber of area 6 $\mathrm{m^2}$ and thickness 0.01 m (for rubber, $k = 0.15\ \mathrm{W\ m^{-1}\ K^{-1}}$);
   (b) a rectangular sheet of perspex of length 2.5 m, width 1.5 m and thickness 0.03 m, (for perspex, $k = 0.2\ \mathrm{W\ m^{-1}\ K^{-1}}$).

6.3 The thermal capacitor

Figure 6.2 shows an object of volume V and density ρ. For simplicity we assume that the material has a uniform temperature T. Heat is being 'pumped' into the material at an energy rate q. This leads to an increase in the tempe-

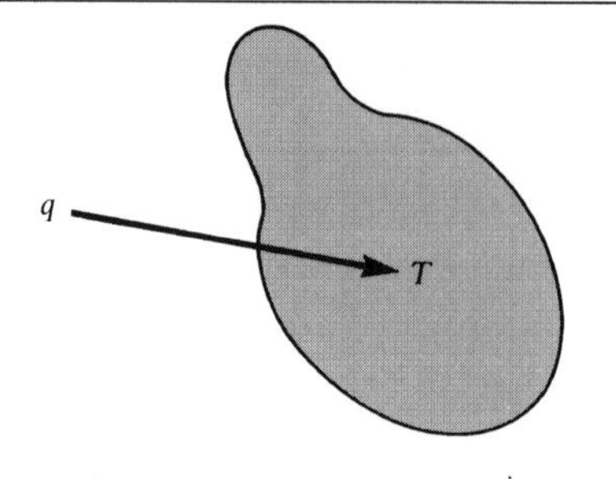

Figure 6.2
An object of volume V and density ρ

rature of the material. The material is therefore storing heat and acts as a **thermal capacitor**.

KEY POINT

Recall that mass = (density) × (volume), that is, $M = \rho V$.

For the thermal capacitor we can state:

$$q = \rho c V \frac{dT}{dt} = Mc\frac{dT}{dt} \qquad \text{Eqn. [6.4]}$$

where c is a constant known as the **specific heat capacity** of the material and M is the mass of the thermal capacitor.

The specific heat capacity of a material is the energy required to raise the temperature of 1 kilogram of the material by 1 kelvin.

The SI unit of specific heat capacity is joule.kilogram^{-1}.kelvin^{-1}. The quantity $\rho c V$ is a constant for a particular object. It is convenient to define

$$C = \rho c V = Mc \qquad \text{Eqn. [6.5]}$$

where C is a constant known as the **thermal capacitance** of the object. The SI unit of thermal capacitance is joule.kelvin^{-1}. Combining Eqns [6.4] and [6.5] gives

KEY POINT

$$q = C\frac{dT}{dt} \qquad \text{Eqn. [6.6]}$$

Water has a surprisingly high specific heat capacity compared with many materials. This is one reason why kettles always seem to take so long to boil!

Note from Eqn. [6.6] that the rate of increase in the temperature of the object is proportional to the rate of heat flow into the object. Also, an object with a high thermal capacitance requires a high rate of heat flow to raise the object temperature at a given rate. The thermal capacitance is proportional to the volume of the object. This is the reason why a kettle full of water takes longer to heat up than a kettle that is half full of water. In both cases the rate of heat input is the same but the thermal capacitance of a full kettle of water is twice that of a half full kettle of water.

Example

6.3 Calculate the thermal capacitance of 0.1 m^3 of water. For water, $\rho = 998$ kg m^{-3} and $c = 4190$ J kg^{-1} K^{-1}.

Solution Recalling Eqn. [6.5] we have

$$C = \rho c V$$

Substituting in the given values yields

$$C = 998 \times 4190 \times 0.1 = 418162 = 4.182 \times 10^5 \text{ J K}^{-1}$$

Example

6.4 Calculate the rate of heat flow needed to raise the temperature of 3 kg of mild steel at a rate of 0.1 K s^{-1}. For mild steel, $\rho = 7860$ kg m^{-3} and $c = 420$ J kg^{-1} K^{-1}.

Solution First we need to calculate the thermal capacitance of the steel. Using Eqn. [6.5] we have

$$C = Mc$$
$$= 3 \times 420$$
$$= 1260 \text{ J K}^{-1}$$

Recalling Eqn. [6.6] we have

$$q = C\frac{dT}{dt}$$

Now we require $dT/dt = 0.1$ K s^{-1}. So,

$$q = C\frac{dT}{dt} = 1260 \times 0.1 = 126.0 \text{ W}$$

Self-assessment questions 6.3

1. What effect does increasing the volume of an object have on its thermal capacitance?
2. Two objects with different thermal capacitances each have a Bunsen burner placed under them. The height of the Bunsen burner flames are adjusted until they are equal. Which object will heat up the quickest, the one with the highest thermal capacitance or the one with the lowest thermal capacitance?

Exercises 6.3

1. Calculate the thermal capacitance of the following objects:

   (a) 6.3 kg of aluminium (for aluminium, $c = 913$ J kg^{-1} K^{-1});
   (b) 9.6 kg of copper (for copper, $c = 385$ J kg^{-1} K^{-1}).
2. Calculate the heat rate needed to raise the temperature of the following lumps of material at a rate of 0.2 K s^{-1}:

   (a) 2.4 kg of brass (for brass, $c = 370$ J kg^{-1} K^{-1});
   (b) 0.34 kg of glass (for glass, $c = 670$ J kg^{-1} K^{-1}).

6.4 Thermal systems

One example of a complicated thermal model is a model of a large building.

Modelling thermal systems can be extremely complicated, especially if radiation and convection effects are taken into account. However, it is possible to make progress with the use of simplified models. These allow some of the more basic features of thermal systems to be analysed. More refined models can be produced if required but their development is beyond the scope of this book. Consider the following example.

Example

6.5 Heating a living room

Figure 6.3 shows a living room with a source of heat which pumps energy into the room at a rate q_i. For simplicity assume the air and furniture in the room are all at the same temperature T. The outside temperature Θ is a constant and less than T so heat escapes from the room to the outside. Furthermore, the interior of the room can be considered to have a combined thermal capacitance C. For convenience, assume that the floor, walls and ceiling have an overall thermal resistance to the heat escaping of R. Develop a differential equation to model this system. Assume the system input is q_i and the system output is T.

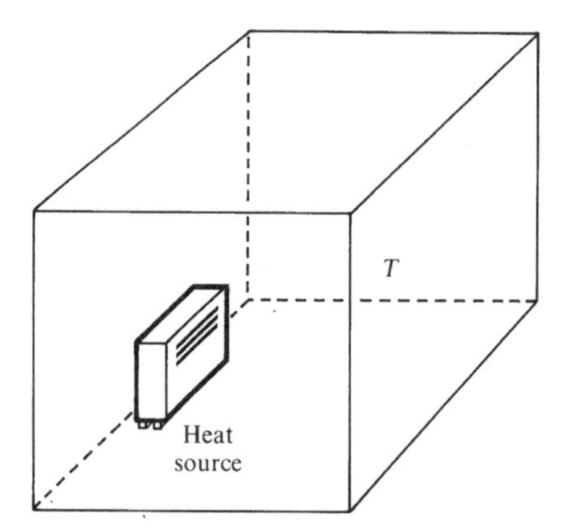

Figure 6.3
A living room containing a source of heat

Solution The temperature difference between the interior and exterior of the room is $T - \Theta$. If we define the total energy rate with which heat is lost from the room to the outside to be q_o then we can write

$$T - \Theta = Rq_o \qquad \text{Eqn. [6.7]}$$

Now as the air and furniture in the room are assumed to be at the same temperature T and have a combined thermal capacitance C we can state

energy rate into room − energy rate out of room
= energy rate to heat the room

In symbols we have

$$q_i - q_o = C\frac{dT}{dt} \qquad \text{Eqn. [6.8]}$$

We require a relationship between q_i and T and so need to eliminate q_o from Eqns [6.7] and [6.8]. This gives

$$q_i - \frac{T - \Theta}{R} = C\frac{dT}{dt}$$

Multiplying by R gives

$$Rq_i - (T - \Theta) = RC\frac{dT}{dt}$$

Rearranging we have

$$Rq_i = RC\frac{dT}{dt} + T - \Theta$$

This is the system differential equation. Recall earlier that we assumed Θ was constant and so this differential equation relates the room temperature T to the heat input to the room, q_i, as required.

In the previous example, if we had not made the assumption that the outside temperature was constant, then the room temperature would have depended on q_i and Θ. This is a more complicated case requiring more advanced analysis.

Example

6.6 Electrical water heater

These sorts of heaters are found in many houses, particularly in shower systems. They have the advantage of not requiring the storage of hot water as hot water is only produced when required.

Figure 6.4 shows an electrical water heater. Electrical power p is supplied to the heater and this causes water entering the heater to be heated from a temperature θ_i to a temperature θ_o. The volume flow rate of water through the

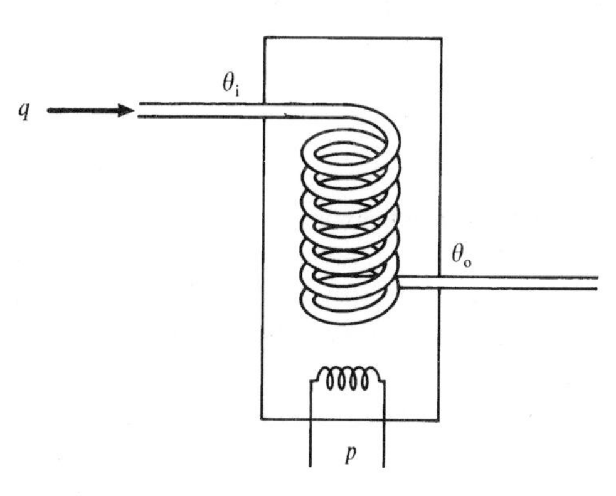

Figure 6.4
Electrical water heater with input power p

heater is q. The heater coil and surrounding parts, together with any water, can be assumed to have a thermal capacitance C_h. The density of the water is ρ and the specific heat capacity of the water is c. Derive a differential equation for the system with input p and output the increase in water temperature θ which is equal to $\theta_o - \theta_i$.

Solution The first stage is to write an energy balance equation for the system. As the water enters the system it has energy due to its temperature θ_i. The volume flow rate of water is q. Now (mass) = (density) × (volume) and so the mass flow rate of water is ρq. Recall the definition of specific heat capacity. It is the amount of energy required to heat 1 kg of material by a temperature of 1 K. This energy is stored in the material. Here the material has a temperature of θ_i and so the energy flow rate is $c\rho q\theta_i$.

Recall Eqn. [6.6]. This gives the energy flow rate required to heat an object at a certain temperature rate. For the water this energy flow rate is $C_h d\theta/dt$. Note the use of θ here because it is the increase in temperature relative to the input temperature that is relevant.

We are now in a position to write an energy balance equation for the system. In words, we have

energy flow rate into heater − energy flow rate out of heater = energy flow rate to increase the temperature of the water heater and the water it contains

In symbols we have,

$$(p + c\rho q\theta_i) - (c\rho q\theta_o) = C_h \frac{d\theta}{dt}$$

Collecting terms gives

$$p = C_h \frac{d\theta}{dt} + c\rho q(\theta_o - \theta_i)$$

Finally, noting that $\theta = \theta_o - \theta_i$ we have

$$p = C_h \frac{d\theta}{dt} + c\rho q\theta$$

This is the system differential equation. Note that in effect θ is a relative temperature measurement. It is the temperature of the hot water relative to the unheated water temperature.

Example

6.7 Mercury in glass thermometer

Figure 6.5 shows a representation of a mercury in glass thermometer. The thermometer is used to measure the temperature θ of various liquids. For simplicity assume the glass has a single temperature θ_g and has thermal capacitance C_g. Furthermore the interface between the glass and the outside is equivalent to a thermal resistance R_g. Also, assume the mercury has a single temperature θ_m and a thermal capacitance C_m. The interface between the mercury and the glass is equivalent to a thermal resistance R_m. Develop a differential equation for the system with input the actual temperature θ and output the mercury temperature θ_m.

Solution Figure 6.6 shows an expanded view of part of the thermometer. For convenience we assume that the outside temperature is greater than the mercury temperature and so there is an energy flow rate q_1 from the outside to

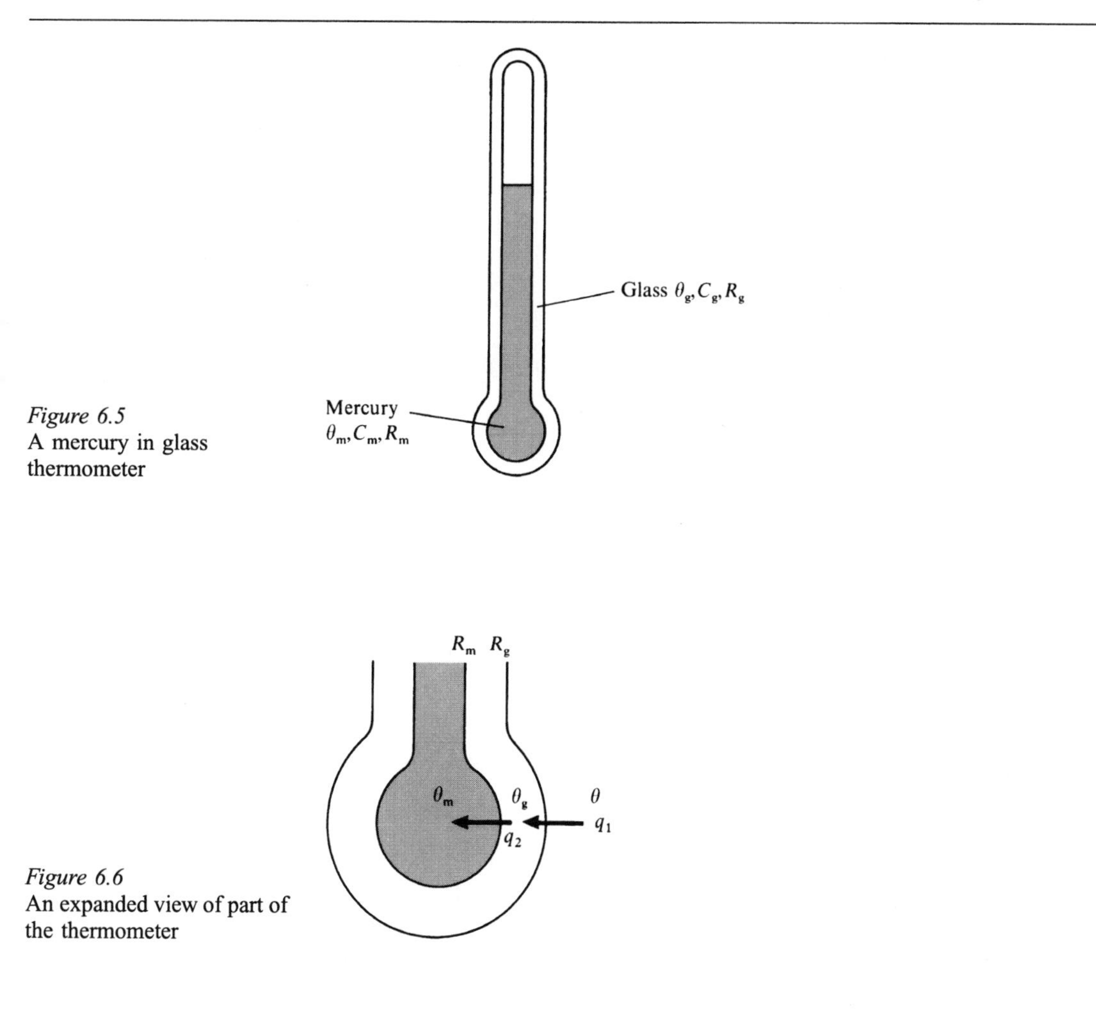

Figure 6.5
A mercury in glass thermometer

Figure 6.6
An expanded view of part of the thermometer

the glass and an energy flow rate q_2 from the glass to the mercury. The equations will still remain valid for the case when the mercury is hotter than the outside temperature. For the mercury we can write

$$q_2 = C_m \frac{d\theta_m}{dt} \qquad \text{Eqn. [6.9]}$$

For the mercury–glass interface we can write

$$\theta_g - \theta_m = q_2 R_m \qquad \text{Eqn. [6.10]}$$

For the glass we can write

$$q_1 - q_2 = C_g \frac{d\theta_g}{dt} \qquad \text{Eqn. [6.11]}$$

For the glass–surroundings interface we can write

$$\theta - \theta_g = q_1 R_g \qquad \text{Eqn. [6.12]}$$

We now have four equations and three intermediate variables that we wish to eliminate, namely, q_1, q_2 and θ_g, and so this is possible.

First we combine Eqns [6.11] and [6.12] to eliminate q_1. This gives

$$\frac{\theta - \theta_g}{R_g} - q_2 = C_g \frac{d\theta_g}{dt} \qquad \theta - \theta_g - R_g q_2 = R_g C_g \frac{d\theta_g}{dt} \qquad \text{Eqn. [6.13]}$$

We now have three equations, namely, Eqns [6.9], [6.10] and [6.13], and two variables to eliminate.

Next we eliminate q_2. Combining Eqns [6.9] and [6.10] gives

$$\theta_g - \theta_m = R_m C_m \frac{d\theta_m}{dt} \qquad \text{Eqn. [6.14]}$$

Combining Eqns [6.9] and [6.13] gives

$$\theta - \theta_g - R_g C_m \frac{d\theta_m}{dt} = R_g C_g \frac{d\theta_g}{dt} \qquad \text{Eqn. [6.15]}$$

We now have two equations, that is, Eqns [6.14] and [6.15], and one variable to eliminate. If we examine Eqn. [6.15] we see that θ_g appears on its own and in differentiated form. We therefore need to differentiate Eqn. [6.14] before we can proceed. This gives

$$\frac{d\theta_g}{dt} - \frac{d\theta_m}{dt} = R_m C_m \frac{d^2\theta_m}{dt^2} \qquad \text{Eqn. [6.14a]}$$

Eqns [6.14] and [6.14a] can now be substituted into Eqn. [6.15] to eliminate θ_g. So,

$$\theta - \left(\theta_m + R_m C_m \frac{d\theta_m}{dt}\right) - R_g C_m \frac{d\theta_m}{dt} = R_g C_g \left(\frac{d\theta_m}{dt} + R_m C_m \frac{d^2\theta_m}{dt^2}\right)$$

Rearranging terms gives

$$\theta = R_m C_m R_g C_g \frac{d^2\theta_m}{dt^2} + (R_m C_m + R_g C_m + R_g C_g)\frac{d\theta_m}{dt} + \theta_m$$

This is the system differential equation.

Self-assessment questions 6.4

1. Give reasons why models of thermal systems tend to be complicated.
2. Explain what is meant by an energy balance equation.

Exercises 6.4

1. An electronic temperature measurement system has a temperature probe with thermal capacitance C and an equivalent thermal resistance R at the probe interface with the liquid whose temperature is being measured. The arrangement is shown in Figure 6.7. The size of the probe has been exaggerated for convenience.

   Assuming the probe has a uniform temperature T_p and the fluid has a uniform temperature T_f, derive a system differential equation for the probe with T_f as the system input and T_p as the system output.

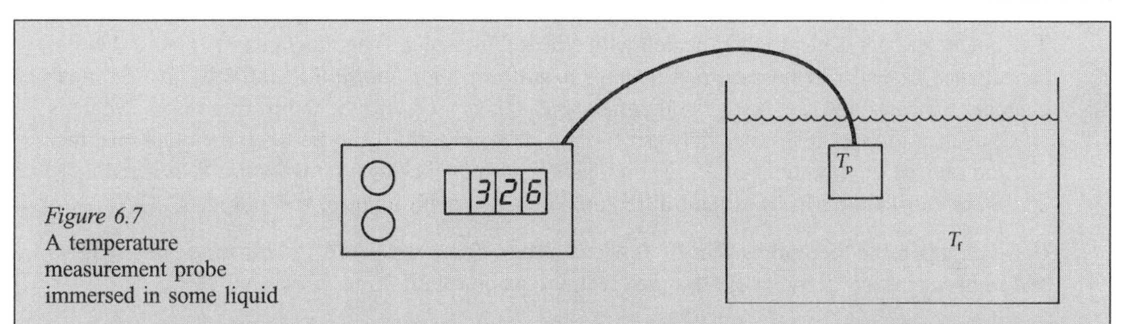

Figure 6.7
A temperature measurement probe immersed in some liquid

2. A building consisting of two rooms is shown in Figure 6.8. Room 1 has a thermal capacitance of C_1 and can be assumed to have a uniform temperature T_1. Room 2 has a thermal capacitance of C_2 and can be assumed to have a uniform temperature T_2. Rooms 1 and 2 are heated electrically and the heat flow rates are q_1 and q_2 respectively. The dividing wall between the two rooms has a thermal resistance R. The rest of the wall, floor and ceiling have an equivalent resistance of R_1 for room 1 and R_2 for room 2. There is a uniform outside temperature T_e. Assuming q_2 and T_e are constants, derive a system differential equation with input q_1 and output T_1. The heat flow rate between room 1 and room 2 is q_a. The heat flow rate out of room 1 is q_b and that out of room 2 is q_c.

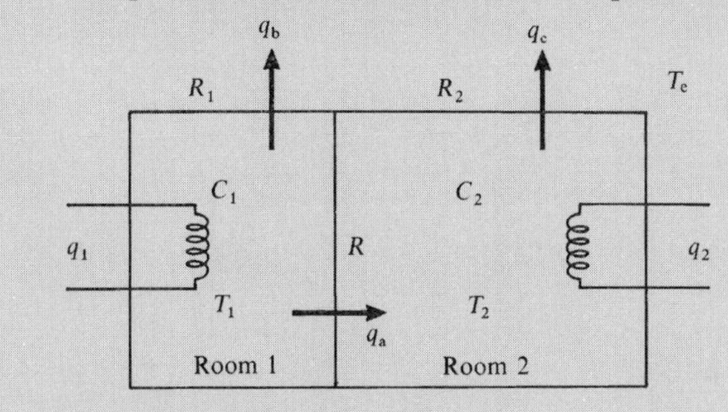

Figure 6.8
A building with two rooms

Test and assignment exercises 6

1. Consider again the two-roomed building of Exercise 6.4.2. This time, assuming q_1 and q_2 are constant, derive a system differential equation with input T_e and output T_2.
2. A thermal system consists of two chambers, one of which is completely contained within the other. The arrangement is shown in Figure 6.9.

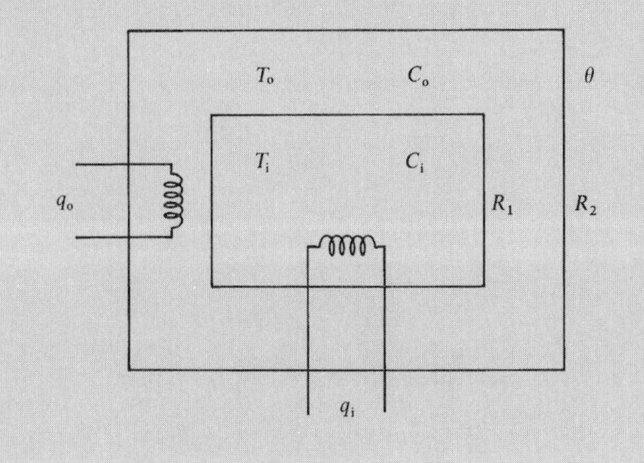

Figure 6.9
A thermal system consisting of two chambers

The inner chamber is electrically heated with a heat flow rate q_i. The inner chamber has a thermal capacitance C_i and can be assumed to have a uniform temperature T_i. Similarly, for the outer chamber the heat flow rate is q_o, the thermal capacitance is C_o and the temperature is T_o. There is an equivalent thermal resistance between the two chambers of R_1 and between the outer chamber and the outside environment of R_2. The outside environment has a temperature θ. Assuming q_o and θ are constant, derive a system differential equation with input q_i and output T_o.

3. Consider again the thermal system of Test and assignment exercises 6.2. This time, assuming q_o and q_i are constant, derive a system differential equation with input θ and output T_i.

7 Simulation

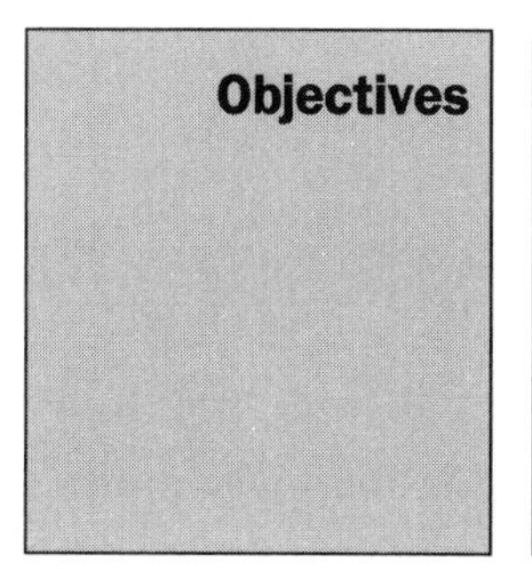

This chapter:

- explains why it is desirable to simulate a system
- explains the difference between analog and digital simulation
- describes how to construct simulation diagrams
- describes how to construct analog computer circuits
- describes how digital simulation is carried out by a computer

7.1 Introduction

We have now examined how to obtain mathematical models for a variety of systems. This chapter looks at how to represent and solve these mathematical models by means of a computer. The main emphasis is on the use of analog computers but there is also a brief discussion of the use of digital computers. Engineers refer to these techniques as **simulation**.

7.2 Simulation diagrams

A **simulation diagram** is a mechanism that allows a mathematical model of a system to be represented in block diagram form. Recall from Chapter 1 that a block diagram consists of a number of blocks together with summing junctions and take-off points. For a simulation diagram there are two types of blocks. The first type of block is a **gain block**. This is illustrated in Figure 7.1.

If an input $r(t)$ is applied to the block then the output is $Kr(t)$. The block has a gain K where K is a constant. For example, if $K=2$ then the output signal is double the input signal.

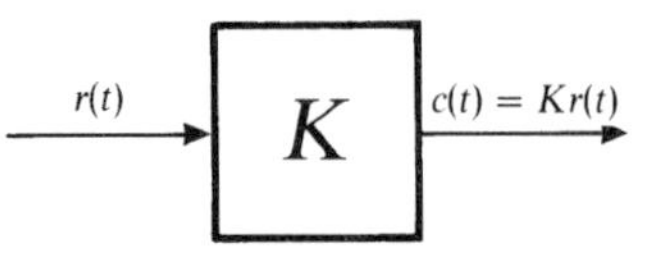

Figure 7.1
A gain block

KEY POINT

For a gain block, the input signal $r(t)$ is multiplied by a gain K to form the output signal $c(t)$. We have, $c(t)=Kr(t)$.

The second type of block is the **integrator block**. This is illustrated in Figure 7.2. If an input $r(t)$ is applied to the block then the output from the block, $c(t)$, is given by

$$c(t) = \int_{-\infty}^{t} r(t)\mathrm{d}t$$

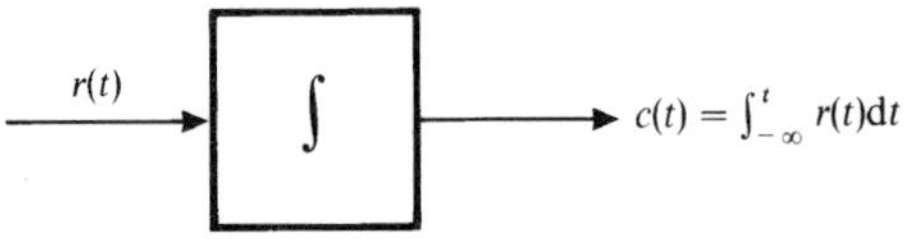

Figure 7.2
An integrator block

Note the form of this integral. It is a summation from the furthest possible point back in time through to the current time, t. In other words, it takes account of the whole history of the values of $r(t)$.

KEY POINT

For an integrator block, the input signal $r(t)$ is integrated to form the output signal $c(t)$. We have

$$c(t) = \int_{-\infty}^{t} r(t)\mathrm{d}t$$

Only these two blocks are needed to represent linear systems. Further types of blocks are needed to represent nonlinear systems.

Using these blocks it is possible to represent a large variety of linear systems.

It is worth remarking at this stage that there is no unique simulation diagram for a particular system and a number of different diagrams produce equivalent results. We now illustrate the method of developing simulation diagrams by means of some examples.

Example

7.1 Draw a simulation diagram for the system shown in Figure 7.3. Note that v_i is the system input and v_o is the system output.

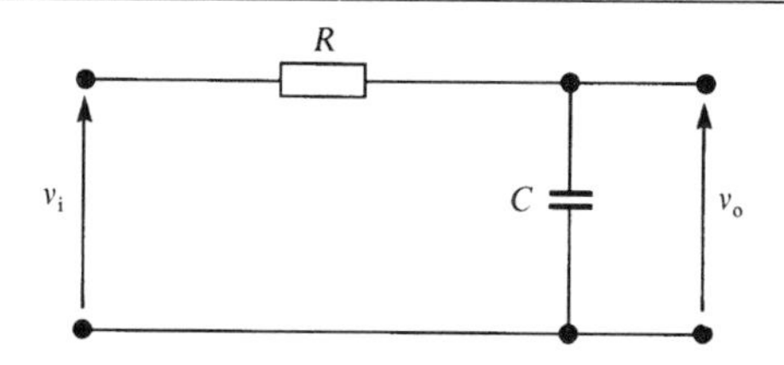

Figure 7.3
An *RC* circuit

We obtained a system differential equation for this circuit in Example 4.4. This is

$$v_i = RC\frac{dv_o}{dt} + v_o$$

The easiest way to construct a simulation diagram is to rearrange the system differential equation so that the term containing the highest derivative is on its own on the left-hand side of the equation. So we have

$$RC\frac{dv_o}{dt} = v_i - v_o$$

$$\frac{dv_o}{dt} = \frac{1}{RC}(v_i - v_o) \qquad \text{Eqn. [7.1]}$$

The number of integrators required is equal to the order of the system differential equation. The term 'order' was defined in Section 1.4.

The idea behind constructing a simulation diagram is to assume that we already have a point in the diagram corresponding to the highest derivative term on the left-hand side of the equation and then to arrange to join this point to a point corresponding to the terms on the right-hand side of the equation.

Examining Eqn. [7.1] we see that there is a v_o term and so we know that we will need v_o. By using an integrator block we can obtain v_o. This is shown in Figure 7.4. This stage is known as constructing the **integrator chain**. In this case there is only one integrator in the chain but with more complicated systems there may be several.

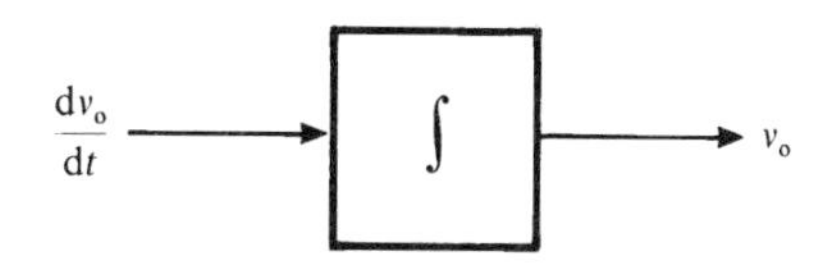

Figure 7.4
The integrator chain for the *RC* circuit

Examining the right-hand side of Eqn. [7.1] again we see that we need $v_i - v_o$. This can be obtained using a summing junction with the v_o term being fed in negatively. The term $v_i - v_o$ then needs scaling by a factor $1/RC$ which can be achieved by use of a gain block. So finally, joining together the two points corresponding to the left-hand side and right-hand side of Eqn. [7.1], we obtain the completed simulation diagram shown in Figure 7.5.

A final point to note is that there is only one integrator in the integrator chain because the system differential equation is first order.

Example

7.2 Draw a simulation diagram for the mass–spring–damper system shown in Figure 7.6, where f is the system input and x is the system output.

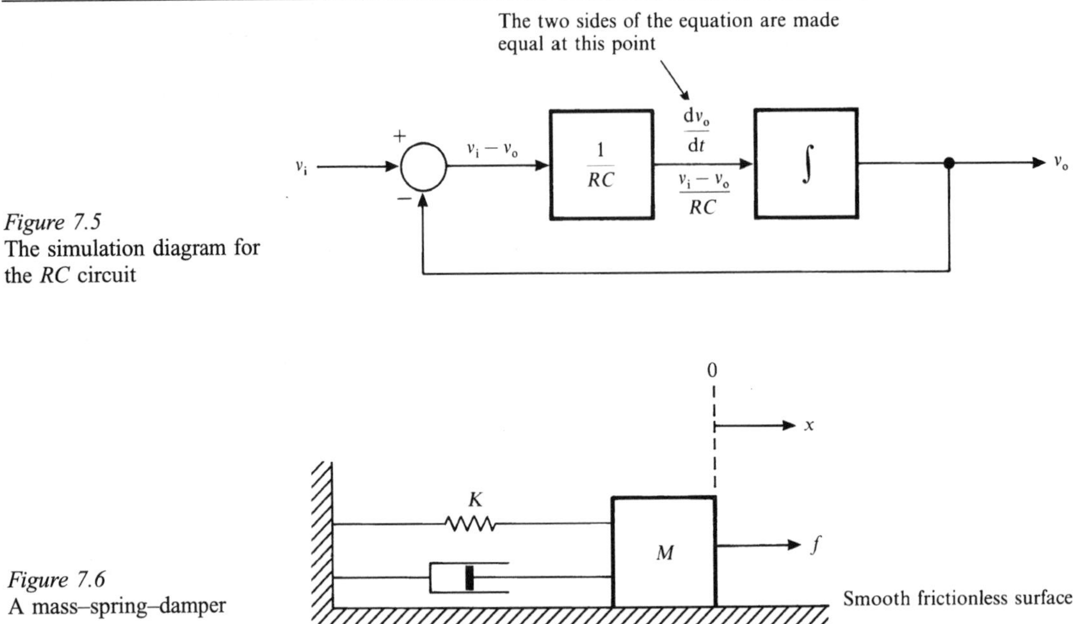

Figure 7.5
The simulation diagram for the *RC* circuit

Figure 7.6
A mass–spring–damper system

Solution We obtained the differential equation for this system in Example 2.3. This is

$$f = M\frac{d^2x}{dt^2} + B\frac{dx}{dt} + Kx$$

Rearranging the system differential equation to obtain the highest derivative term on its own we have

$$M\frac{d^2x}{dt^2} = f - B\frac{dx}{dt} - Kx$$

$$\frac{d^2x}{dt^2} = \frac{1}{M}\left(f - B\frac{dx}{dt} - Kx\right) \qquad \text{Eqn. [7.2]}$$

Next we form the integrator chain. This is shown in Figure 7.7.

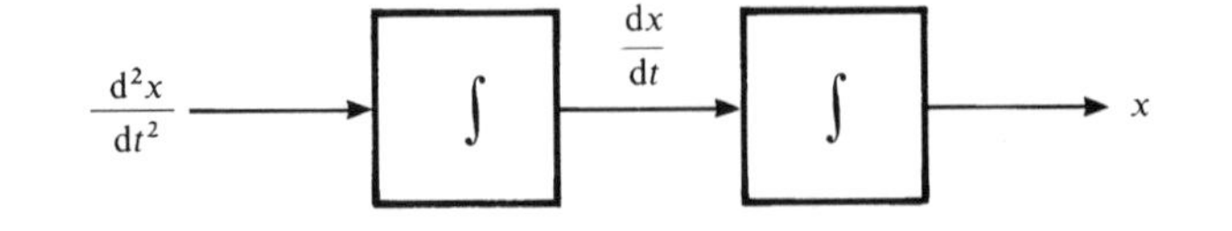

Figure 7.7
Integrator chain for the mass–spring–damper system

There are two integrators in the integrator chain because the system differential equation is second order. Examining the right-hand side of Eqn. [7.2] we see that we need to multiply dx/dt by B using a gain block. We also need Kx and so we need to multiply x by K using a gain block. The input signal f is readily available. We can combine the right-hand side terms using a summing junction with the $B\mathrm{d}x/\mathrm{d}t$ and Kx terms being fed in negatively. Finally, we need to scale the output from this summing junction by $1/M$. The completed simulation diagram for the mass–spring–damper system is shown in Figure 7.8.

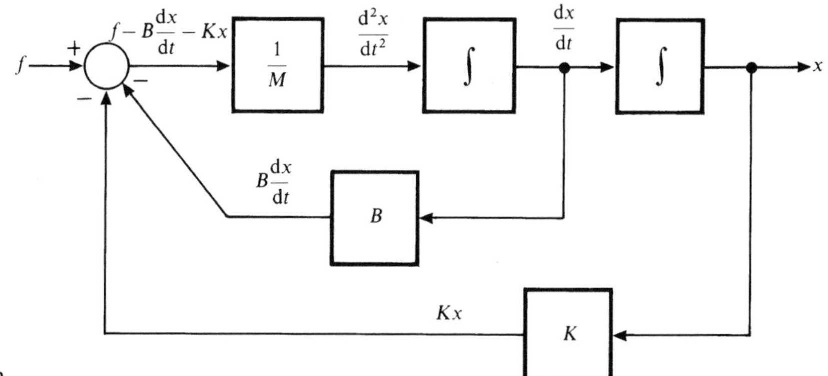

Figure 7.8
Simulation diagram for the mass–spring–damper system

Sometimes problems can occur when constructing a simulation diagram if the system differential equation has a term containing a derivative of the input variable. The following example shows how to deal with this case.

Example

7.3 Draw a simulation diagram for the system shown in Figure 7.9 where v_i is the system input and v_o is the system output.

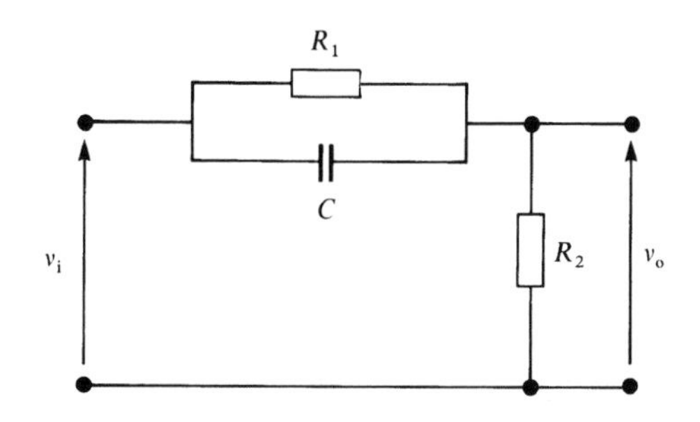

Figure 7.9
Circuit for Example 7.3

Solution We obtained the differential equation for this system in Example 4.5. It is

$$R_1R_2C\frac{\mathrm{d}v_i}{\mathrm{d}t} + R_2v_i = R_1R_2C\frac{\mathrm{d}v_o}{\mathrm{d}t} + (R_1 + R_2)v_o$$

Examining this equation we see that there is a term containing a derivative of the input. We need to dispose of this term before we proceed because we only have access to v_i and the process of differentiation is not available to obtain $\mathrm{d}v_i/\mathrm{d}t$. The reason why we cannot use differentiation will be explained later in the chapter. The solution to the problem is to integrate the system differential equation in order to remove the dv_i/dt term. This gives

$$R_1R_2Cv_i + R_2\int_{-\infty}^{t} v_i dt = R_1R_2Cv_o + (R_1+R_2)\int_{-\infty}^{t} v_o dt \qquad \text{Eqn. [7.3]}$$

We see that this introduces some integral terms but these are easily dealt with. We can now form the integrator chain. The highest derivative of the output variable is now v_o and so this term must form the start point of the integrator chain. Rearranging Eqn. [7.3] to obtain this term on its own we get

$$R_1R_2Cv_o = R_1R_2Cv_i + R_2\int_{-\infty}^{t} v_i dt - (R_1+R_2)\int_{-\infty}^{t} v_o dt$$

Dividing by R_1R_2C gives

$$v_o = v_i + \frac{1}{R_1C}\int_{-\infty}^{t} v_i dt - \frac{R_1+R_2}{R_1R_2C}\int_{-\infty}^{t} v_o dt \qquad \text{Eqn. [7.4]}$$

The integrator chain for this system is shown in Figure 7.10.

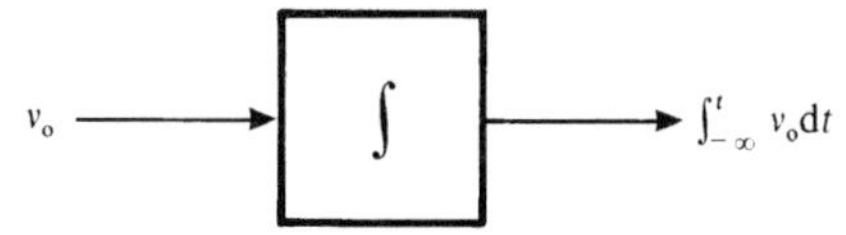

Figure 7.10
The integrator chain for the circuit of Figure 7.9

We see that the integrator chain contains one integrator as the system is first order. We can now form the terms of the right-hand side of Eqn. [7.4]. Note that an integrated version of v_i is required and so an integrator is needed to obtain this term. The terms can be summed using a summing junction. The simulation diagram for the system is shown in Figure 7.11. Note that the output signal v_o has been tapped off at a point before the end of the integrator chain. This is a result of the action of integrating the system differential equation prior to developing the simulation diagram.

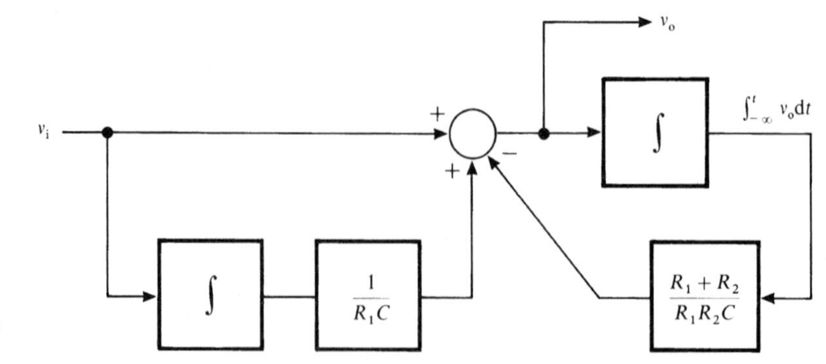

Figure 7.11
Simulation diagram for the circuit of Figure 7.9

Recall that there is no unique simulation diagram for a particular system. As this system is first order then it should be possible to construct a simulation diagram with only one integrator. Recall Eqn. [7.4]:

$$v_o = v_i + \frac{1}{R_1 C}\int_{-\infty}^{t} v_i \mathrm{d}t - \frac{R_1 + R_2}{R_1 R_2 C}\int_{-\infty}^{t} v_o \mathrm{d}t$$

Gathering the terms under a common integral sign gives

$$v_o = v_i + \int_{-\infty}^{t} \frac{1}{R_1 C} v_i - \frac{R_1 + R_2}{R_1 R_2 C} v_o \mathrm{d}t$$

Figure 7.12 shows the simulation diagram corresponding to this form of the system differential equation. Note that only one integrator has been used.

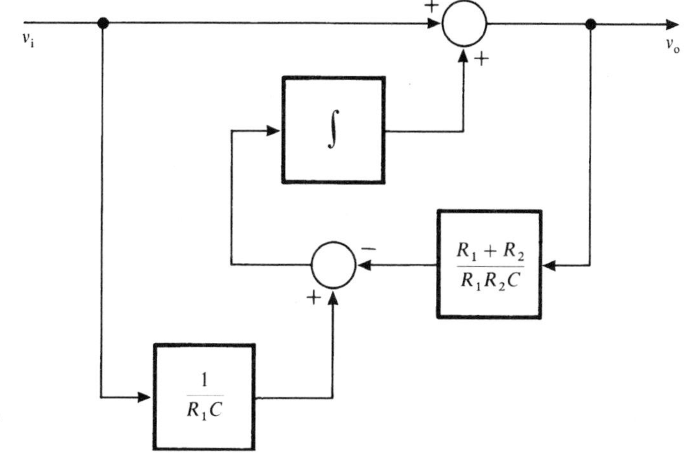

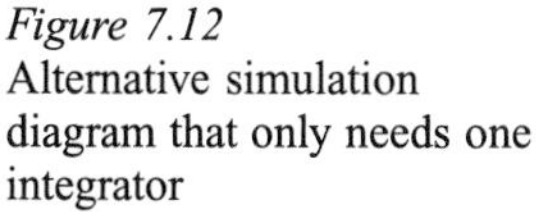
Figure 7.12
Alternative simulation diagram that only needs one integrator

We mentioned in the previous example that differentiation was a process that was not allowed in a simulation diagram. The reason for this will now be discussed. We shall see in Section 7.3 that simulation diagrams can be implemented electronically in order to allow solution of a mathematical model to be obtained with relative ease. Now all electronic circuits generate some electrical noise which becomes superimposed on signals as they propagate through a circuit. The process of integration reduces this noise and so is a process to be encouraged because, in general, noise is not welcome in most electronic circuits. To see why this is so, examine Figure 7.13.

Engineers refer to integration as a **virtuous** process in this context.

Figure 7.13(a) shows a sinusoidal signal. Figure 7.13(b) shows the same signal when noise has been superimposed on it. We see that the noise causes deviations in the value of the sinusoidal signal. If this signal is presented to an integrator then the process of integration corresponds to calculating the area under the curve of the signal against time. We see that, when calculating the area of the noisy signal, sometimes the signal value is slightly less than that of the true signal and at other points it is slightly more. This is indicated in region 1 of the signal of Figure 7.13(b). The result is that positive and negative errors tend to cancel each other out and the value of the area under the curve of the noisy signal will be almost the same as that of the true signal. Therefore the

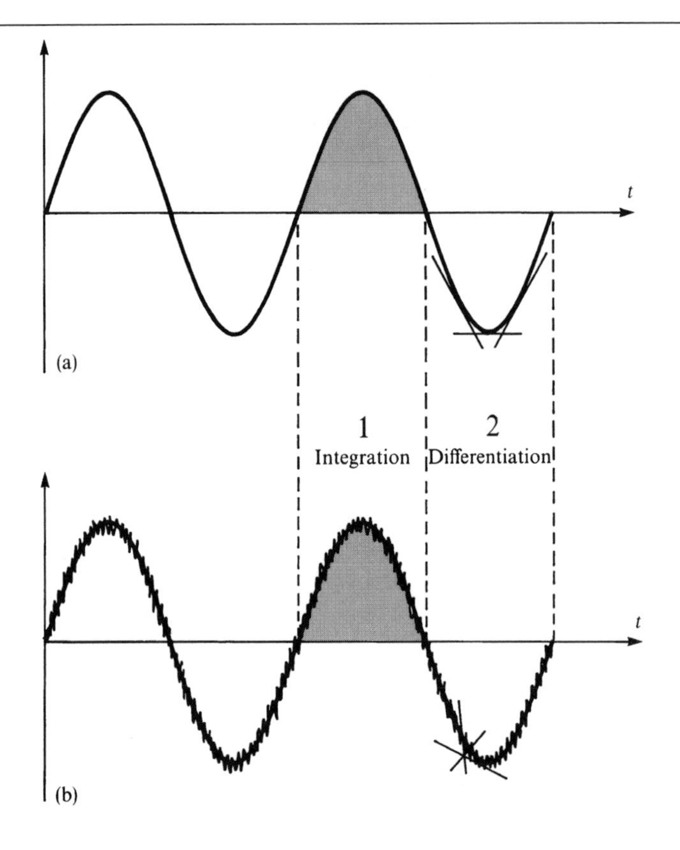

Figure 7.13
(a) A sinusoidal signal
(b) the same sinusoidal signal with noise superimposed on it

process of integrating the noisy signal has reduced the importance of the noise.

Let us now examine the effect of passing the noisy signal through a differentiator. Recall that differentiation corresponds to calculating the slope of the curve. We see from examining region 2 of Figure 7.13(a) that the slope of the true signal changes slowly from negative to zero to positive values with increasing time. However, if we examine the slope of the signal contaminated with noise we see that it has positive and negative slope values that alternate very quickly with time. Therefore the effect of differentiating the noisy signal is to amplify the noise, thus increasing its importance. This is clearly a process to be discouraged.

Engineers refer to differentiation as a **vicious** process in this context.

Self-assessment questions 7.2

1. Explain why the process of differentiation is not used in a simulation diagram.
2. State the two types of blocks needed to construct a simulation diagram and explain their operation.
3. Explain what is meant by the term 'integrator chain'.

Exercises 7.2

1. Construct a simulation diagram for the coupled tanks of Example 5.2. Assume that the system input is q_i and the system output is q_o.
2. Construct a simulation diagram for the torsional pendulum of Example 3.2. Assume the system input is T and the system output is θ.
3. Construct a simulation diagram for the mercury in glass thermometer in Example 6.7. Assume the system input is the outside temperature and the system output is the mercury temperature.
4. Construct a simulation diagram for the *RLC* circuit of Example 4.6. Assume the system input is v_i and the system output is v_o.
5. Construct a simulation diagram for the armature-controlled d.c. motor shown in Example 4.8. Assume the system input is v_a and the system output is θ.

7.3 Analog computer circuits

Analog computers are no longer as popular as they used to be. However, they still have an important role to play in engineering. For example, they can be used to simulate industrial systems in a laboratory, thus allowing controllers to be tested before being put into service.

In the last section we saw how to construct simulation diagrams. These provide a means of pictorially representing an engineering system and are often used for just this purpose. By drawing a simulation diagram it is possible to discern more easily the structure of an engineering system. However, it is also possible to use a simulation diagram as the basis for constructing an electronic circuit that will simulate the behaviour of a system. It is a simple matter to produce circuits that can multiply, integrate and sum signals, as we shall shortly see. Often these building block circuits are collected together in order to provide a convenient means of simulating systems. The device containing this collection of building block circuits is known as an **analog computer**. **Analog computation** or **analog simulation** is a phrase used to describe the simulation of an engineering system by means of electronic circuits.

Before designing any analog computer circuits we first need to examine the various building block circuits.

7.3.1 The summer

A **summer** is a circuit that can add together signals and additionally has the facility to multiply them by a fixed gain. The heart of this circuit is an **operational amplifier** which is a device that has a very large negative voltage gain, typically 10 000, and is represented by a triangular symbol. In order for it to operate in a useful manner several passive electronic components need to be connected to the amplifier. The circuit that turns the operational amplifier into a summer is shown in Figure 7.14.

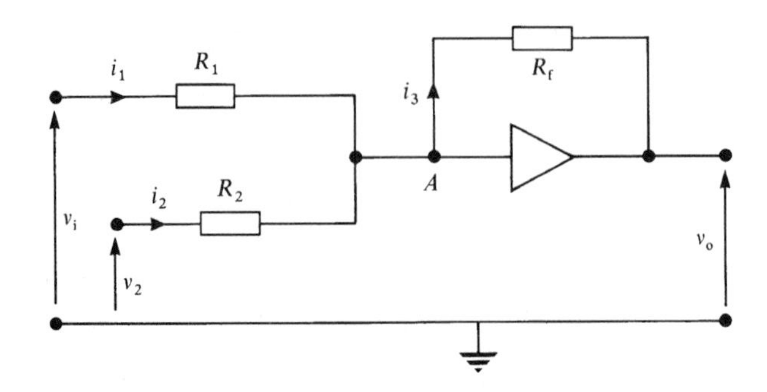

Figure 7.14
The circuit for a summer

The circuit is easy to analyse if it is noted that the input point to the amplifier, point A, is at a very low voltage in order to prevent the output voltage v_o being impossibly high; recall that the gain of an operational amplifier is very large. This point is said by engineers to be a **virtual earth**. Also the current flowing into the amplifier is virtually zero for similar reasons. Figure 7.14 shows a circuit with two input voltages v_1 and v_2, but it is possible to have more if desired. Now, because the current flowing into the amplifier is so small we have

$$i_1 + i_2 = i_3 \qquad \text{Eqn. [7.5]}$$

by Kirchhoff's current law.

Also, because A is virtually at earth potential, we have

$$i_1 = \frac{v_1}{R_1} \qquad \text{Eqn. [7.6]}$$

$$i_2 = \frac{v_2}{R_2} \qquad \text{Eqn. [7.7]}$$

$$i_3 = \frac{-v_o}{R_f} \qquad \text{Eqn. [7.8]}$$

Note the negative sign in Eqn. [7.8] due to the direction chosen for i_3. Combining Eqns [7.5], [7.6], [7.7] and [7.8] to eliminate i_1, i_2 and i_3 we obtain

$$\frac{v_1}{R_1} + \frac{v_2}{R_2} = \frac{-v_o}{R_f}$$

$$v_o = -\left(\frac{R_f}{R_1}v_1 + \frac{R_f}{R_2}v_2\right) \qquad \text{Eqn. [7.9]}$$

KEY POINT

For a summer we have

$$v_o = -\left(\frac{R_f}{R_1}v_1 + \frac{R_f}{R_2}v_2\right)$$

where v_1 and v_2 are the input voltages and v_o is the output voltage.

We see that if we choose $R_1 = R_2 = R_f$ then this reduces to

$$v_o = -(v_1 + v_2)$$

For this case the circuit merely sums the two input voltages. Note the sign change which is a feature of all these operational amplifier circuits and is easily dealt with. However, it is possible to scale the input voltages by a judicious choice of the values of R_f, R_1 and R_2. For example, if we choose $R_f = R_1 = 10R_2$ then

$$\begin{aligned} v_o &= -\left(\frac{R_1}{R_1}v_1 + \frac{10R_2}{R_2}v_2\right) \\ &= -(v_1 + 10v_2) \end{aligned}$$

If we only have a single input voltage then Eqn. [7.9] reduces to

$$v_o = -\frac{R_f}{R_1}v_1$$

At first the sign changes appear to be an awkward complication to deal with but they can be an advantage as often as they are a disadvantage once familiarity is gained.

The circuit is acting in an equivalent manner to a gain block which we discussed in the previous section, although there is a sign reversal. However, we often find that this sign reversal can be used advantageously as we shall see when we come to design some analog computer circuits. In order to avoid the clutter of putting in components each time a summer is drawn it is convenient to use a symbol to represent it in the knowledge that the resistor values are easily included to achieve a particular configuration. The symbol is shown in Figure 7.15. Note the placing of a number against each input line. This represents the gain by which that particular input voltage is scaled. For the summer of Figure 7.15 we have

$$v_o = -(v_1 + 10v_2 + v_3)$$

Figure 7.15
The symbol for a summer

It is conventional to use only resistor values that give a gain of 1 or 10. Typically the feedback resistor R_f is 1 MΩ and the input resistors are 1 MΩ or 100 KΩ to give a gain of either 1 or 10. This cuts down on the number of resistors needed on the analog computer. Gains between 1 and 10 are achieved by using an amplifier and a potentiometer, as we shall see shortly. Gains above 10 are best achieved by using two amplifiers in series.

7.3.2 The inverter

An **inverter** is essentially a special case of a summer with one input and with the feedback resistor equal to the input resistor. However, because it is so common, it is given a special name. Its main use is to invert the polarity of signals to correct for the inversion the other circuits produce. The symbol for an inverter is shown in Figure 7.16.

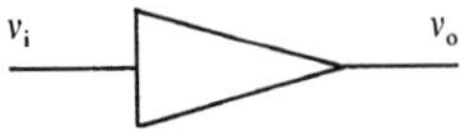

Figure 7.16
The symbol for an inverter

For the inverter, we have

$$v_o = -v_i$$

where v_i is the input voltage and v_o is the output voltage.

7.3.3 The potentiometer

The **potentiometer** is used to provide gains between 0 and 1. The circuit and symbol are shown in Figure 7.17.

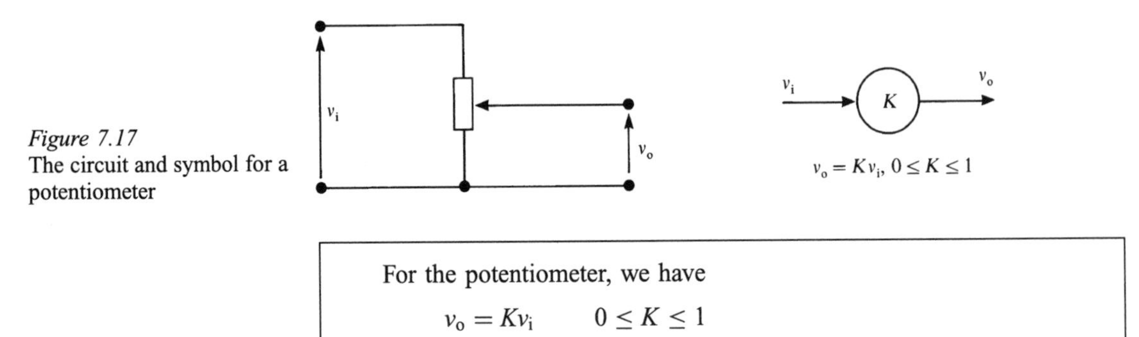

Figure 7.17
The circuit and symbol for a potentiometer

For the potentiometer, we have

$$v_o = Kv_i \qquad 0 \leq K \leq 1$$

where v_i is the input voltage and v_o is the output voltage.

It is possible to achieve any gain that is required by connecting a potentiometer in series with one or more one input summers acting as gain blocks. However, care must be taken not to connect two potentiometers in series as one will **load** the other. What we mean by this is that one potentiometer will draw current from the other, thus effectively changing the gain value. The solution is to place an operational amplifier circuit in between as these draw negligible current and so do not load a potentiometer. On analog computers the potentiometers are precision wire-wound resistors to improve their accuracy.

7.3.4 The integrator

An **integrator** has the same circuit as a summer but the feedback resistor is replaced by a capacitor. This is shown with three input voltages in Figure 7.18. Analysis of this circuit is also simple because A is a virtual earth point and the

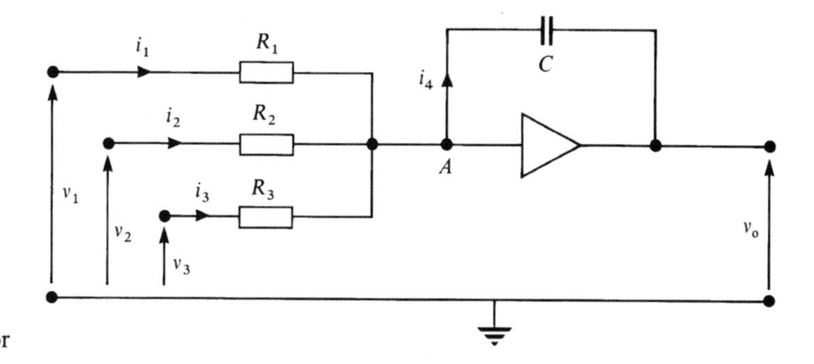

Figure 7.18
The circuit for an integrator

amplifier draws negligible current. So we have

$$i_1 + i_2 + i_3 = i_4 \qquad \text{Eqn. [7.10]}$$

Also,

$$i_1 = \frac{v_1}{R_1} \qquad \text{Eqn. [7.11]}$$

$$i_2 = \frac{v_2}{R_2} \qquad \text{Eqn. [7.12]}$$

$$i_3 = \frac{v_3}{R_3} \qquad \text{Eqn. [7.13]}$$

$$i_4 = -C\frac{\mathrm{d}v_o}{\mathrm{d}t} \qquad \text{Eqn. [7.14]}$$

Combining Eqns [7.10], [7.11], [7.12], [7.13] and [7.14] to eliminate i_1, i_2, i_3, i_4 we obtain

$$\frac{v_1}{R_1} + \frac{v_2}{R_2} + \frac{v_3}{R_3} = -C\frac{\mathrm{d}v_o}{\mathrm{d}t}$$

Dividing by $-C$ gives

$$\frac{\mathrm{d}v_o}{\mathrm{d}t} = -\left(\frac{v_1}{R_1C} + \frac{v_2}{R_2C} + \frac{v_3}{R_3C}\right)$$

Integrating gives

$$v_o = -\int_{-\infty}^{t} \frac{v_1}{R_1C} + \frac{v_2}{R_2C} + \frac{v_3}{R_3C}\mathrm{d}t \qquad \text{Eqn. [7.15]}$$

KEY POINT

For an integrator we have

$$v_o = -\int_{-\infty}^{t} \frac{v_1}{R_1C} + \frac{v_2}{R_2C} + \frac{v_3}{R_3C}\mathrm{d}t$$

where v_1, v_2 and v_3 are the input voltages and v_o is the output voltage.

We therefore see that this circuit can both integrate and provide a voltage gain depending on the choice of the values of R_1, R_2, R_3 and C. It is common practice to assume $t=0$ when a simulation is started. The integration in Eqn. [7.15] can also be started at $t=0$ provided that the initial value of v_o, denoted $v_o(0)$, is included. So we can write

$$v_o = -\int_0^t \frac{v_1}{R_1C} + \frac{v_2}{R_2C} + \frac{v_3}{R_3C}\,dt + v_o(0)$$

It is quite common for a system to have nonzero initial conditions. For example, a water storage tank may contain water at $t=0$, rather than being empty.

The symbol for an integrator is shown in Figure 7.19. Note that the gain values are written against the relevant input lines. Again it is usual to use only gains of 1 and 10. Note, also, the facility to set up an initial output voltage in the integrator at the start of a simulation. The equation for the integrator of Figure 7.19 is

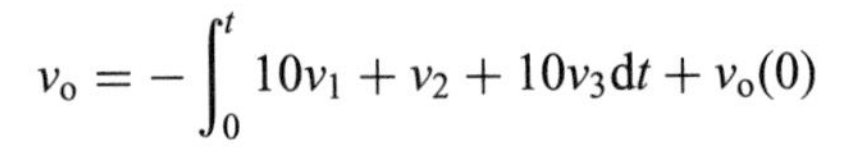

$$v_o = -\int_0^t 10v_1 + v_2 + 10v_3\,dt + v_o(0)$$

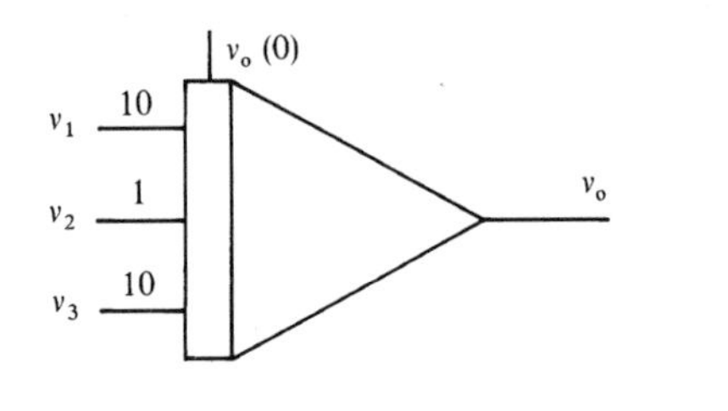

Figure 7.19
The symbol for an integrator

7.3.5 Analog computer circuits

We are now in a position to construct some simple analog computer circuits. For convenience we only use the symbols for the various building blocks rather than the full circuits. Also we work from the examples developed in Section 7.2 and use the simulation diagrams already obtained.

Example

7.4 Obtain an analog computer circuit for the *RC* circuit analysed in Examples 4.4 and 7.1. For convenience the simulation diagram, together with the intermediate signals, are shown in Figure 7.20.

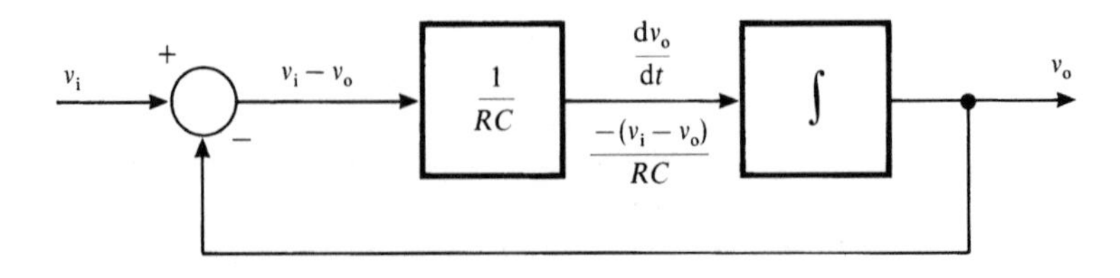

Figure 7.20
Simulation diagram for an *RC* circuit

Solution We see immediately that we need an integrator. Assuming that the gain block $1/RC$ has a gain between 0 and 1 then this can be obtained by means of a potentiometer. If the value is greater than 1 then the input gains on the amplifiers can be adjusted or further amplifiers can be introduced. The summing junction can be obtained by means of a summer. The only problem now is to take care of any sign changes needed. The output voltage needs to be fed back negatively into the summer and so an inverter will be needed to take care of this. The output from the summer will be negated but, conveniently, this will be corrected when it is fed into the integrator and so therefore no more inverters are needed. The final analog computer circuit is shown in Figure 7.21 together with all the intermediate signals to aid comprehension.

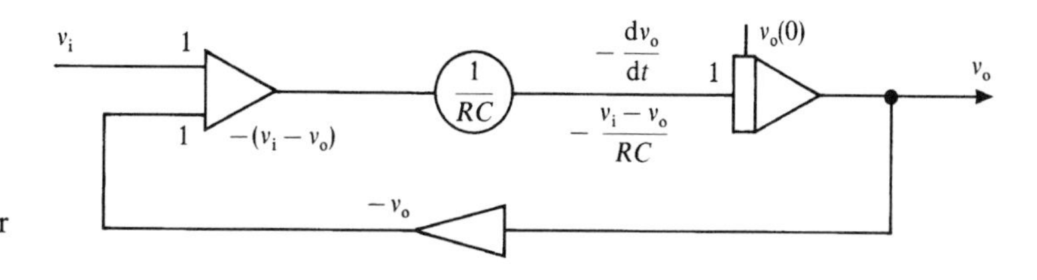

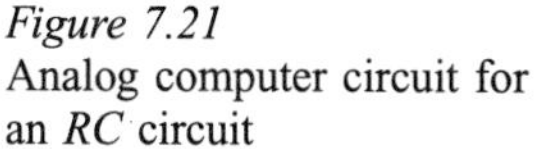
Figure 7.21 Analog computer circuit for an RC circuit

Example

7.5 Obtain an analog computer circuit for the spring–mass–damper system analysed in Examples 2.3 and 7.2. The simulation diagram for this system is shown in Figure 7.8.

Solution Examining Figure 7.8, we see that two integrators are needed. We shall assume that potentiometers can provide the necessary gain values $1/M$, B and K. If not, then the input gains in the amplifiers can be adjusted or further amplifiers can be introduced. A three-input summer is also required. The final analog computer circuit is shown in Figure 7.22. Careful examination of the intermediate signals will reveal why the various inverters are needed.

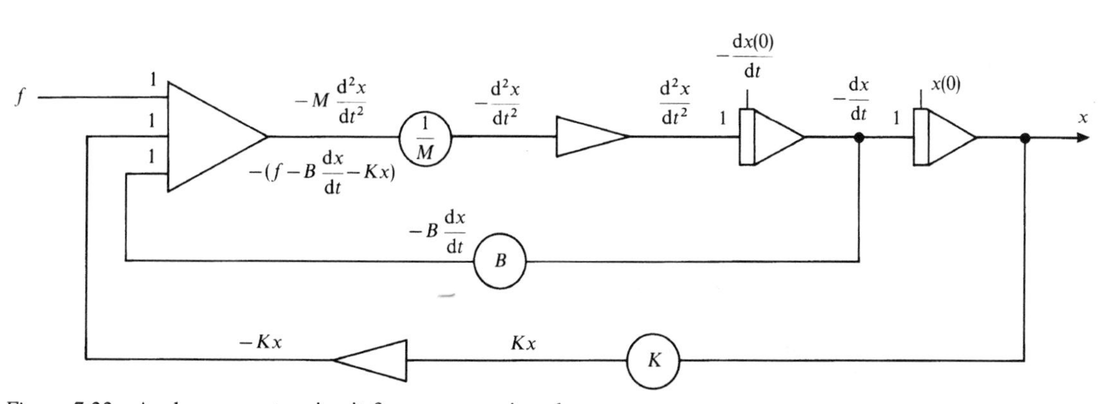

Figure 7.22 Analog computer circuit for a mass–spring–damper system

We have now seen how to construct analog computer circuits. Carrying out a simulation is straightforward once the circuit has been connected up. The potentiometers are set to give the correct values of the system parameters. Any initial conditions are placed on the integrators and then the simulation is started ensuring that the correct input signal is applied. The output and any intermediate variables can be observed using voltmeters, or more conveniently, by displaying them on an oscilloscope with a slow time base. The values of these signals correspond to the values that would be present in the actual system that is being simulated. In Example 7.5 we obtained an analog computer circuit for a spring–mass–damper. The values of the mass, damping coefficient and spring stiffness can be set by adjusting the three potentiometers. Any initial nonzero values of velocity and displacement can be set on the integrators. As the simulation proceeds, the output voltage from the circuit corresponds to the displacement of the mass as an input force f is applied.

Self-assessment questions 7.3

1. Explain why it is necessary to take account of any initial conditions on a system when carrying out a simulation.
2. Why is it important not to connect two potentiometers in series in an analog computer circuit?
3. State the equation for a summer and give its symbol.
4. State the equation for an integrator and give its symbol.
5. State the equation for a potentiometer and give its symbol.
6. State the equation for an inverter and give its symbol.

Exercises 7.3

1. Obtain an analog computer circuit for the electrical system of Example 7.3.
2. Obtain an analog computer circuit for the coupled tanks of Example 5.2.
3. Obtain an analog computer circuit for the electrical water heater of Example 6.6.
4. Obtain an analog computer circuit for the torsional pendulum of Example 3.3.
5. Obtain an analog computer circuit for the d.c. motor of Example 4.5.

7.4 Digital simulation

Digital simulation is the solution of a mathematical model of the system by means of a digital computer. The heart of the process is the solution of the system differential equations using numerical methods. A variety of methods exist, the most popular of which are the Runge–Kutta methods. Space prevents these methods being discussed but details are available in many engineering mathematics textbooks. Nowadays many of the digital simulation computer packages have a graphical interface thus allowing engineering systems to be represented in a pictorial manner. This greatly improves the ease with which an engineering system can be simulated by a computer. Also, the many packages tend to have a variety of ways in which the results of the simulation can be represented, ranging from a conventional two-dimensional graph through to a three-dimensional 'animated movie'.

Test and assignment exercises 7

1. Draw a simulation diagram for the following system differential equation:

$$2x = \frac{dy}{dt} + 3y$$

where x is the system input and y is the system output.

2. Draw a simulation diagram for the following system differential equation:

$$3r = \frac{d^2c}{dt^2} + 6\frac{dc}{dt} + 4c$$

where r is the system input and c is the system output.

3. Draw a simulation diagram for the following system differential equation:

$$5\frac{dp}{dt} + 5p = 8\frac{d^2q}{dt^2} + 5\frac{dq}{dt} + 3q$$

where p is the system input and q is the system output.

4. Construct an analog computer circuit for the system of Test and assignment exercises 7.1.
5. Construct an analog computer circuit for the system of Test and assignment exercises 7.2.
6. Construct an analog computer circuit for the system of Test and assignment exercises 7.3.

8 Transfer functions and the Laplace transform

Objectives

This chapter:

- introduces the Laplace transform
- describes some theorems associated with the Laplace transform
- solves some linear differential equations using the Laplace transform
- explains the concept of a transfer function
- shows how to calculate the poles and zeros of a transfer function

8.1 Introduction

We have seen in earlier chapters that engineering systems can be modelled by means of differential equations. In order to calculate how a system will respond to various inputs it is necessary to solve the system differential equation. The Laplace transform provides a way of doing this by converting the system differential equation into an algebraic equation which is then easier to solve. A development of this is to represent a system algebraically by means of a transfer function rather than by means of a differential equation. This approach considerably simplifies the analysis of engineering systems.

8.2 The Laplace transform

Suppose we have a function $f(t)$ which depends on time. It is possible to convert this function into a new function $F(s)$ using the **Laplace transform**.

KEY POINT

The Laplace transform $F(s)$ of a function $f(t)$ is given by

$$F(s) = \int_0^\infty e^{-st} f(t)\mathrm{d}t$$

Several assumptions are involved in this process. First we assume that $f(t)$ is zero for $t < 0$. This is the reason we can start the integral at $t = 0$ without any loss of generality. We say that the function is **switched on** at $t = 0$. This is not a limitation as many engineering functions can be conveniently thought of as commencing at $t = 0$. Second, we assume that this integral can be calculated. In practice, this does not present a problem for most of the functions met in engineering but it is important to be aware that there are restrictions on the existence of the integral. It is common practice to write $\mathscr{L}\{f(t)\}$ to denote the Laplace transform of a function and so we can write $F(s) = \mathscr{L}\{f(t)\}$. Note also that $F(s)$ is a function of s, and s is a complex variable. It is usual for engineers to talk of transforming a function from the **time domain** to the ***s* domain**. It is conventional to use a lower case letter to denote a time domain function and to use the same letter, but upper case to denote the s domain equivalent function.

Before discussing the Laplace transform in more detail let us first evaluate the Laplace transforms of some simple functions.

Example

8.1 Find the Laplace transforms of the following functions:

(a) $f(t) = 1 \quad t \geq 0$
(b) $f(t) = e^{-at} \quad t \geq 0$

Solution (a) For $f(t) = 1$ we have

$$F(s) = \mathscr{L}\{f(t)\} = \int_0^\infty e^{-st} f(t)\mathrm{d}t = \int_0^\infty e^{-st} 1\mathrm{d}t$$

$$F(s) = \left[\frac{e^{-st}}{-s}\right]_0^\infty$$

Assuming $s > 0$, then $e^{-st} \to 0$ as $t \to \infty$ and so

$$F(s) = 0 - \frac{1}{-s}$$

Finally,

$$F(s) = \frac{1}{s}$$

(b) For $f(t) = e^{-at}$ we have

$$F(s) = \mathscr{L}\{f(t)\} = \int_0^\infty e^{-st} f(t)\mathrm{d}t = \int_0^\infty e^{-st} e^{-at}\mathrm{d}t$$

$$F(s) = \int_0^\infty e^{-(s+a)t}\mathrm{d}t$$

$$F(s) = \left[\frac{e^{-(s+a)t}}{-(s+a)}\right]_0^\infty$$

Assuming $s+a>0$, then $e^{-(s+a)t} \to 0$ as $t \to \infty$ and so

$$F(s) = 0 - \frac{-1}{(s+a)}$$

Finally we have

$$F(s) = \frac{1}{s+a}$$

Note that in each case the function $f(t)$ has been transformed into a new function $F(s)$.

In the previous example restrictions were put on the value of s in order to evaluate the integrals. In practice, these restrictions are not a limitation for the type of engineering systems we shall analyse. However, it is important to be aware that such restrictions do exist. A more detailed mathematical treatment would explore these ideas further but they are beyond the scope of this book. We have carried out these calculations in order to gain familiarity with the Laplace transform. In practice, engineers use a table of Laplace transforms of the common engineering functions to avoid doing these calculations, which can be very tedious for all but the simplest functions. Table 8.1 gives the Laplace transforms of some common engineering functions.

Example

8.2 Use Table 8.1 to evaluate the Laplace transform of the following functions:

(a) $f(t)=t$
(b) $f(t)=t^5$
(c) $f(t)=\cos(3t)$
(d) $f(t)=e^{-6t}$

Solution (a) We have $f(t)=t$ and so directly from Table 8.1 we deduce

$$F(s) = \frac{1}{s^2}$$

(b) Using the entry for t^n we note that $n=5$ and so we have

$$F(s) = \frac{5!}{s^{5+1}} = \frac{5 \times 4 \times 3 \times 2 \times 1}{s^6} = \frac{120}{s^6}$$

(c) Using the entry for cos (bt) and noting that $b=3$ we have

$$F(s) = \frac{s}{s^2+3^2} = \frac{s}{s^2+9}$$

(d) Using the entry for e^{-at} and noting that $a=6$ we have

$$F(s) = \frac{1}{s+6}$$

Table 8.1
The Laplace transforms of some common engineering functions

Function $f(t)$	Laplace transform $F(s)$
1	$1/s$
t	$1/s^2$
t^n	$n!/s^{n+1}$
$\sin(bt)$	$b/(s^2+b^2)$
$\cos(bt)$	$s/(s^2+b^2)$
e^{-at}	$1/(s+a)$
$t^n e^{-at}$	$n!/(s+a)^{n+1}$
damped sine, $e^{-at}\sin(bt)$	$b/[(s+a)^2+b^2]$
damped cosine, $e^{-at}\cos(bt)$	$(s+a)/[(s+a)^2+b^2]$
$t\sin(bt)$	$2bs/(s^2+b^2)^2$
$t\cos(bt)$	$(s^2-b^2)/(s^2+b^2)^2$
unit step, $u(t)$	$1/s$
delayed unit step, $u(t-d)$	e^{-sd}/s
unit impulse, $\delta(t)$	1
delayed unit impulse, $\delta(t-d)$	e^{-sd}

It is sometimes necessary to obtain a time domain function given its Laplace transform. This is easily done by using a table of Laplace transforms in reverse. This process is known as taking the **inverse Laplace transform** of a function.

Example

8.3 Use Table 8.1 to find the inverse Laplace transform of the following functions:

(a) $\dfrac{1}{s+2}$ (b) $\dfrac{s}{s^2+4}$ (c) $\dfrac{3}{s^2+9}$ (d) $\dfrac{2}{(s+3)^3}$

Solution (a) Examining the entry for e^{-at} we note that $a=2$ and so

$$f(t) = e^{-at} = e^{-2t}$$

(b) We can write

$$F(s) = \frac{s}{s^2+4} = \frac{s}{s^2+2^2}$$

Examining the entry for $\cos(bt)$ we note that $b=2$ and so

$$f(t) = \cos(bt) = \cos(2t)$$

(c) We can write

$$F(s) = \frac{3}{s^2+9} = \frac{3}{s^2+3^2}$$

Examining the entry for $\sin(bt)$ we note that $b=3$ and so

$$f(t) = \sin(bt) = \sin(3t)$$

(d) We can write

$$F(s) = \frac{2}{(s+3)^3} = \frac{1\times 2}{(s+3)^{2+1}}$$

Examining the entry for $t^n e^{-at}$ we note that $n = 2$ and $a = 3$ and so

$$F(s) = t^n e^{-at} = t^2 e^{-3t}$$

If a function is too complicated to appear in the table then it may be possible to break it down into simpler functions using partial fractions. We examine this process in more detail in Section 8.4.

Self-assessment questions 8.2

1. Explain what is meant by the terms 's domain' and 'time domain'.
2. What do we mean when we say a function is switched on at $t = 0$?
3. State the formula for the Laplace transform of a function $f(t)$.

Exercises 8.2

1. Calculate the Laplace transform of the following functions:
   (a) t^2
   (b) $\sin(6t)$
   (c) $\cos(4t)$
   (d) e^{5t}
   (e) $u(t-3)$ where $u(t)$ is the unit step function
   (f) $t^2 e^{-3t}$
   (g) $t\cos(2t)$
   (h) $t\sin(8t)$
2. Using Table 8.1, find the inverse Laplace transform of the following functions:
   (a) $\frac{1}{s}$ (b) $\frac{1}{s^2}$ (c) $\frac{6}{s^4}$ (d) $\frac{4}{s^2+16}$ (e) $\frac{24}{(s+6)^5}$

8.3 Properties of the Laplace transform

There are many properties of the Laplace transform. We shall only examine those necessary for developing our work on engineering systems.

8.3.1 Linearity

The first property we shall examine is that of **linearity**. It is convenient to split the property of linearity into two parts.

The Laplace transform of the sum of two time domain functions can be obtained by adding together the Laplace transforms of the individual functions.

KEY POINT

> For two functions $f(t)$ and $g(t)$ we have
>
> $$\mathscr{L}\{f(t)+g(t)\} = \mathscr{L}\{f(t)\} + \mathscr{L}\{g(t)\} = F(s) + G(s)$$
>
> where $F(s) = \mathscr{L}\{f(t)\}$ and $G(s) = \mathscr{L}\{g(t)\}$

The Laplace transform of a scaled version of a function can be obtained by scaling the Laplace transform of the original function by the same amount.

KEY POINT

> For a function $f(t)$ we have
>
> $$\mathscr{L}\{af(t)\} = a\mathscr{L}\{f(t)\} = aF(s)$$
>
> where a is a constant and $F(s) = \mathscr{L}\{f(t)\}$.

Example

8.4 Calculate the Laplace transforms of the following functions:

(a) $3t + 8t^2$
(b) $2\sin(2t) + 3\cos(3t)$

Solution (a) We have, using linearity,

$$\begin{aligned}\mathscr{L}\{3t + 8t^2\} &= \mathscr{L}\{3t\} + \mathscr{L}\{8t^2\} \\ &= 3\mathscr{L}\{t\} + 8\mathscr{L}\{t^2\}\end{aligned}$$

So, using Table 8.1, we have

$$\begin{aligned}\mathscr{L}\{3t + 8t^2\} &= \frac{3}{s^2} + 8 \times \frac{2}{s^3} \\ &= \frac{3}{s^2} + \frac{16}{s^3}\end{aligned}$$

(b) We have

$$\begin{aligned}\mathscr{L}\{2\sin(2t) + 3\cos(3t)\} &= \mathscr{L}\{2\sin(2t)\} + \mathscr{L}\{3\cos(3t)\} \\ &= 2\mathscr{L}\{\sin(2t)\} + 3\mathscr{L}\{\cos(3t)\} \\ &= 2\frac{2}{s^2+2^2} + 3\frac{s}{s^2+3^2} \\ &= \frac{4}{s^2+4} + \frac{3s}{s^2+9}\end{aligned}$$

8.3.2 Final value theorem

The next property we examine is the **final value theorem**. This only applies if a function $f(t)$ has a limit as $t \to \infty$. For this case we have

KEY POINT

$$\lim_{t\to\infty} f(t) = \lim_{s\to 0} sF(s)$$

Example

8.5 Verify that the final value of $f(t) = \mathrm{e}^{-3t}$ as $t \to \infty$ is 0 using the final value theorem.

Solution We know that

$$F(s) = \mathscr{L}\{f(t)\} = \frac{1}{s+3}$$

So we have

$$\lim_{s\to 0} sF(s) = \lim_{s\to 0} \frac{s}{s+3} = 0$$

Therefore the final value theorem is verified in this case.

The final value theorem is particularly useful for calculating the steady state values of system outputs when they have settled. We shall see this in Chapter 9.

8.3.3 Laplace transform of a derivative

The next property we examine is the Laplace transform of the derivative of a function. This is given by

KEY POINT

$$\mathscr{L}\left\{\frac{\mathrm{d}^n f(t)}{\mathrm{d}t^n}\right\} = s^n F(s) - s^{n-1} f(0) - s^{n-2}\frac{\mathrm{d}f(0)}{\mathrm{d}t} - \ldots - \frac{\mathrm{d}^{n-1} f(0)}{\mathrm{d}t^{n-1}}$$

The most common cases are when $n = 1$ and $n = 2$.

KEY POINT

$$\mathscr{L}\left\{\frac{\mathrm{d}f(t)}{\mathrm{d}t}\right\} = sF(s) - f(0)$$

$$\mathscr{L}\left\{\frac{\mathrm{d}^2 f(t)}{\mathrm{d}t^2}\right\} = s^2 F(s) - sf(0) - \frac{\mathrm{d}f(0)}{\mathrm{d}t}$$

Example

8.6 Calculate the Laplace transform of the following expression given that $f(0) = 2$ and $\mathrm{d}f(0)/\mathrm{d}t = 1$:

$$2\frac{\mathrm{d}^2 f(t)}{\mathrm{d}t^2} + \frac{\mathrm{d}f(t)}{\mathrm{d}t}$$

Solution We have, using linearity

$$\mathscr{L}\left\{2\frac{\mathrm{d}^2f(t)}{\mathrm{d}t^2}+\frac{\mathrm{d}f(t)}{\mathrm{d}t}\right\}=2\mathscr{L}\left\{\frac{\mathrm{d}^2f(t)}{\mathrm{d}t^2}\right\}+\mathscr{L}\left\{\frac{\mathrm{d}f(t)}{\mathrm{d}t}\right\}$$

First of all we note that

$$\mathscr{L}\left\{\frac{\mathrm{d}^2f(t)}{\mathrm{d}t^2}\right\}=s^2F(s)-sf(0)-\frac{\mathrm{d}f(0)}{\mathrm{d}t}$$

and

$$\mathscr{L}\left\{\frac{\mathrm{d}f(t)}{\mathrm{d}t}\right\}=sF(s)-f(0)$$

So, using the formula for the Laplace transform of a derivative, we have

$$\begin{aligned}\mathscr{L}\left\{2\frac{\mathrm{d}^2f(t)}{\mathrm{d}t^2}+\frac{\mathrm{d}f(t)}{\mathrm{d}t}\right\}&=2\left[s^2F(s)-sf(0)-\frac{\mathrm{d}f(0)}{\mathrm{d}t}\right]+sF(s)-f(0)\\&=2[s^2F(s)-s\times 2-1]+sF(s)-2\\&=2s^2F(s)-4s-2+sF(s)-2\\&=(2s^2+s)F(s)-4s-4\end{aligned}$$

8.3.4 Laplace transform of an Integral

Another result which can be useful is the Laplace transform of the integral of a function. This is

KEY POINT

$$\mathscr{L}\left\{\int_0^t f(t)\mathrm{d}t\right\}=\frac{F(s)}{s}$$

where

$$F(s)=\mathscr{L}\{f(t)\}.$$

Example

8.7 Calculate the Laplace transform of the following expression:

$$6\int_0^t f(t)\mathrm{d}t+3f(t)$$

Solution We have

$$\mathscr{L}\left\{6\int_0^t f(t)\mathrm{d}t+3f(t)\right\}=6\mathscr{L}\left\{\int_0^t f(t)\mathrm{d}t\right\}+3\mathscr{L}\{f(t)\}$$

by linearity. So

$$\mathscr{L}\left\{6\int_0^t f(t)\mathrm{d}t + 3f(t)\right\} = 6\frac{F(s)}{s} + 3F(s)$$
$$= F(s)\left(\frac{6}{s} + 3\right)$$

Self-assessment questions 8.3

1. What condition is necessary for the final value theorem to be valid?
2. State the formula for the Laplace transform of the nth derivative of a function, $f(t)$.

Exercises 8.3

1. Using linearity, calculate the Laplace transform of the following expressions:
   (a) $3\mathrm{e}^{-t} + 5\mathrm{e}^{-5t}$
   (b) $4\sin t + 2\cos(2t)$
   (c) $8t\,\mathrm{e}^{-2t} - 5t^2$
   (d) $6\mathrm{e}^{-4t}\sin(2t) + 5\mathrm{e}^{-2t}\cos(4t)$
2. Calculate the Laplace transform of the following expressions:
   (a) $6\frac{\mathrm{d}f(t)}{\mathrm{d}t} + 2f(t)$ given $f(0) = 2$
   (b) $8\frac{\mathrm{d}^2f(t)}{\mathrm{d}t^2} + 3\frac{\mathrm{d}f(t)}{\mathrm{d}t}$ given $f(0) = 3, \frac{\mathrm{d}f(0)}{\mathrm{d}t} = 2$
   (c) $8\int_0^t f(t)\mathrm{d}t + 4\frac{\mathrm{d}^2f(t)}{\mathrm{d}t^2}$ given $f(0) = 2, \frac{\mathrm{d}f(0)}{\mathrm{d}t} = 1$
3. Verify the final value theorem in the following cases:
   (a) $\lim_{t\to\infty} \mathrm{e}^{-8t}$
   (b) $\lim_{t\to\infty} 2 - \mathrm{e}^{-3t}$
4. State the formula for the Laplace transform of $\mathrm{d}^3f(t)/\mathrm{d}t^3$ and $\mathrm{d}^4f(t)/\mathrm{d}t^4$.

8.4 Solving linear differential equations using the Laplace transform

One of the major uses of the Laplace transform is in obtaining a solution to a linear differential equation. We shall introduce this technique by means of an example.

Example

8.8 Obtain the solution of the following differential equation:

$$\frac{\mathrm{d}f(t)}{\mathrm{d}t} + 2f(t) = 1$$

given the initial condition $f(0) = 0$.

Solution In order to solve this differential equation we take the Laplace transform of both sides and solve the problem in the s domain. Taking Laplace transforms of the differential equation gives

$$\mathscr{L}\left\{\frac{\mathrm{d}f(t)}{\mathrm{d}t} + 2f(t)\right\} = \mathscr{L}\{1\}$$

We note that

$$\mathscr{L}\left\{\frac{\mathrm{d}f(t)}{\mathrm{d}t}\right\} = sF(s) - f(0)$$

and so we have

$$sF(s) - f(0) + 2F(s) = \frac{1}{s}$$

Now $f(0) = 0$ and so

$$sF(s) - 0 + 2F(s) = \frac{1}{s}$$

$$F(s)(s + 2) = \frac{1}{s}$$

$$F(s) = \frac{1}{s(s + 2)}$$

Thus far we have obtained the solution of the differential equation but this solution is given in the s domain. In order to obtain $f(t)$ we need to take the inverse Laplace transform of this expression. Examining Table 8.1 we see that no entry exists for an expression of this form. Therefore we first need to split the expression into simpler components using partial fractions (Appendix 2 contains a brief introduction to partial fractions). This gives

$$\frac{1}{s(s + 2)} = \frac{A}{s} + \frac{B}{s + 2}$$

Placing the partial fractions over a common denominator gives

$$\frac{1}{s(s + 2)} = \frac{A(s + 2) + Bs}{s(s + 2)}$$

$$\frac{1}{s(s + 2)} = \frac{(A + B)s + 2A}{s(s + 2)}$$

Comparing numerators we have

$$1 = (A + B)s + 2A \qquad \text{Eqn. [8.1]}$$

where A and B are constants to be determined.

Comparing the constant terms on the left- and right-hand sides of Eqn. [8.1] we have

$$2A = 1$$

$$A = \frac{1}{2}$$

Comparing the coefficient of s on the left- and right-hand sides of Eqn. [8.1] we have

$$A + B = 0$$

$$\frac{1}{2} + B = 0$$

$$B = -\frac{1}{2}$$

So

$$F(s) = \frac{1}{s(s+2)} = \frac{1}{2s} - \frac{1}{2(s+2)} = \frac{1}{2}\left(\frac{1}{s}\right) - \frac{1}{2}\left(\frac{1}{s+2}\right)$$

Now we can use Table 8.1 to invert $F(s)$. This gives

$$f(t) = \frac{1}{2} \times 1 - \frac{1}{2} \times \mathrm{e}^{-2t}$$

$$f(t) = \frac{1}{2} - \frac{\mathrm{e}^{-2t}}{2}$$

$$f(t) = \frac{1}{2}(1 - \mathrm{e}^{-2t}) \quad t \geq 0$$

This is the required solution of the differential equation.

The procedure is the same for solving differential equations of order 2 or higher. The only difference is that the algebra tends to be more complicated.

Self-assessment questions 8.4

1. Describe the procedure for solving a linear differential equation using the Laplace transform.

Exercises 8.4

1. Solve the following differential equations using the Laplace transform:

   (a) $\frac{\mathrm{d}y(t)}{\mathrm{d}t} + 3y(t) = 2 \quad y(0) = 1$

   (b) $\frac{\mathrm{d}f(t)}{\mathrm{d}t} + 8f(t) = 5 \quad f(0) = 3$

(c) $\dfrac{dc(t)}{dt} + 12c(t) = 4 + e^{-2t} \quad c(0) = 3$

(d) $\dfrac{d^2f(t)}{dt^2} + 6\dfrac{df(t)}{dt} + 9f(t) = 2e^{-t} \quad \dfrac{df(0)}{dt} = 1, f(0) = 1$

(e) $\dfrac{d^2y(t)}{dt^2} + 3\dfrac{dy(t)}{dt} + 2y(t) = 6 + e^{-3t} \quad \dfrac{dy(0)}{dt} = 2, y(0) = 4$

8.5 Transfer functions

If we assume that the initial conditions on a system are zero when taking the Laplace transform of a differential equation then a particularly simple relationship exists between the system input and the system output. This relationship is known as the **transfer function** of the system. The transfer function of a system is the ratio of the Laplace transform of the output signal to the Laplace transform of the input signal. It is usually denoted by $G(s)$.

KEY POINT

For a system with input $X(s)$ and output $Y(s)$ we have

$$Y(s) = G(s)X(s)$$

where $G(s)$ is the transfer function of the system.

Note that in order to obtain the output signal it is merely necessary to multiply the input signal by the transfer function. This algebraic simplicity is of great value when analysing complicated systems.

KEY POINT

A transfer function is an algebraic relationship between the input signal and the output signal of a system.

Example

8.9 A system with input signal $x(t)$ and output signal $y(t)$ has the following differential equation:

$$\frac{d^2y(t)}{dt^2} + 6\frac{dy(t)}{dt} + 8y(t) = x(t)$$

Calculate a transfer function for this system.

Solution First we take the Laplace transform of both sides of the equation. So we have

$$\mathscr{L}\left\{\frac{d^2y(t)}{dt^2} + 6\frac{dy(t)}{dt} + 8y(t)\right\} = \mathscr{L}\{x(t)\}$$

$$s^2Y(s) - sy(0) - \frac{dy(0)}{dt} + 6sY(s) - 6y(0) + 8Y(s) = X(s)$$

where

$$Y(s) = \mathscr{L}\{y(t)\} \qquad X(s) = \mathscr{L}\{x(t)\}$$

However, as we intend to obtain a transfer function for the system, we assume zero initial conditions, that is, we assume $y(0)=0$ and $\mathrm{d}y(0)/\mathrm{d}t=0$, and so we have

$$s^2Y(s) + 6sY(s) + 8Y(s) = X(s)$$

$$Y(s)(s^2 + 6s + 8) = X(s)$$

Rearranging gives

$$\frac{Y(s)}{X(s)} = \frac{1}{s^2 + 6s + 8}$$

The function $Y(s)/X(s)$ is the transfer function for the system. It relates the input signal and the output signal in the s domain.

We saw in the previous example that by assuming zero initial conditions the relationship between the input signal and the output signal became reasonably straightforward. If we had not made this assumption then the relationship would have been considerably more complicated with terms involving $f(0)$ and $\mathrm{d}f(0)/\mathrm{d}t$. Engineers find the concept of a transfer function particularly useful because it allows a simple algebraic relationship between the input signal and the output signal to be obtained. They can live with the limitation of assuming zero initial conditions because in practice many systems are **relaxed**, that is, contain no stored energy, before an input is applied. Also, the Laplace transform is formed on the assumption that a signal is zero prior to $t=0$. If initial conditions need to be taken into account then there are ways of dealing with this. We shall not need to use them.

It is common practice to represent a transfer function pictorially using a block diagram. This is shown in Figure 8.1. In this case we have $G(s) = Y(s)/X(s)$.

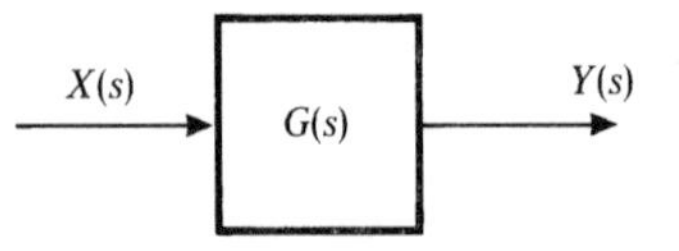

Figure 8.1
A transfer function $G(s)$

We shall see in later chapters that two key features of transfer functions are important in determining the type of system that exists. The first of these is the **system poles**.

KEY POINT

A pole is any value of s that makes the denominator of the transfer function equal to zero. Equivalently, it is any value of s that makes the transfer function tend to infinity.

For example the transfer function

$$G(s) = \frac{6(s+2)}{(s+1)(s+3)}$$

has poles at $s = -1$ and $s = -3$ because, for both these values of s, the denominator of $G(s)$ is zero. The other key feature is the **system zeros**.

KEY POINT

> A zero is any value of s that makes the numerator of the transfer function equal to zero. Equivalently, it is any value of s that makes the transfer function equal to zero.

We shall see just how useful a pole–zero plot of a system can be in later chapters.

In this example there is a system zero at $s = -2$. We shall examine the significance of poles and zeros later. It is usual to draw the system poles and zeros on an s plane diagram. Such a diagram is known as a **pole–zero** plot. Poles are represented by small crosses and zeros are represented by small circles. A pole–zero plot for $G(s)$ is shown in Figure 8.2

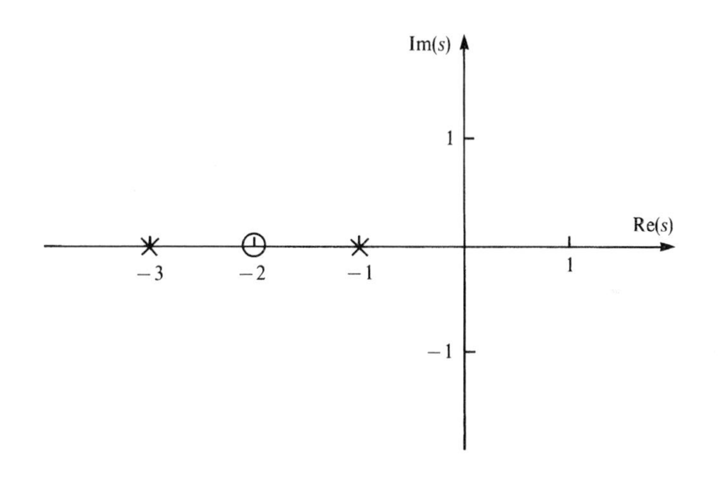

Figure 8.2

A pole–zero plot for $G(s) = \frac{6(s+2)}{(s+1)(s+3)}$

Sometimes a point has more than one pole. For this case it is usual to space the crosses slightly apart to indicate there is more than one. There may also be more than one zero at a point. This is indicated by two or more circles slightly separated.

Example

8.10 Calculate the poles and zeros for the systems with the following transfer functions:

(a) $G(s) = \frac{s+5}{s^2+3s+2}$

(b) $G(s) = \frac{6(s^2+5s+6)}{s^2+2s+2}$

In each case draw a pole–zero plot for the system.

Solution (a) The zeros are obtained by solving $s+5=0$. This gives $s=-5$. The poles are obtained by solving $s^2+3s+2=0$. By inspection we have $(s+1)(s+2)=0$ and so the poles are $s=-1$ and $s=-2$. A pole–zero plot for the system is shown in Figure 8.3.

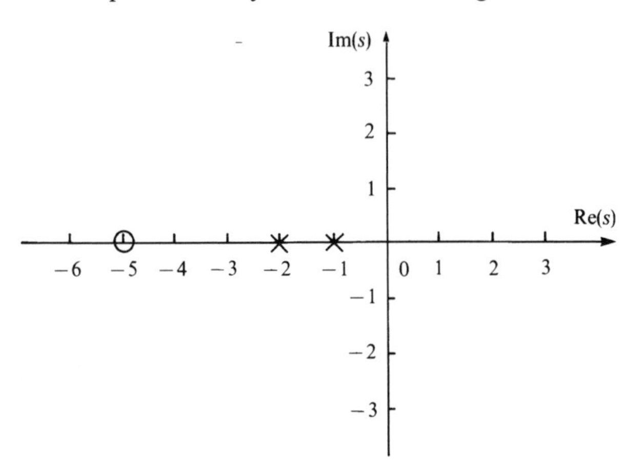

Figure 8.3
A pole–zero plot for the system of Example 8.10(a)

In general, the solution of the quadratic equation $ax^2+bx+c=0$ is

$$x=\frac{-b\pm\sqrt{b^2-4ac}}{2a}$$

(b) The zeros are given by solving $6(s^2+5s+6)=0$. This factorises to $6(s+2)(s+3)=0$. So $s=-2$ and $s=-3$ are the system zeros. The poles are given by solving $s^2+2s+2=0$. There are no obvious factors in this case and so we use the formula for solving a quadratic equation. This gives

$$\begin{aligned}s&=\frac{-2\pm\sqrt{2^2-4\times 2}}{2}\\&=\frac{-2\pm\sqrt{-4}}{2}\\&=-1\pm\frac{2}{2}\mathrm{j}\\&=-1\pm\mathrm{j}\end{aligned}$$

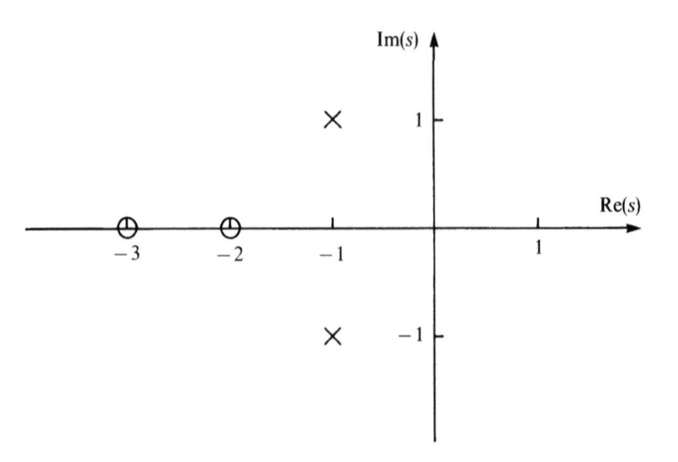

Figure 8.4
A pole–zero plot for the system of Example 8.10(b)

Note that in this case the poles are complex. A pole–zero plot for the system is shown in Figure 8.4.

Self-assessment questions 8.5

1. Explain what is meant by a transfer function of a system.
2. How are the poles and zeros of a system obtained?

Exercises 8.5

1. Calculate transfer functions for the following systems:

(a) $\dfrac{d^2y}{dt^2} + 6y = 4x \quad x$, input; y, output

(b) $\dfrac{d^3x}{dt^3} + 8\dfrac{d^2x}{dt^2} + 7\dfrac{dx}{dt} + x = 6\dfrac{du}{dt} + 2u$
u, input; x, output

2. Calculate the poles and zeros of the following transfer functions and hence draw a pole–zero plot in each case:

(a) $\dfrac{s+1}{(s+4)(s+5)}$

(b) $\dfrac{s^2+s+1}{(s+3)(s^2+3s+6)}$

(c) $\dfrac{(s+1)(s^2+3s+6)}{(s^2+10s+25)(s^2-3s+2)}$

Test and assignment exercises 8

1. Solve the following differential equations:

(a) $\dfrac{dy(t)}{dt} + 6y(t) = 3 \quad y(0) = 0$

(b) $3\dfrac{dy(t)}{dt} + 6y(t) = 2 - e^{-3t} \quad y(0) = 3$

(c) $2\dfrac{d^2f(t)}{dt^2} + 3df(t) + 6f(t) = 3\sin t$

$f(0) = 1, \quad \dfrac{df(0)}{dt} = 1$

(d) $\dfrac{d^3x(t)}{dt^3} + \dfrac{dx(t)}{dt} + 6x(t) = \cos(4t)$

$x(0) = 2, \quad \dfrac{dx(0)}{dt} = 0, \quad \dfrac{d^2x(0)}{dt^2} = 1$

2. Calculate the poles and zeros of the following transfer functions and hence draw a pole–zero plot in each case:

(a) $\dfrac{s(s+2)}{(s+3)(s+4)}$

(b) $\dfrac{(s+1)(s+0.5)}{(s+0.1)(s^2+5s+8)}$

(c) $\dfrac{s^2+3s-6}{(s-0.1)(s^2-3s+6)}$

(d) $\dfrac{(s+0.4)(s^2+3s+10)}{(s^2+3s+7)(s^2-3s+5)}$

9 First order systems

Objectives

This chapter:

- describes the type of test signals used to analyse a system
- describes the main characteristics of a first order system
- defines the time constant of a first order system
- defines the d.c. gain of a first order system
- calculates the time response of a first order system
- explains how the position of the system poles and zeros affect the time response

9.1 Introduction

We have already seen in earlier chapters that several different types of system have similar system differential equations. For this reason engineers find it convenient to classify systems in terms of the type of system differential equation they have. The simplest type of system is one which is modelled by a first order differential equation. Such a system is known as a **first order system**. We shall analyse these systems in this chapter. Before doing so we first look at the types of test signals that are commonly used when analysing the response of any engineering system.

9.2 Test signals

In order to determine the **response** of a system, engineers make use of a variety of **test signals**. The general arrangement is shown in Figure 9.1. A test signal $r(t)$ is applied to the system. This results in a response $c(t)$. Note that

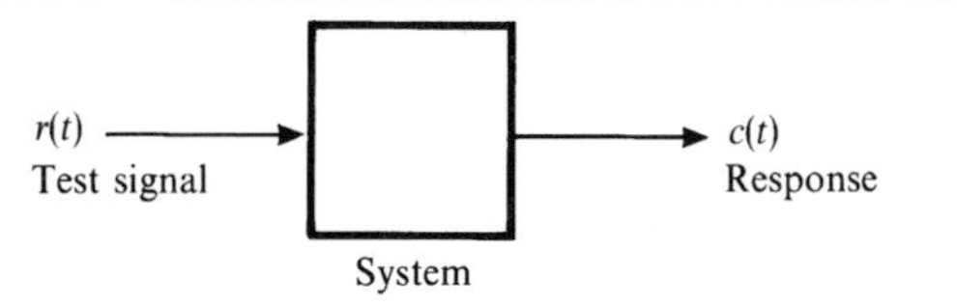

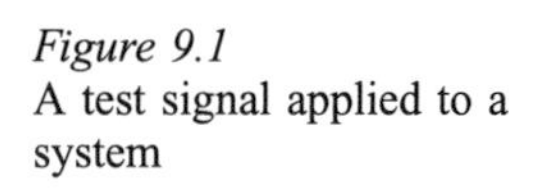

Figure 9.1
A test signal applied to a system

both the input to the system and the output from the system are functions that vary with time. We shall now examine some of the more common test signals.

9.2.1 Step signal

The **step signal** is a signal that has zero value which changes to a constant value at a certain point in time. The most common type of step signal is the **unit step**. This is shown in Figure 9.2. It is common practice to denote the unit step signal by $u(t)$.

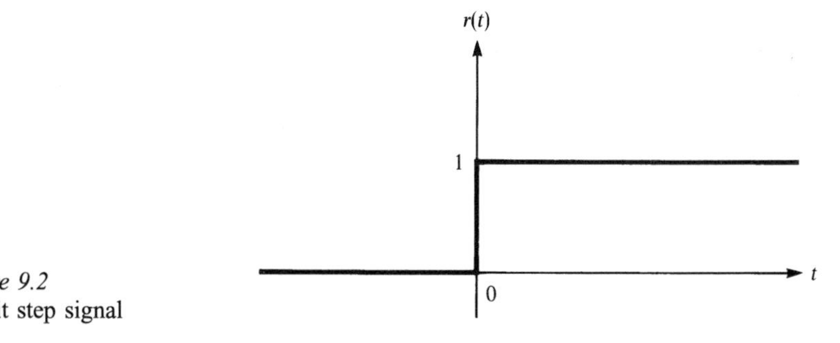

Figure 9.2
A unit step signal

KEY POINT

A unit step signal is given by

$$r(t) = u(t) = \begin{cases} 0 & t < 0 \\ 1 & t \geq 0 \end{cases}$$

In general, if a function $f(t)$ changes its value at $t = 0$ then the function $f(t - t_0)$ changes its value at $t = t_0$. This is easily seen by noting that if $t - t_0 = 0$ then $t = t_0$

The brace notation is used by engineers to define a signal that has different mathematical descriptions in different regions of the time domain. Note that the unit step switches on at $t = 0$ and has a height of 1 after the switch. A more general step signal has a height k and starts at $t = t_0$. This signal is shown in Figure 9.3. It is known as a **delayed step signal**.

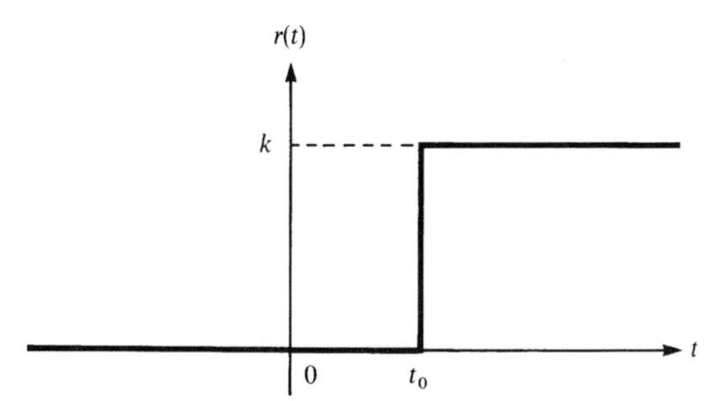

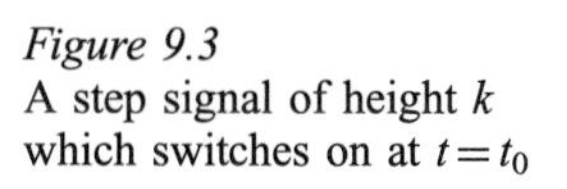

Figure 9.3
A step signal of height k which switches on at $t = t_0$

KEY POINT

A delayed step signal of height k is given by

$$r(t) = ku(t - t_0) = \begin{cases} 0 & t < t_0 \\ k & t \geq t_0 \end{cases}$$

The step signal is one of the most common signals used by engineers because it corresponds to so many different situations found in real life. For example, switching on a cooker, opening a pipe valve and closing an electrical switch are all processes that can be modelled by a step signal.

9.2.2 Ramp signal

The **ramp signal** has a value of zero until a certain point in time at which it changes to a steadily rising value. The most common type of ramp signal is the **unit ramp**, so called because the rising part of its graph has a slope of 1. This is shown in Figure 9.4. There is no commonly agreed symbol to denote a unit ramp. Note, however, that it is possible to define a unit ramp signal in terms of a unit step signal $u(t)$.

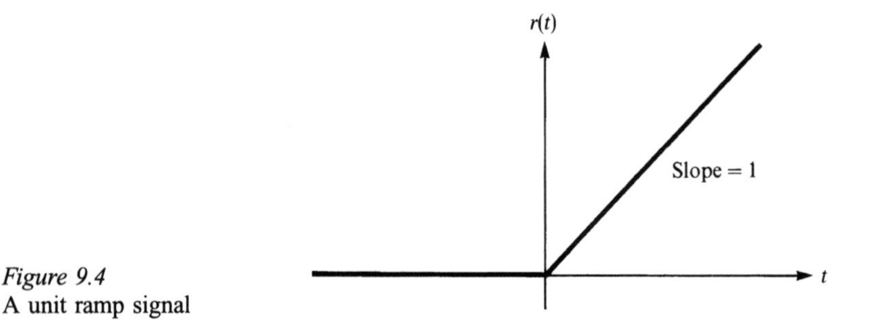

Figure 9.4
A unit ramp signal

A unit ramp signal is given by

$$r(t) = tu(t) = \begin{cases} 0 & t < 0 \\ t & t \geq 0 \end{cases}$$

A more general ramp signal is shown in Figure 9.5.

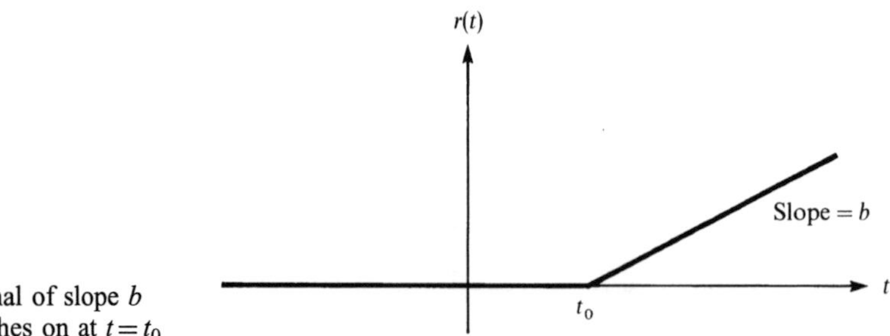

Figure 9.5
A ramp signal of slope b which switches on at $t = t_0$

KEY POINT

A delayed ramp signal with slope b is given by

$$r(t) = btu(t - t_0) = \begin{cases} 0 & t < t_0 \\ bt & t \geq t_0 \end{cases}$$

The delayed ramp signal changes its behaviour at $t = t_0$ and then has a slope of b. The ramp signal is useful for modelling the behaviour of quantities that change gradually. For example, the angular movement of a radio telescope to follow the changing position of the stars as the Earth rotates can be modelled by a ramp signal.

9.2.3 Sinusoidal signal

Sometimes it is useful to apply a **sinusoidal** signal to a system. The term sinusoid is a general one which covers both sine and cosine signals.

KEY POINT

A sinusoidal signal is given by

$$r(t) = A\ \sin(\omega t + \phi)$$

or

$$r(t) = A\ \cos(\omega t + \phi)$$

where A is the **amplitude**, ω is the **angular frequency** and ϕ is the phase angle.

The sine signal and the cosine signal have the same shape. The only difference between the two is that one is 'shifted' in time relative to the other. Which one is used is a matter of convenience. A sinusoidal signal is shown in Figure 9.6.

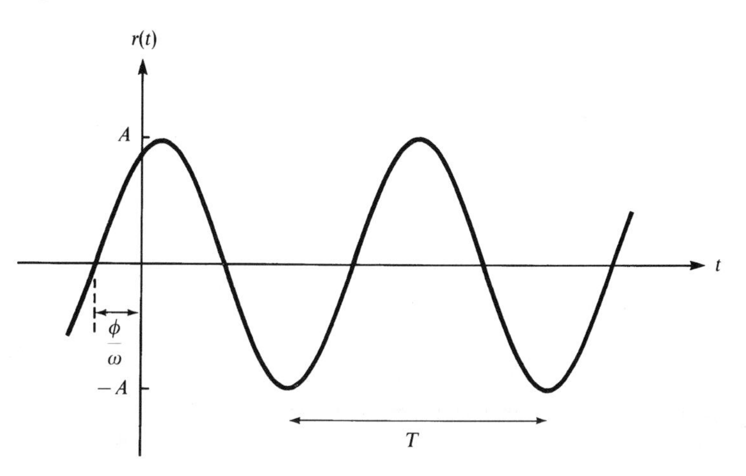

Figure 9.6
The sinusoidal signal $r(t) = A\ \sin\ (\omega t + \phi)$

The **time period** T of the sinusoidal signal is shown in Figure 9.6. This is the time to complete one cycle. The angular frequency is related to the time period by the following:

$$\omega = \frac{2\pi}{T}$$

Examining Figure 9.6 we see that the signal has a nonzero value at $t = 0$. The signal has been **shifted** along the time axis. We can determine the amount of this time shift by noting that

$$\begin{aligned} r(t) &= A \ \sin(\omega t + \phi) \\ &= A \ \sin \omega\left(t + \frac{\phi}{\omega}\right) \end{aligned}$$

Therefore the time shift is ϕ/ω.

9.2.4 Impulse signal

The easiest way of understanding the impulse signal is by first considering a pulse signal $p(t)$ of height $1/h$ and width h. This is shown in Figure 9.7. As h

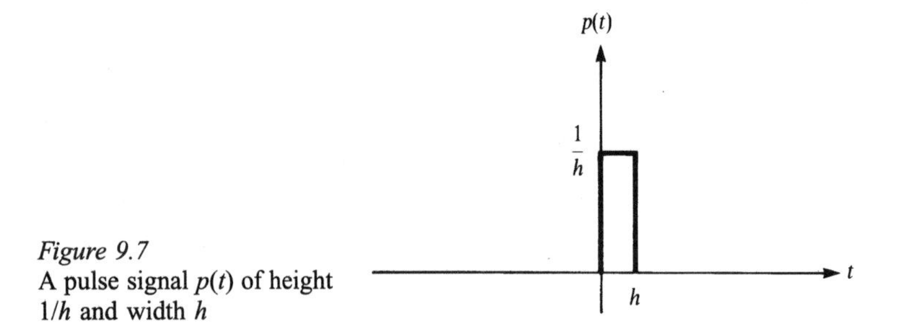

Figure 9.7
A pulse signal $p(t)$ of height $1/h$ and width h

becomes smaller the base of the pulse becomes smaller but the height of the pulse increases. If this process is continued then in the limit the base becomes infinitesimally small and the height becomes infinitely large. We then call the pulse signal the **impulse signal** commonly denoted $\delta(t)$.

KEY POINT

Mathematically we write

$$r(t) = \delta(t) = \lim_{h \to 0} p(t)$$

A useful way of thinking about an impulse signal is to compare it with a hammer blow. This produces an extremely large force for a very short length of time.

The impulse signal is drawn as an arrow, as shown in Figure 9.8. The height of the arrow denotes the **strength** of the impulse. This corresponds to the area of the rectangle before h is shrunk to zero. In the above example the area of the rectangle is $(1/h) \times h = 1$ and so the impulse has strength 1 and is known as a **unit impulse**. An impulse of strength b occurring at time t_0 is given by $b\delta(t - t_0)$. It is known as a **delayed impulse signal**.

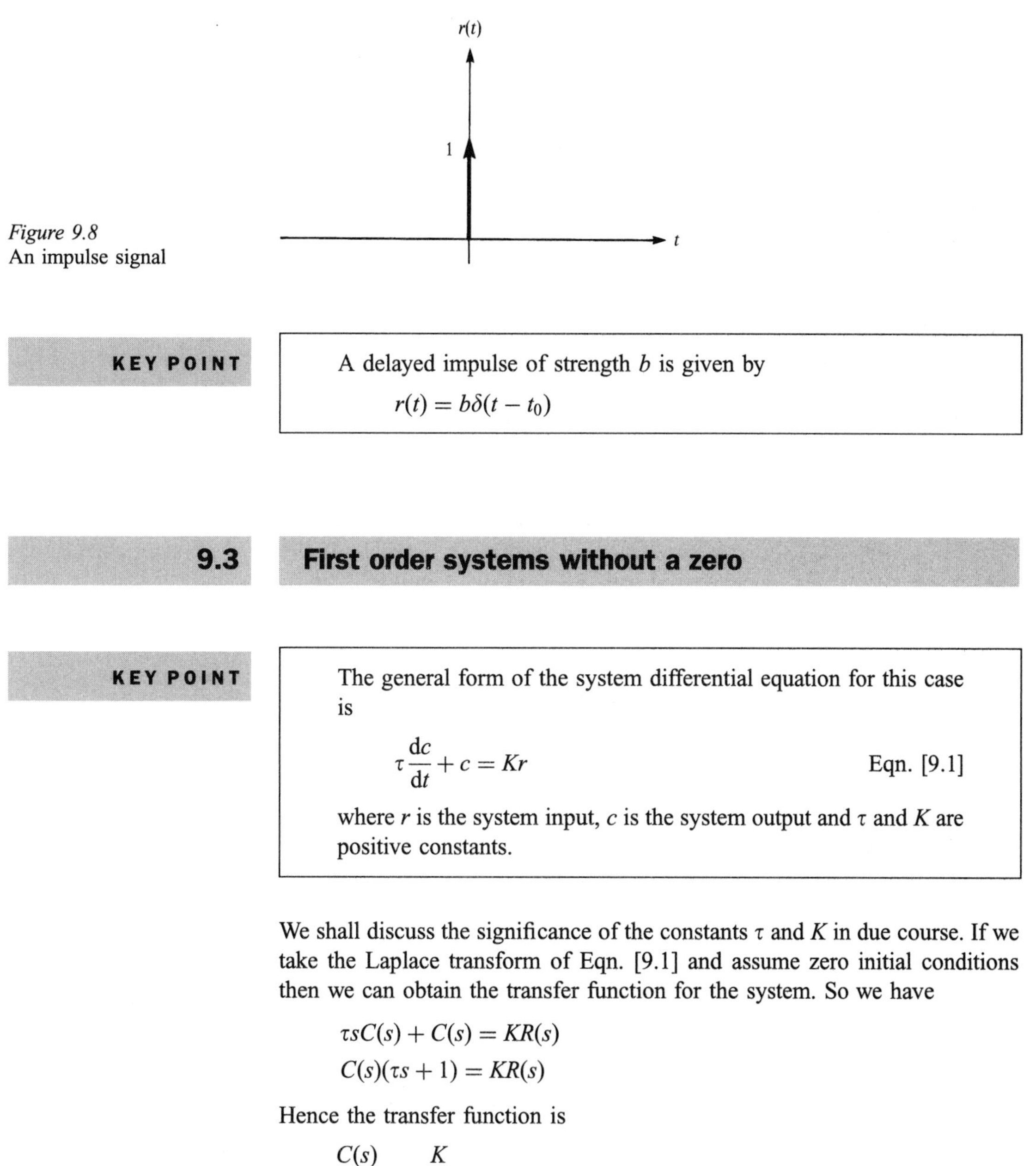

Figure 9.8
An impulse signal

KEY POINT

A delayed impulse of strength b is given by

$$r(t) = b\delta(t - t_0)$$

9.3 First order systems without a zero

KEY POINT

The general form of the system differential equation for this case is

$$\tau \frac{dc}{dt} + c = Kr \qquad \text{Eqn. [9.1]}$$

where r is the system input, c is the system output and τ and K are positive constants.

We shall discuss the significance of the constants τ and K in due course. If we take the Laplace transform of Eqn. [9.1] and assume zero initial conditions then we can obtain the transfer function for the system. So we have

$$\tau sC(s) + C(s) = KR(s)$$
$$C(s)(\tau s + 1) = KR(s)$$

Hence the transfer function is

$$\frac{C(s)}{R(s)} = \frac{K}{1 + \tau s}$$

Recall from Chapter 8 that in order to calculate the poles for a system we need to equate the denominator of the transfer function to zero. This equation is known as the **characteristic equation** of the system. So we have

$$1 + \tau s = 0$$
$$s = -\frac{1}{\tau}$$

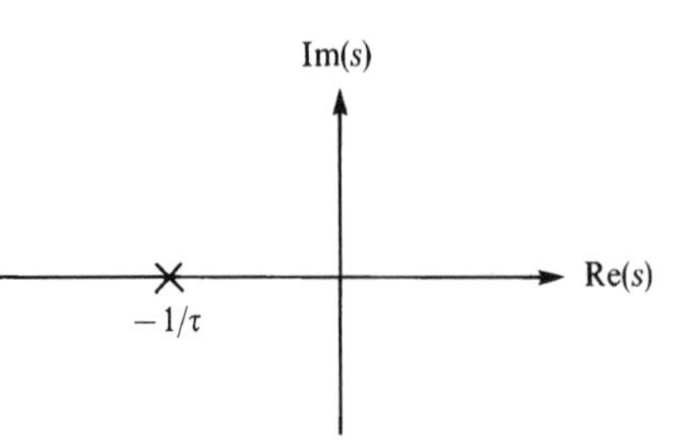

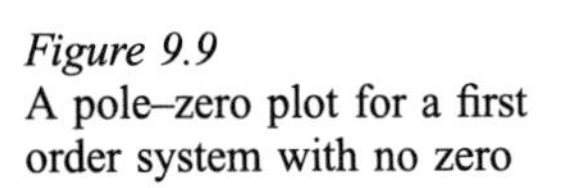
Figure 9.9
A pole–zero plot for a first order system with no zero

Recall that a cross is used to indicate a pole and a circle is used to indicate a zero on a pole–zero plot.

A pole–zero plot for this system is shown in Figure 9.9. Note that the system has one pole and no zeros.

Let us now see how this system behaves with a variety of inputs.

Example

9.1 Calculate the response of the first order system defined by Eqn. [9.1] to a unit step input. Assume the initial conditions are zero for the system when the step is applied.

Solution In order to work out the system response we make use of the transfer function. Recall that

$$\frac{C(s)}{R(s)} = \frac{K}{1 + \tau s}$$

Rearranging we have

$$C(s) = \frac{K}{1 + \tau s} R(s)$$

From Table 8.1 we see that the Laplace transform of a unit step is $1/s$. So,

$$R(s) = \frac{1}{s}$$

and so we can write

$$C(s) = \frac{K}{s(1 + \tau s)}$$

This expression contains two poles. The pole at $s = 0$ is due to the input signal and the pole at $s = -1/\tau$ is due to the system. In order to solve this system we need to separate the expression for $C(s)$ into partial fractions (see Appendix 2). In this case the denominator contains two linear factors. So we have

$$C(s) = \frac{K}{s(1 + \tau s)} = \frac{A}{s} + \frac{B}{1 + \tau s}$$

where A and B are constants to be determined. Placing the partial fractions over a common denominator gives

$$\frac{K}{s(1 + \tau s)} = \frac{A(1 + \tau s) + Bs}{s(1 + \tau s)}$$

Comparing the numerator polynomials gives

$$K = A(1 + \tau s) + Bs$$

Collecting terms we have

$$K = s(A\tau + B) + A$$

Comparing the constant terms of the numerator polynomials we have

$$K = A$$

and so

$$A = K$$

Comparing the coefficients of s gives

$$0 = A\tau + B$$

$$B = -A\tau = -K\tau$$

Finally we can write

$$C(s) = \frac{K}{s} - \frac{K\tau}{1 + \tau s}$$

In order to be able to invert the partial fractions of $C(s)$ we need to manipulate them into the standard forms of Table 8.1. So we have

$$C(s) = \frac{K}{s} - \frac{K}{(1/\tau) + s}$$

We can now invert this expression using Table 8.1. This gives

$$c(t) = K - Ke^{-t/\tau}$$

$$c(t) = K(1 - e^{-t/\tau}) \qquad \text{Eqn. [9.2]}$$

We can generalise an observation from the previous example.

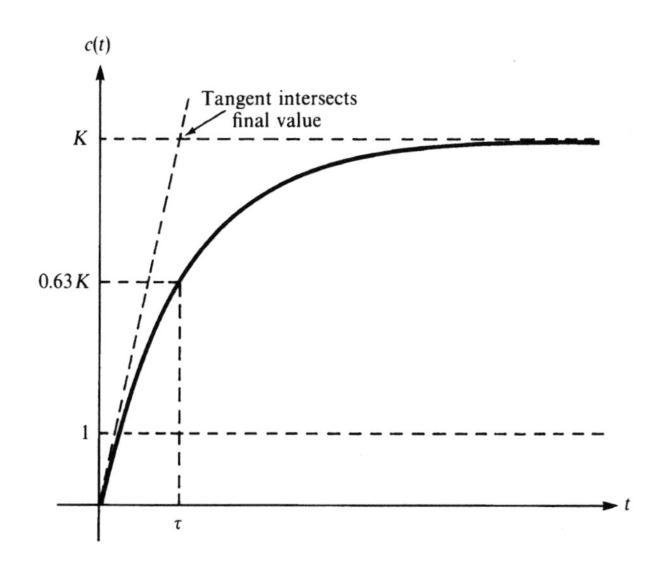

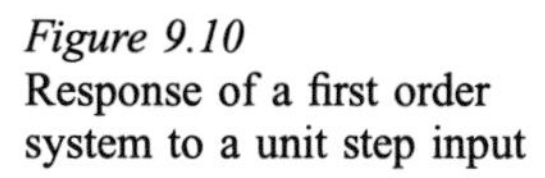
Figure 9.10
Response of a first order system to a unit step input

KEY POINT

The expression for the Laplace transform of a time response contains two types of poles. The poles due to the input are known as **signal poles** and the poles due to the system are known as **system poles**.

Let us now examine the response calculated in the previous example in a little more detail. This response is sketched in Figure 9.10. Note that when the output of the system has settled down it has a steady state value of K. For systems that settle down to a constant value when a step input is applied it is convenient to define a quantity known as the **d.c. gain**. This is the ratio between the steady state output value and the steady state input value.

KEY POINT

The d.c. gain of a system is the ratio between the steady state value of the output signal and the steady state value of the input signal, when the system has settled.

The use of K to represent d.c. gain is common. Unfortunately it clashes with its use to represent string stiffness. Notation clashes often occur in engineering and so it is important to watch out for them. There are not enough symbols in the various alphabets used by engineers to allow a unique symbol to be assigned to every variable and constant of interest.

Dynamic systems are not always changing. Quite often they are in a settled state until disturbed.

The term d.c. is borrowed from electrical engineering because a constant signal is similar to a direct current in that it has a constant value. As we can see, for this system the d.c. gain is K.

The d.c. gain is an important quantity, as engineers are often interested in knowing the signal levels in a system after it has settled down. The units of K depend on the units of the system input and system output. For example, for an armature-controlled d.c. motor the input signal is a voltage (V) and the output signal may be an angular velocity (rad s^{-1}). In this case K has units of rad s^{-1} V^{-1}.

Further examination of Figure 9.10 shows that we can draw a tangent to the response curve at the time origin. Recalling Eqn. [9.2] we can calculate the slope of this tangent. We have

$$c(t) = K(1 - \mathrm{e}^{-t/\tau})$$

Therefore,

$$\frac{\mathrm{d}c(t)}{\mathrm{d}t} = K\left(-\frac{1}{\tau}\right)(-\mathrm{e}^{-t/\tau}) = \frac{K}{\tau}\mathrm{e}^{-t/\tau}$$

At the origin we have

$$\left.\frac{\mathrm{d}c(t)}{\mathrm{d}t}\right|_{t=0} = \frac{K}{\tau}\mathrm{e}^{0} = \frac{K}{\tau}$$

Therefore the tangent has equation

$$c_{\mathrm{T}}(t) = \frac{K}{\tau}t$$

This intersects the final value of the response when $c_{\mathrm{T}}(t) = K$. This gives

$$K = \frac{K}{\tau}t$$

and so

$$t = \tau$$

Clearly the quantity τ is significant. It is known as the **time constant** of the system. It is a measure of how quickly the system responds to an input.

KEY POINT

> The time constant is a measure of how quickly a first order system responds to a step input. The time constant has units of seconds.

If we subsitute $t = \tau$ into Eqn. [9.2] we obtain

$$\begin{aligned} c(\tau) &= K(1 - e^{-\tau/\tau}) \\ &= K(1 - e^{-1}) \\ &= 0.6321K \end{aligned}$$

A common rule of thumb used by engineers is that a first order system reaches 63% of its final value after one time constant. When engineers try to measure the time constant of a system from tests they prefer the 63% rule rather than draw a tangent to the curve at the origin, as it is often difficult to draw the tangent accurately, especially if noise is present on the response curve.

KEY POINT

> A first order system reaches 63% of its final value after one time constant.

Another rule of thumb is that a first order system has reached its final value after four time constants. We can see why this is so by putting $t = 4\tau$ into Eqn. [9.2]. We obtain

$$\begin{aligned} K(1 - e^{-4\tau/\tau}) &= K(1 - e^{-4}) \\ &= 0.9817K \end{aligned}$$

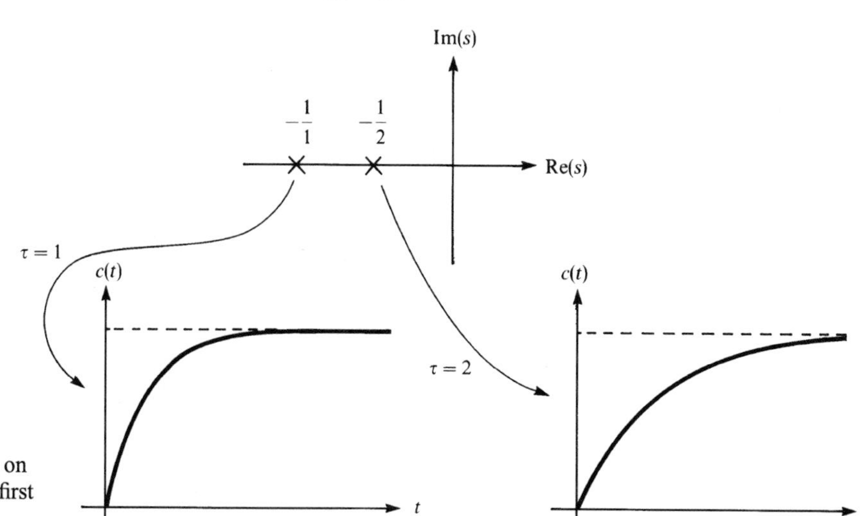

Figure 9.11
Effect of pole position on the step response of a first order system

We see that this is 98% of the final value, which is usually good enough for engineers. Engineers say the system **settles** after four time constants. Note that, theoretically, the response never quite reaches K because the exponential transient only decays to zero as $t \to \infty$.

KEY POINT

A first order system has settled after four time constants.

It is interesting to examine the effect of moving the system pole. Recall that the system pole is at $s = -1/\tau$ as shown in Figure 9.9. Increasing τ corresponds to decreasing $1/\tau$ and so the pole moves closer to the imaginary axis. Similarly, decreasing τ corresponds to moving the system pole further from the imaginary axis. The effect of moving the system pole on the step response of a first order system is shown in Figure 9.11 for time constants of 1 s and 2 s.

KEY POINT

Moving a system pole away from the imaginary axis corresponds to decreasing the response time of a first order system. Moving a system pole toward the imaginary axis corresponds to increasing the response time of a first order system.

To summarise, the d.c. gain tells an engineer the ratio between the output signal level and the input signal level once the system has settled down. The time constant provides an indication of how quickly the system responds to a change in the input. These two quantities are the key to understanding a first order system.

KEY POINT

A first order system with no zero has a transfer function

$$\frac{K}{1 + \tau s}$$

where τ is the time constant of the system and K is the d.c. gain of the system.

It is important that the constant in the denominator is 1 if K and τ are to be correctly identified. Let us now revisit some of the first order systems we met in earlier chapters and identify the d.c. gain and time constant in each case.

Example

9.2 Identify the d.c. gain and time constant, stating their units, for the following first order systems:

(a) the RC circuit examined in Example 4.4;
(b) the spring–damper system examined in Example 2.5.

Solution (a) Recall that for an RC circuit the system differential equation is

$$v_i = RC\frac{dv_o}{dt} + v_o$$

where v_i is the input voltage and v_o is the output voltage.

The first thing we need to do is obtain a transfer function for this system. So taking the Laplace transform and assuming zero initial conditions we have

$$V_i(s) = RCsV_o(s) + V_o(s)$$

$$\frac{V_o(s)}{V_i(s)} = \frac{1}{RCs + 1}$$

The coefficient in the denominator is 1 and so we can proceed to identify the d.c. gain and the time constant. The coefficient of the s term is RC and so this is the time constant of the circuit. It has units of seconds. We can write $\tau = RC$. By varying the product RC we can adjust the time it takes for the circuit to respond to a step input.

The numerator constant is 1 and so the d.c. gain is 1, that is, $K = 1$. We note that both the input signal and the output signal are voltages and so in this case K has no units and is purely a number. Because the d.c. gain is 1, when the system has settled down the input voltage and the output voltage will be the same.

Here we see a clash of notation in the use of K. This sometimes happens and so it is instructive to learn how to deal with it.

(b) Recall for the spring–mass–damper the system differential equation is

$$f = B\frac{dy}{dt} + Ky$$

where f is the input force and y is the output displacement. The constant K is the spring stiffness and the constant B is the damping coefficient.

Taking the Laplace transform and assuming zero initial conditions we obtain

$$F(s) = BsY(s) + KY(s)$$

$$\frac{Y(s)}{F(s)} = \frac{1}{K + Bs}$$

Before we can identify the d.c. gain and the time constant we first need to reorganise the transfer function. Recall that the standard form is

$$\frac{\text{d.c.gain}}{1 + \text{time constant} \times s}$$

That is, there is a 1 in the denominator polynomial. If we divide the top and bottom of the transfer function by K we can obtain the standard form. So we have

$$\frac{Y(s)}{F(s)} = \frac{1/K}{(K + Bs)/K} = \frac{1/K}{1 + sB/K}$$

It is always useful to imagine how systems change as their parameters are varied.

We now see that the system time constant is B/K. This has units of seconds. In this case we see that if the damping coefficient B of the damper is reduced then the time constant is reduced and the system responds more quickly. Increasing the spring stiffness K also reduces the system time constant.

The d.c. gain of the system is $1/K$. The input signal is a force with units of newtons and the output signal is a displacement with units of metres. Therefore, the d.c. gain has units of m N^{-1}.

Engineers have developed a quick way of finding the d.c. gain of a transfer function known as the **cover-up rule**. It is worth spending some time dealing with this rule as it is applicable to both first and higher order systems. Consider Figure 9.12 which shows a signal $R(s)$ being applied to a transfer function $G(s)$ to produce an output $C(s)$. Let us now assume that the input is a unit step and so $R(s) = 1/s$. We now wish to calculate the steady state output from the system as a result of a step input, that is, $c(t)$ as $t \to \infty$. Recall from Chapter 8 the final value theorem. This states that, provided a final value exists,

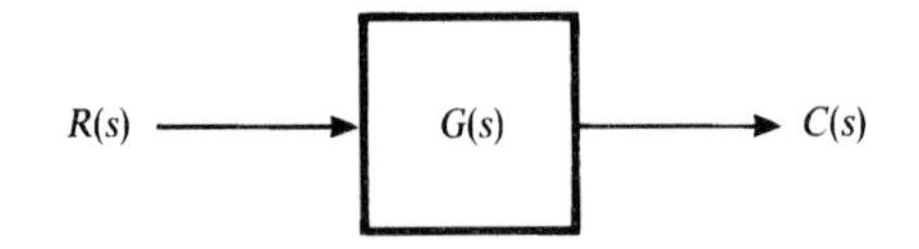

Figure 9.12
A system with transfer function $G(s)$

$$\lim_{t\to\infty} c(t) = \lim_{s\to 0} sC(s)$$

Now $R(s) = 1/s$ and so $C(s) = G(s)/s$. Therefore,

$$\lim_{t\to\infty} c(t) = \lim_{s\to 0} \frac{sG(s)}{s} = \lim_{s\to 0} G(s)$$

Note that all this is based on the assumption that the system settles down to a steady output; some systems do not.

We see that the s in the numerator due to the theorem has cancelled with the s in the denominator due to the step input. This is very convenient because it allows us to calculate the final value of $c(t)$ merely by letting s tend to zero in the transfer function. Now as a unit step has been applied to the system its height is unity and so the final value of $c(t)$ corresponds to the d.c. gain of the system.

We see now why engineers call this the cover-up rule. If we cover up any terms that have s in them then what is left gives the d.c. gain. For example, $3/(2 + 4s)$ has a d.c. gain of $3/(2 + 0) = 1.5$. We have covered up the $4s$ term because as s tends to zero the term disappears. The usefulness of the cover-up rule will become clearer when we examine more complicated transfer functions in later chapters.

KEY POINT

The cover-up rule states that the d.c. gain of a system can be obtained by putting $s = 0$ in the system transfer function and then calculating the resultant gain. The rule is only valid if the system output settles down to a constant value when subjected to a step input.

Example

9.3 Calculate the d.c. gain for each of the following transfer functions using the cover-up rule.

(a) $\dfrac{3}{1+2s}$ (b) $\dfrac{4}{2+4s}$ (c) $\dfrac{9}{4+s}$ (d) $\dfrac{18.2}{3.2+7.3s}$

Solution (a) $\dfrac{3}{1+0}=3$ (b) $\dfrac{4}{2+0}=2$ (c) $\dfrac{9}{4+0}=2.25$ (d) $\dfrac{18.2}{3.2+0}=5.69$

We shall now examine how a first order system responds to a unit ramp input.

Example

9.4 Calculate the response of a first order system to a unit ramp input.

Solution Recall from Table 8.1 that the Laplace transform of a unit ramp is $R(s) = 1/s^2$. So we have

$$C(s) = \frac{KR(s)}{1+\tau s} = \frac{K}{(1+\tau s)s^2}$$

Note that this expression has a double pole at $s = 0$ due to the input signal and a pole at $s = -1/\tau$ which is the system pole.

We now need to split this expression into partial fractions. So we have

$$C(s) = \frac{K}{(1+\tau s)s^2} = \frac{A}{1+\tau s} + \frac{B}{s^2} + \frac{C}{s}$$

where A, B and C are constants to be determined.

Placing the partial fractions over a common denominator gives

$$\frac{K}{(1+\tau s)s^2} = \frac{As^2 + B(1+\tau s) + C(1+\tau s)s}{(1+\tau s)s^2}$$

Rearranging we have

$$\frac{K}{(1+\tau s)s^2} = \frac{(A+C\tau)s^2 + (B\tau + C)s + B}{(1+\tau s)s^2}$$

Comparing numerators we have

$$K = (A + C\tau)s^2 + (B\tau + C)s + B$$

Comparing constant terms of the numerator polynomials, we have

$$B = K$$

Comparing the coefficients of the s terms of the numerator polynomials, we have

$$0 = B\tau + C$$

and so

$$C = -B\tau = -K\tau$$

Comparing the coefficients of the s^2 terms of the numerator polynomials, we have

$$0 = A + C\tau$$

and so

$$A = -C\tau = K\tau^2$$

Substituting the values of A, B and C into the expression for $C(s)$ gives

$$C(s) = \frac{K\tau^2}{1 + \tau s} + \frac{K}{s^2} - \frac{K\tau}{s}$$

Arranging this expression to fit the standard forms of Table 8.1 gives

$$C(s) = \frac{K\tau}{(1/\tau) + s} + \frac{K}{s^2} - \frac{K\tau}{s}$$

We can now invert this expression using Table 8.1. This gives

$$c(t) = K\tau e^{-t/\tau} + Kt - K\tau$$

Let us now examine the ramp response in a little more detail. Note from the previous example that the response consists of three parts. There is a ramp term Kt, which is the same as the input ramp except that it is multiplied by K. There is an exponential term which decays to zero with time. Such a term is known as a **transient** term because it decays to zero with time.

KEY POINT

> Any term in a system response that decays to zero with time is known as a transient term.

Finally, there is a fixed offset term $-K\tau$. It is easier to visualise these three parts if we examine the response of a system with a d.c. gain of 1. The input and output signals for such a case are shown in Figure 9.13.

Note that after a short time, corresponding to the decay of the exponential transient term, the output lags the input by a fixed amount τ. There are many practical examples of this sort of behaviour. One such example is the reading presented by a car speedometer as a car is brought to a halt. In this case a negative ramp is being applied to the car speedometer and the reading tends to lag behind the actual value of the car speed. So for a short time after the car stops the speedometer is reading a positive speed. This is only observed on dial speedometers. Electronic displays do not have an appreciable time constant and hence there is no lag in the speed displayed.

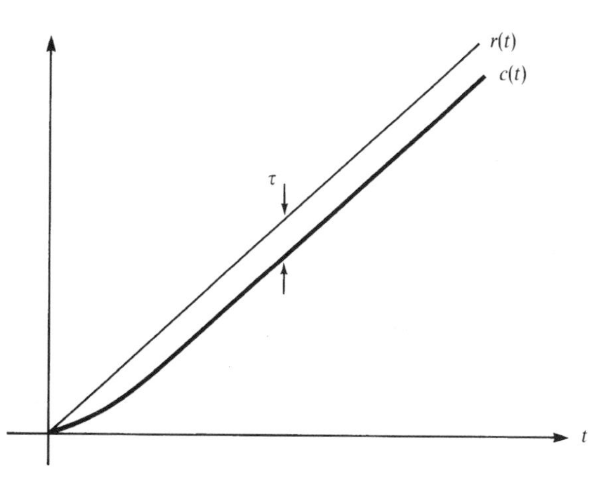

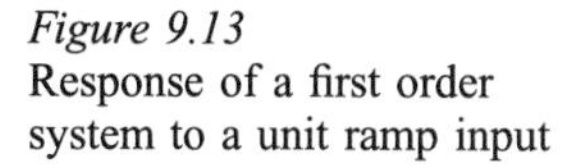
Figure 9.13
Response of a first order system to a unit ramp input

Self-assessment questions 9.3

1. Explain what is meant by the time constant of a first order system.
2. Explain what is meant by the d.c. gain of a first order system.
3. Describe the cover-up rule.
4. Explain the effect on the step response of a first order system of varying the position of the system pole.

Exercises 9.3

1. Identify the d.c. gain and time constant, stating their units, of the water storage tank system described in Example 5.1.
2. Identify the d.c. gain and time constant of the systems with the following transfer functions:
   (a) $\dfrac{5}{1+4s}$ (b) $\dfrac{6}{3+8s}$ (c) $\dfrac{7}{2+3s}$ (d) $\dfrac{9.2}{3.6+7.1s}$
3. Consider the RC circuit of Example 4.4. Calculate the response of this system to a step input voltage signal of height 10 V. Sketch this response. Use component values of $R = 10\ \mathrm{k\Omega}$ and $C = 1\ \mu\mathrm{F}$.
4. Consider the liquid storage tank system of Example 5.1. Calculate the response of this system to a step input flow rate signal of height $5\ \mathrm{m^3\ s^{-1}}$. Sketch this response. Use system parameters of $C = 1 \times 10^{-3}\ \mathrm{m^5\ N^{-1}}$ and $R = 1 \times 10^4\ \mathrm{N\ s\ m^{-5}}$.

9.4 First order systems with a zero

This type of first order system is less common but does exist and is worth examining because it allows the effect of a zero on a pole to be examined.

KEY POINT

The general form for the system differential equation is

$$\tau \frac{dc}{dt} + c = K\left(\alpha \frac{dr}{dt} + r\right) \quad \text{Eqn. [9.3]}$$

where r is the system input, c is the system output and α, τ and K are positive constants.

First we need to obtain a transfer function for this system. Taking the Laplace transform and assuming zero initial conditions we have

$$\tau sC(s) + C(s) = K[\alpha sR(s) + R(s)]$$

Factorising we have

$$C(s)(\tau s + 1) = KR(s)(\alpha s + 1)$$

Rearranging to obtain a transfer function gives

$$\frac{C(s)}{R(s)} = \frac{K(1 + \alpha s)}{1 + \tau s}$$

Again we see that the characteristic equation is $1 + \tau s = 0$ and so the system pole is $s = -1/\tau$.

However, we now have a system zero which can be calculated by equating the numerator of the transfer function to zero. So,

$$K(1 + \alpha s) = 0$$

and so we have

$$s = -\frac{1}{\alpha}$$

If we plot a pole–zero plot for this system we see that there are two possibilities. These are shown in Figure 9.14. Either the pole is further from the imaginary axis or the zero is. Let us now examine the step response for both cases.

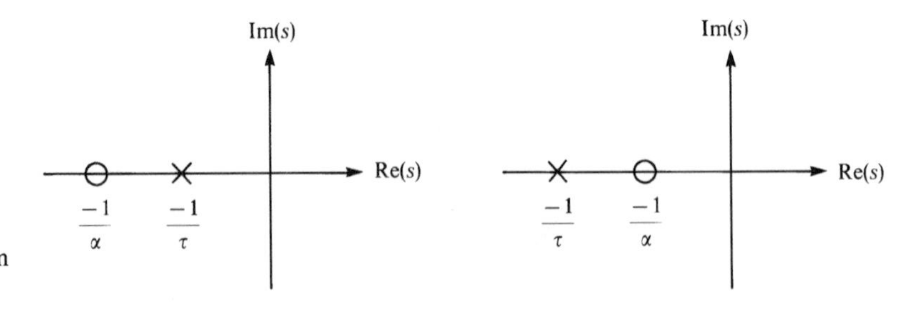

Figure 9.14
Pole–zero plots for a first order system with a system zero

Example

9.5 Calculate the unit step response for the system described by Eqn. [9.3]. Sketch the response.

Solution We have already obtained the transfer function for this system. It is

$$\frac{C(s)}{R(s)} = \frac{K(1+\alpha s)}{1+\tau s}$$

Multiplying by $R(s)$ gives

$$C(s) = \frac{K(1+\alpha s)}{1+\tau s}R(s)$$

Now $R(s) = 1/s$ for a unit step input and so

$$C(s) = \frac{K(1+\alpha s)}{(1+\tau s)s}$$

Splitting this expression into partial fractions gives

$$C(s) = \frac{A}{1+\tau s} + \frac{B}{s}$$

where A and B are constants to be determined. Placing the partial fractions over a common denominator gives

$$\frac{K(1+\alpha s)}{(1+\tau s)s} = \frac{As + B(1+\tau s)}{(1+\tau s)s}$$

Rearranging we have

$$\frac{K(1+\alpha s)}{(1+\tau s)s} = \frac{(A+B\tau)s + B}{(1+\tau s)s}$$

Comparing numerator polynomials we have

$$K(1+\alpha s) = (A+B\tau)s + B$$

Collecting terms gives

$$\alpha Ks + K = (A+B\tau)s + B$$

Comparing constant terms of the numerator polynomials we obtain

$$B = K$$

Comparing coefficients of s of the numerator polynomials gives

$$\alpha K = A + B\tau$$

and so

$$A = K\alpha - B\tau$$

Substituting for B gives

$$A = K\alpha - K\tau = K(\alpha - \tau)$$

Substituting the values of A and B into the expression for $C(s)$ gives

$$C(s) = \frac{K(\alpha - \tau)}{1+\tau s} + \frac{K}{s}$$

Rearranging to put the partial fraction into the standard forms of Table 8.1 gives

$$C(s) = \frac{K(\alpha - \tau)/\tau}{1/\tau + s} + \frac{K}{s}$$

Inverting this expression using Table 8.1 we have

$$c(t) = \frac{K(\alpha - \tau)}{\tau} \mathrm{e}^{-t/\tau} + K$$

Rearranging gives

$$c(t) = K\left(1 + \frac{\alpha - \tau}{\tau} \mathrm{e}^{-t/\tau}\right)$$

Note that putting $\alpha = 0$ reduces the response to that obtained in the previous section, that is, a first order system with no zero. Figure 9.15 shows the unit step response for the two cases described earlier, $\alpha > \tau$ and $\alpha < \tau$. $\alpha = 0$ is shown for comparison. The case $\alpha = \tau$ is trivial because then the pole and zero cancel and the system becomes a pure gain.

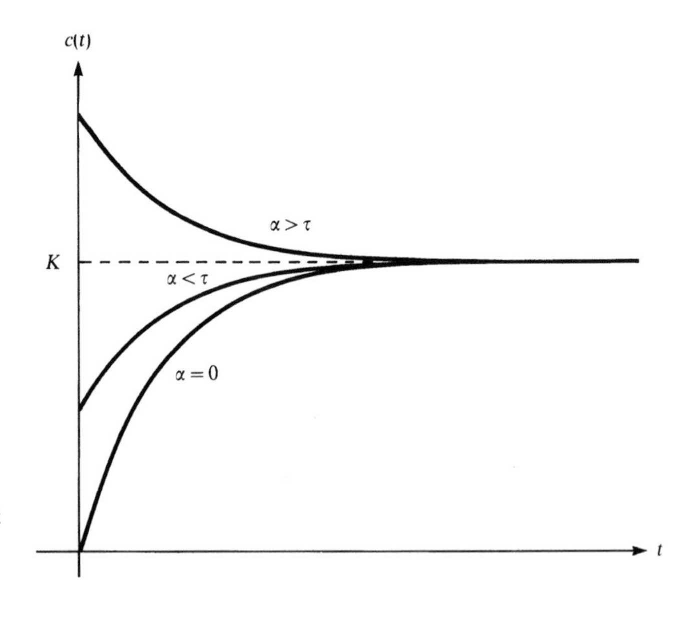

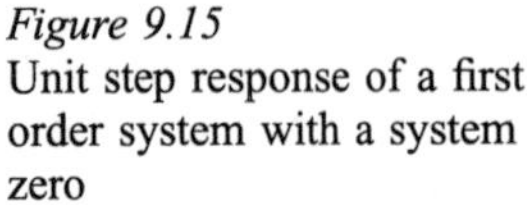
Figure 9.15
Unit step response of a first order system with a system zero

The interesting point to note from the previous example is that introducing the zero reduces the size of the coefficient of the exponential term associated with the system pole. In other words, it reduces the effect of the transient term due to the pole. In fact, if $\alpha = \tau$ then the zero sits directly on top of the pole and effectively cancels it out. As mentioned earlier, engineers often introduce zeros to reduce the effect of unwanted poles when designing systems. The nearer the zero is to the pole, the less the effect that the pole has on the system response.

KEY POINT

Placing a zero near to a system pole reduces the size of the transient term associated with that pole. The nearer the zero to the pole, the less effect the pole has on the system response.

Let us now examine a practical first order system that has a zero.

Example

9.6 Carry out the following for the system shown in Figure 9.16:

(a) Calculate the system differential equation. The input signal is v_i and the output signal is v_o.
(b) Obtain a transfer function for the system and hence determine the system poles and zeros.
(c) Calculate the response of the system to a unit step input when the component values are $R_1 = 5$ kΩ, $R_2 = 2$ kΩ and $C = 100$ μF.

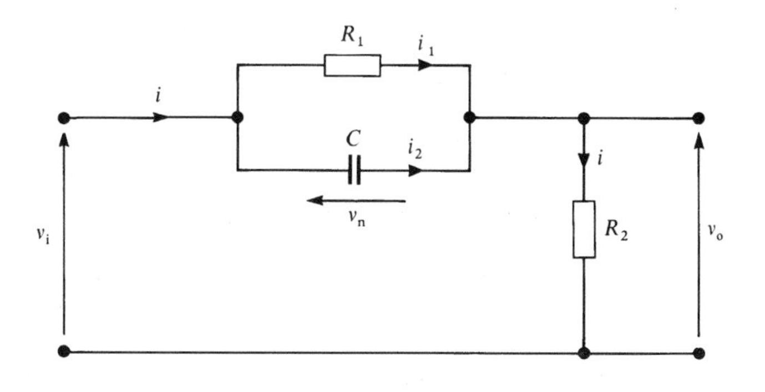

Figure 9.16
Circuit for Example 9.6

Solution (a) In order to solve this circuit we need to introduce the intermediate variables v_n, i, i_1 and i_2 shown in Figure 9.16.

Using Kirchhoff's current law, we have

$$i = i_1 + i_2 \qquad \text{Eqn. [9.4]}$$

Using Kirchhoff's voltage law gives

$$v_i = v_n + v_o \qquad \text{Eqn. [9.5]}$$

For R_1 we have

$$v_n = i_1 R_1 \qquad \text{Eqn. [9.6]}$$

For R_2 we have

$$v_o = iR_2 \qquad \text{Eqn. [9.7]}$$

For C we have

$$i_2 = C\frac{dv_n}{dt} \qquad \text{Eqn. [9.8]}$$

We have five equations and four intermediate variables to eliminate and so this is possible. By substituting Eqn. [9.6] and Eqn. [9.8] into Eqn. [9.4] we can eliminate i_1 and i_2. So,

$$i = \frac{v_n}{R_1} + C\frac{dv_n}{dt} \qquad \text{Eqn. [9.9]}$$

We now have three unused equations, namely, Eqns [9.5], [9.7] and [9.9], and two intermediate variables left to eliminate, which are i and v_R. Substituting Eqn. [9.7] into Eqn. [9.9] allows i to be eliminated. So,

$$\frac{v_o}{R_2} = \frac{v_n}{R_1} + C\frac{dv_n}{dt} \qquad \text{Eqn. [9.10]}$$

Finally, we can eliminate v_n by substituting Eqn. [9.5] into Eqn. [9.10]. In order to do this we need to use the differentiated form of Eqn. [9.5] as well. We will label this Eqn. [9.5a] as it is essentially the same. So,

$$\frac{dv_i}{dt} = \frac{dv_n}{dt} + \frac{dv_o}{dt} \qquad \text{Eqn. [9.5a]}$$

Finally, substituting Equations [9.5] and [9.5a] into Eqn. [9.10] gives

$$\frac{v_o}{R_2} = \frac{v_i - v_o}{R_1} + C\left(\frac{dv_i}{dt} - \frac{dv_o}{dt}\right)$$

Multiplying each term by R_1R_2 gives

$$R_1 v_o = R_2(v_i - v_o) + R_1R_2C\left(\frac{dv_i}{dt} - \frac{dv_o}{dt}\right)$$

Finally collecting terms gives

$$R_1R_2C\frac{dv_o}{dt} + (R_1 + R_2)v_o = R_1R_2C\frac{dv_i}{dt} + R_2v_i$$

This is the system differential equation with input v_i and output v_o.

(b) Taking Laplace transforms and assuming zero initial conditions to form a transfer function we have

$$R_1R_2CsV_o(s) + (R_1 + R_2)V_o(s) = R_1R_2CsV_i(s) + R_2V_i(s)$$

Rearranging we have

$$\frac{V_o(s)}{V_i(s)} = \frac{R_2 + R_1R_2Cs}{R_1 + R_2 + R_1R_2Cs}$$

Factorising the numerator gives

$$\frac{V_o(s)}{V_i(s)} = \frac{R_2(1 + R_1Cs)}{R_1 + R_2 + R_1R_2Cs}$$

The characteristic equation is

$$R_1 + R_2 + R_1R_2Cs = 0$$

Hence the system pole is

$$s = -\frac{R_1 + R_2}{R_1R_2C}$$

The system zero is obtained by equating the numerator of the transfer function to zero. We have

$$R_2(1 + R_1 Cs) = 0$$

Hence,

$$s = -\frac{1}{R_1 C}$$

We see that the circuit has a system pole and a system zero. By adjusting the component values it is possible to change the position of the pole and the zero.

(c) Substituting the component values into the transfer function gives

$$\frac{V_o(s)}{V_i(s)} = \frac{2 \times 10^3(1 + 5 \times 10^3 \times 1 \times 10^{-4}s)}{5 \times 10^3 + 2 \times 10^3 + 5 \times 10^3 \times 2 \times 10^3 \times 1 \times 10^{-4}s}$$

$$\frac{V_o(s)}{V_i(s)} = \frac{2 \times 10^3(1 + 0.5s)}{7 \times 10^3 + 10^3 s}$$

Cancelling a common factor of 10^3 gives

$$\frac{V_o(s)}{V_i(s)} = \frac{2(1 + 0.5s)}{7 + s}$$

Now the system input is a unit step and so $V_i(s) = 1/s$. Therefore,

$$V_o(s) = \frac{2(1 + 0.5s)}{s(7 + s)}$$

Splitting into partial fractions we have

$$\frac{2(1 + 0.5s)}{s(7 + s)} = \frac{A}{s} + \frac{B}{7 + s}$$

Putting the partial fractions over a common denominator gives

$$\frac{2(1 + 0.5s)}{s(7 + s)} = \frac{A(7 + s) + Bs}{s(7 + s)}$$

Comparing numerators we have

$$2(1 + 0.5s) = A(7 + s) + Bs$$

Collecting terms gives

$$s + 2 = (A + B)s + 7A$$

Comparing constant terms gives

$$2 = 7A$$

and so

$$A = \frac{2}{7} = 0.2857$$

Comparing coefficients of s gives

$$1 = A + B$$

Substituting in the value of A we have

$$B = 1 - 0.2857 = 0.7143$$

Using the values of A and B obtained we can write

$$V_o(s) = \frac{0.2857}{s} + \frac{0.7143}{7+s}$$

Finally, inverting this expression using Table 8.1 we have

$$v_o(t) = 0.2857 + 0.7143e^{-7t}$$

This is the unit step response of the circuit.

Self-assessment questions 9.4

1. How does introducing a zero modify the step response of a first order system?
2. Why do engineers sometimes deliberately introduce a zero into a system?

Exercises 9.4

1. Consider again the circuit of Figure 9.16. The component values are $R_1 = 5$ kΩ, $R_2 = 2$ kΩ and $C = 100$ μF. Calculate the response of the system to a unit step input.

Test and assignment exercises 9

1. Calculate the d.c. gain of the following transfer functions using the cover-up rule:

   (a) $\dfrac{45}{6+4s}$ (b) $\dfrac{23.3}{12.2+6.5s}$

   (c) $\dfrac{78.45}{12.34+34.5s}$

2. Identify the d.c. gain and time constant, stating their units, of the space heating system described in Example 6.5.

3. Identify the d.c. gain and time constant of the following transfer functions:

   (a) $\dfrac{5}{2+4s}$ (b) $\dfrac{12}{32+6s}$ (c) $\dfrac{43.4}{12.8+5.2s}$

4. Consider the space heating system described in Example 6.5. Calculate the response of the system to a unit step input.

10 Second order systems

Objectives

This chapter:

- describes the main characteristics of a second order system
- calculates the d.c. gain of a second order system
- calculates the damping ratio of a second order system
- calculates the natural frequency of a second order system
- derives the time response of a second order system
- describes how the position of the system poles affects the time response
- examines some practical second order systems

10.1 Introduction

A common type of engineering system is one that is modelled by a second order differential equation. Such a system is known as a **second order system**. We shall examine this type of system in this chapter.

10.2 General form of a second order system

KEY POINT

The general form of the system differential equation for a second order system is

$$\frac{\mathrm{d}^2c}{\mathrm{d}t^2} + 2\zeta\omega_\mathrm{n}\frac{\mathrm{d}c}{\mathrm{d}t} + \omega_\mathrm{n}^2c = K\omega_\mathrm{n}^2r \qquad \text{Eqn. [10.1]}$$

where c is the system output, r is the system input and K, ζ, ω_n are positive constants.

Before examining the significance of the constants K, ζ and ω_n we first obtain the transfer function for this system by taking the Laplace transform of Eqn. [10.1] and assuming zero initial conditions. So we have

$$s^2C(s) + 2\zeta\omega_n sC(s) + \omega_n^2 C(s) = K\omega_n^2 R(s)$$

Factorising gives

$$C(s)(s^2 + 2\zeta\omega_n s + \omega_n^2) = K\omega_n^2 R(s)$$

Rearranging gives the transfer function for the system. So,

$$\frac{C(s)}{R(s)} = \frac{K\omega_n^2}{s^2 + 2\zeta\omega_n s + \omega_n^2}$$

KEY POINT

The transfer function for a second order system is

$$\frac{C(s)}{R(s)} = \frac{K\omega_n^2}{s^2 + 2\zeta\omega_n s + \omega_n^2} \qquad \text{Eqn. [10.2]}$$

10.2.1 The d.c. gain of a second order system

The concept of d.c. gain is the same for a second order system as it is for a first order system. It is a general concept that is applicable to any order system.

If we put $s = 0$ into the transfer function then we can obtain the d.c. gain of the system using the cover-up rule discussed in Chapter 9. So we have

$$\text{d.c. gain} = \left.\frac{K\omega_n^2}{s^2 + 2\zeta\omega_n s + \omega_n^2}\right|_{s=0} = \frac{K\omega_n^2}{\omega_n^2} = K$$

We see that the d.c. gain of a second order system is K. It is the ratio between the output signal value and the input signal value when the system has settled. The units of K depend on the system being modelled.

10.2.2 The damping ratio of a second order system

Recall the spring–mass–damper system examined in Example 2.3. This is a second order system. The value of the damping coefficient in this system is crucial in determining how the system will respond to a step input, assuming the mass M and spring stiffness K remain constant. This is true in general for second order systems. If the damping coefficient is large then the system will respond slowly to a step input. For this case the system is said to be **over-damped**. If the damping coefficient is small then the system will respond rapidly to a step input. For this case the system is said to be **underdamped**. When a system is underdamped it overreacts to the step input and the system is said to **overshoot** its final value. There is also a special case when the system responds the fastest that is possible without overshooting the final value. For this case the system is said to be **critically damped**. If we denote the damping coefficient corresponding to critical damping by B_c, then we can define a **damping ratio** ζ as

$$\zeta = \frac{B_s}{B_c}$$

where B_s is the damping coefficient of the system.

We see that when the system is critically damped then $\zeta = 1$. When the system is underdamped then $B_s < B_c$ and so $\zeta < 1$. When the system is overdamped then $B_s > B_c$ and so $\zeta > 1$. The damping ratio has no units as it is a ratio between two quantities that have the same units. It is known as a **dimensionless** quantity.

The poles of a second order system depend on the value of the damping ratio. We now examine what effect the damping ratio has on the system poles. The characteristic equation for the system is obtained by equating the denominator of the transfer function to zero. So we have

$$s^2 + 2\zeta\omega_n s + \omega_n^2 = 0$$

Recall that for $ax^2 + bx + c = 0$ we have

$$x = \frac{-b \pm \sqrt{b^2 - 4ac}}{2a}$$

We can solve this equation by using the formula for a quadratic. This gives

$$s = \frac{-2\zeta\omega_n \pm \sqrt{(2\zeta\omega_n)^2 - 4\omega_n^2}}{2}$$

Simplifying the square root term gives,

$$s = \frac{-2\zeta\omega_n \pm \sqrt{4\omega_n^2\zeta^2 - 4\omega_n^2}}{2} = \frac{-2\zeta\omega_n \pm \sqrt{4\omega_n^2(\zeta^2 - 1)}}{2}$$

Taking a common factor outside the square root term gives

$$s = \frac{-2\zeta\omega_n \pm 2\omega_n\sqrt{\zeta^2 - 1}}{2}$$

Finally, dividing by 2 gives

KEY POINT

$$s = -\zeta\omega_n \pm \omega_n\sqrt{\zeta^2 - 1} \qquad \text{Eqn. [10.3]}$$

In order to gain some familiarity with the poles of a second order system, consider the following example.

Example

10.1 Calculate the poles of a second order system for the following values of damping ratio:

(a) $\zeta = 2$
(b) $\zeta = 1$
(c) $\zeta = 0.5$

For convenience, assume $\omega_n = 1$.

Solution (a) Substituting $\zeta = 0.5$ into Eqn. [10.3] we have

$$\begin{aligned} s &= -\zeta\omega_n \pm \omega_n\sqrt{\zeta^2 - 1} \\ &= -2\omega_n \pm \omega_n\sqrt{4 - 1} \\ &= -2\omega_n \pm 1.732\omega_n \\ &= -3.732\omega_n, \ -0.268\omega_n \\ &= -3.732, \ -0.268 \text{ (for } \omega_n = 1) \end{aligned}$$

These poles are plotted in Figure 10.1. The poles are real, negative and unequal.

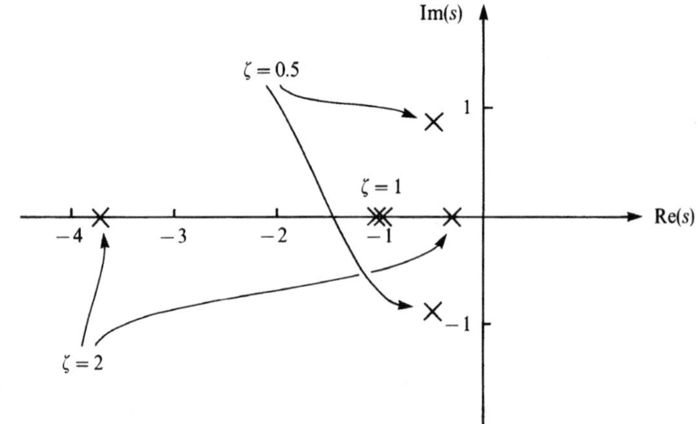

Figure 10.1
The poles of a second order system for various values of damping ratio

(b) Substituting $\zeta = 1$ into Eqn. [10.3] we have

$$\begin{aligned} s &= -\zeta\omega_n \pm \omega_n\sqrt{\zeta^2 - 1} \\ &= -\omega_n \pm \omega_n\sqrt{1 - 1} \\ &= -\omega_n \\ &= -1 \text{ (for } \omega_n = 1) \end{aligned}$$

There is a double pole at $s = -1$. So, the poles are real, negative and equal. These poles are plotted in Figure 10.1.

(c) Substituting $\zeta = 0.5$ into Eqn. [10.3] we have,

$$\begin{aligned} s &= -\zeta\omega_n \pm \omega_n\sqrt{\zeta^2 - 1} \\ &= -0.5\omega_n \pm \omega_n\sqrt{0.5^2 - 1} \\ &= -0.5\omega_n \pm \omega_n\sqrt{-0.75} \\ &= -0.5\omega_n \pm 0.8660\omega_n \text{j} \\ &= -0.5 \pm 0.8660\text{j (for } \omega_n = 1) \end{aligned}$$

The poles form a complex conjugate pair. They are plotted in Figure 10.1.

We see that the damping ratio ζ is crucial in determining the type of system poles a second order system has. If $\zeta > 1$ then the poles are real and unequal. If $\zeta = 1$ then the poles are real and equal. If $\zeta < 1$ then the poles are complex. We find that the time response of a second order system is different for each of these three cases and so we examine each case separately in the next three sections.

KEY POINT

The damping ratio ζ is a measure of the rapidity with which a second order system responds to a step input, assuming other system parameters remain constant. There are three cases:

(i) The system is overdamped corresponding to $\zeta > 1$. Here the system is slow to respond to the step input.
(ii) The system is critically damped corresponding to $\zeta = 1$. This is the fastest a system can respond without overshooting its final value.
(iii) The system is underdamped corresponding to $\zeta < 1$. Here the system is fast to respond to the step input but it does overshoot its final value.

10.2.3 The undamped natural frequency of a second order system

The final quantity of interest is ω_n which is known as the **undamped natural frequency** of the system. This corresponds to the angular frequency with which a second order system would oscillate if there was no damping in the system, that is, if the damping ratio of the system was zero. Recall again the spring–mass–damper of Example 2.3. If the damper has a damping coefficient of zero then this corresponds to there being no damping in the system. If a step input is applied to the system for this case then the mass would simply oscillate and the displacement of the mass would have the shape of a sinusoid. The oscillations would not decay because there is no damping to 'damp' the oscillations. This type of motion is known as **simple harmonic motion**.

Recall the general form of the system differential equation given in Eqn. [10.1]. For $\zeta = 0$ we have

$$\frac{d^2c}{dt^2} + \omega_n^2 c = K\omega_n^2 r$$

It can be shown that the response of the second order system to a step input in this case is a sinusoidal signal of angular frequency ω_n once the system has settled and any transient terms have decayed. The units of ω_n are radian.second^{-1}.

10.2.4 Identifying values of K, ζ and ω_n for a second order system

It is important to be able to identify values of K, ζ and ω_n for a second order system given the transfer function of the system. Consider the following example.

Example

10.2 Calculate the values of K, ζ and ω_n for the following second order system transfer functions:

(a) $\dfrac{C(s)}{R(s)} = \dfrac{4}{s^2 + 4s + 4}$

(b) $\dfrac{C(s)}{R(s)} = \dfrac{16}{s^2 + 18s + 16}$

(c) $\dfrac{C(s)}{R(s)} = \dfrac{10}{s^2 + 6s + 20}$

(d) $\dfrac{C(s)}{R(s)} = \dfrac{8}{2s^2 + 10s + 16}$

Solution (a) Let us compare this transfer function with the general form. We have

$$\frac{4}{s^2 + 4s + 4} = \frac{K\omega_n^2}{s^2 + 2\zeta\omega_n s + \omega_n^2}$$

Comparing the constant terms of the denominator term gives

$$\omega_n^2 = 4$$
$$\omega_n = 2$$

Note that we have only used the positive square root as ω_n was defined to be positive earlier. Comparing the numerator terms gives

$$K\omega_n^2 = 4$$

However, we know $\omega_n^2 = 4$ and so

$$K = \frac{4}{\omega_n^2} = \frac{4}{4} = 1$$

Finally we have

$$2\zeta\omega_n = 4$$
$$\zeta = \frac{4}{2\omega_n} = \frac{4}{2 \times 2} = 1$$

(b) We have

$$\frac{16}{s^2 + 18s + 16} = \frac{K\omega_n^2}{s^2 + 2\zeta\omega_n + \omega_n^2}$$

So,

$$\omega_n^2 = 16$$
$$\omega_n = 4$$

Also,

$$K\omega_n^2 = 16$$
$$K = \frac{16}{\omega_n^2} = \frac{16}{16} = 1$$

Finally we have

$$2\zeta\omega_n = 18$$
$$\zeta = \frac{18}{2\omega_n} = \frac{18}{2 \times 4} = 2.25$$

(c) We have

$$\frac{10}{s^2 + 6s + 20} = \frac{K\omega_n^2}{s^2 + 2\zeta\omega_n s + \omega_n^2}$$

So,

$$\omega_n^2 = 20$$
$$\omega_n = \sqrt{20} = 4.472$$

Also,

$$K\omega_n^2 = 10$$
$$K = \frac{10}{\omega_n^2} = \frac{10}{20} = 0.5$$

Finally,

$$2\zeta\omega_n = 6$$
$$\zeta = \frac{6}{2\omega_n} = \frac{6}{2 \times \sqrt{20}} = 0.6708$$

(d) Here we must be careful. We note that the coefficient of the s^2 term in the denominator is not 1. Therefore before we can compare the transfer function with the general form we must rearrange the transfer function so that the coefficient of s^2 is 1. This is achieved by dividing numerator and denominator by the coefficient of s^2, which in this case is 2. So we have

$$\frac{8}{2s^2 + 10s + 16} = \frac{4}{s^2 + 5s + 8}$$

So,

$$\frac{4}{s^2 + 5s + 8} = \frac{K\omega_n^2}{s^2 + 2\zeta\omega_n s + \omega_n^2}$$

We have

$$\omega_n^2 = 8$$
$$\omega_n = \sqrt{8} = 2.828$$

Also,

$$K\omega_n^2 = 4$$
$$K = \frac{4}{\omega_n^2} = \frac{4}{8} = 0.5$$

Finally,

$$2\zeta\omega_n = 5$$
$$\zeta = \frac{5}{2\omega_n} = \frac{5}{2\sqrt{8}} = 0.8839$$

Self-assessment questions 10.2

1. State the standard form of the transfer function for a second order system.
2. Name the three constants K, ζ and ω_n.
3. Explain what is meant by the d.c. gain of a second order system.
4. Explain what is meant by the damping ratio of a second order system.
5. Explain what is meant by the undamped natural frequency of a second order system.
6. Explain what is meant by the term overshoot.

Exercises 10.2

1. Calculate the values of K, ζ and ω_n for the following transfer functions:

(a) $\dfrac{10}{s^2 + 10s + 10}$ (b) $\dfrac{20}{s^2 + 5s + 10}$ (c) $\dfrac{18}{3s^2 + 9s + 12}$

(d) $\dfrac{6}{6s^2 + 15s + 18}$ (e) $\dfrac{8}{3s^2 + 4s + 9}$ (f) $\dfrac{6.3}{5.6s^2 + 2.9s + 6.4}$

10.3 Overdamped second order system

When $\zeta > 1$, the system is overdamped. Let us consider the unit step response of such a system. For a unit step input $R(s) = 1/s$, and so

$$C(s) = \frac{K\omega_n^2 R(s)}{s^2 + 2\zeta\omega_n s + \omega_n^2}$$

$$C(s) = \frac{K\omega_n^2}{s(s^2 + 2\zeta\omega_n s + \omega_n^2)} \qquad \text{Eqn. [10.4]}$$

The poles for this expression are the solutions of $s(s^2 + 2\zeta\omega_n s + \omega_n^2) = 0$. There is a pole at $s = 0$ which is a signal pole. The system poles are found by solving $s^2 + 2\zeta\omega_n s + \omega_n^2 = 0$.

Recall that because $\zeta > 1$ the system poles are real and unequal. Let them be denoted by p_1 and p_2 so that $s^2 + 2\zeta\omega_n s + \omega_n^2 = (s - p_1)(s - p_2) = 0$.

In order to be able to invert this expression we first need to split it into partial fractions. So we have

$$C(s) = \frac{K\omega_n^2}{s(s-p_1)(s-p_2)} = \frac{A}{s} + \frac{B}{s-p_1} + \frac{C}{s-p_2} \qquad \text{Eqn. [10.5]}$$

where A, B and C are constants to be determined. Putting the partial fractions over a common denominator gives

$$\frac{K\omega_n^2}{s(s^2 + 2\zeta\omega_n s + \omega_n^2)} = \frac{A(s-p_1)(s-p_2) + Bs(s-p_2) + Cs(s-p_1)}{s(s-p_1)(s-p_2)}$$

Equating numerators gives

$$K\omega_n^2 = A(s-p_1)(s-p_2) + Bs(s-p_2) + Cs(s-p_1) \qquad \text{Eqn. [10.6]}$$

Putting $s = 0$ in Eqn. [10.6] gives

$$K\omega_n^2 = A(-p_1)(-p_2)$$

$$A = \frac{K\omega_n^2}{(-p_1)(-p_2)} = \frac{K\omega_n^2}{p_1 p_2} \qquad \text{Eqn. [10.7]}$$

Recall that

$$s^2 + 2\zeta\omega_n s + \omega_n^2 = (s-p_1)(s-p_2)$$

Expanding the right-hand side of this equation gives

$$s^2 + 2\zeta\omega_n + \omega_n^2 = s^2 - (p_1 + p_2)s + p_1 p_2$$

Comparing the constant terms of these two polynomials gives

$$p_1 p_2 = \omega_n^2 \qquad \text{Eqn. [10.8]}$$

So substituting Eqn. [10.8] into [10.7] we have

$$A = \frac{K\omega_n^2}{p_1 p_2} = \frac{K\omega_n^2}{\omega_n^2} = K$$

Thus we have obtained the first of the partial fraction constants. Recall from Eqn. [10.3] that the system poles are given by

$$p_1 = -\zeta\omega_n + \omega_n\sqrt{\zeta^2 - 1}$$

$$p_2 = -\zeta\omega_n - \omega_n\sqrt{\zeta^2 - 1}$$

Putting $s = p_1$ into Eqn. [10.6] gives

$$K\omega_n^2 = A \times 0 + Bp_1(p_1 - p_2) + C \times 0$$

$$B = \frac{K\omega_n^2}{p_1(p_1 - p_2)}$$

Substituting the values of p_1 and p_2 we obtain

$$B = \frac{K\omega_n^2}{(-\zeta\omega_n + \omega_n\sqrt{\zeta^2 - 1})(-\zeta\omega_n + \omega_n\sqrt{\zeta^2 - 1} + \zeta\omega_n + \omega_n\sqrt{\zeta^2 - 1})}$$

$$= \frac{K\omega_n^2}{(-\zeta\omega_n + \omega_n\sqrt{\zeta^2 - 1})(2\omega_n\sqrt{\zeta^2 - 1})}$$

$$= \frac{K\omega_n^2}{-2\zeta\omega_n^2\sqrt{\zeta^2 - 1} + 2\omega_n^2(\zeta^2 - 1)}$$

Cancelling a common factor of ω_n^2 gives

$$B = \frac{K}{-2\zeta\sqrt{\zeta^2 - 1} + 2(\zeta^2 - 1)}$$

It is convenient to manipulate this expression further in order to obtain B in the form that it is usually quoted in most textbooks. Removing the factor $-K/2$ gives

$$B = -\frac{K}{2}\left[\frac{1}{\zeta\sqrt{\zeta^2 - 1} - (\zeta^2 - 1)}\right]$$

Removing the denominator factor $\sqrt{\zeta^2 - 1}$ by noting $\zeta^2 - 1 = (\sqrt{\zeta^2 - 1}) \times (\sqrt{\zeta^2 - 1})$ gives

$$B = -\frac{K}{2\sqrt{\zeta^2 - 1}}\left[\frac{1}{\zeta - \sqrt{\zeta^2 - 1}}\right]$$

Multiplying numerator and denominator by $\zeta + \sqrt{\zeta^2 - 1}$ gives

$$B = -\frac{K}{2\sqrt{\zeta^2 - 1}}\left[\frac{\zeta + \sqrt{\zeta^2 - 1}}{(\zeta - \sqrt{\zeta^2 - 1})(\zeta + \sqrt{\zeta^2 - 1})}\right]$$

The phrase 'difference of two squares' is used to express the mathematical statement

$a^2 - b^2 = (a + b)(a - b)$

Noting that the denominator term inside the brackets is the difference of two squares gives

$$B = -\frac{K}{2\sqrt{\zeta^2 - 1}}\left[\frac{\zeta + \sqrt{\zeta^2 - 1}}{\zeta^2 - (\zeta^2 - 1)}\right]$$

This simplifies to

$$B = -\frac{K}{2\sqrt{\zeta^2 - 1}}\left(\zeta + \sqrt{\zeta^2 - 1}\right)$$

Rearranging gives

$$B = K\left(-\frac{\zeta}{2\sqrt{\zeta^2 - 1}} - \frac{1}{2}\right)$$

Finally in the more usual form for B we have

KEY POINT

$$B = \left(-\frac{1}{2} - \frac{\zeta}{2\sqrt{\zeta^2 - 1}}\right)K \qquad \text{Eqn. [10.9]}$$

This is the second of the partial fraction constants. We can obtain C by comparing the coefficients of s^2 in Eqn. [10.6]. This gives

$$0 = A + B + C$$
$$C = -A - B$$

Noting the value of B in Eqn. [10.9] and that $A = K$ we have

$$C = -K - \left(-\frac{K}{2} - \frac{\zeta K}{2\sqrt{\zeta^2 - 1}}\right)$$
$$= -\frac{K}{2} + \frac{\zeta K}{2\sqrt{\zeta^2 - 1}}$$

Finally in the more usual form for C we have

KEY POINT

$$C = \left(-\frac{1}{2} + \frac{\zeta}{2\sqrt{\zeta^2 - 1}}\right)K \qquad \text{Eqn. [10.10]}$$

The derivation of A, B and C is difficult. Fortunately it is not necessary for engineers to reproduce this derivation in order to be able to analyse second order systems. It is the result that is important.

The expressions for B and C may look complicated but remember that ζ is a constant for a particular system and so B and C are also merely constant numbers. We can obtain an expression for the unit step response by taking the inverse Laplace transform of Eqn. [10.5]. This gives

$$c(t) = K + B\mathrm{e}^{p_1 t} + C\mathrm{e}^{p_2 t}$$

Using the values of B and C in Eqns [10.9] and [10.10] we have

$$c(t) = K + \left(-\frac{1}{2} - \frac{\zeta}{2\sqrt{\zeta^2 - 1}}\right)K\mathrm{e}^{p_1 t} + \left(-\frac{1}{2} + \frac{\zeta}{2\sqrt{\zeta^2 - 1}}\right)K\mathrm{e}^{p_2 t}$$

$$c(t) = K\left[1 + \left(-\frac{1}{2} - \frac{\zeta}{2\sqrt{\zeta^2 - 1}}\right)\mathrm{e}^{p_1 t} + \left(-\frac{1}{2} + \frac{\zeta}{2\sqrt{\zeta^2 - 1}}\right)\mathrm{e}^{p_2 t}\right]$$

For convenience we define the constants

$$B' = -\frac{1}{2} - \frac{\zeta}{2\sqrt{\zeta^2 - 1}} \qquad \text{Eqn. [10.11]}$$

$$C' = -\frac{1}{2} + \frac{\zeta}{2\sqrt{\zeta^2 - 1}} \qquad \text{Eqn. [10.12]}$$

The unit step response then becomes

$$c(t) = K(1 + B'e^{p_1 t} + C'e^{p_2 t}) \qquad \text{Eqn. [10.13]}$$

We see that the system poles give rise to two exponential terms. Recalling Eqn. [10.3] we have

$$p_1 = -\zeta\omega_n + \omega_n\sqrt{\zeta^2 - 1}$$

$$p_2 = -\zeta\omega_n - \omega_n\sqrt{\zeta^2 - 1}$$

Examining the expression for p_2 and noting that ζ and ω_n are positive by definition we see that p_2 must be negative because the expression consists of two negative terms.

Consider the case when $\zeta = 3$. We have $3 > \sqrt{3^2 - 1}$.

Examining the expression for p_1 we note that it consists of a negative term and a positive term. However, $\zeta > \sqrt{\zeta^2 - 1}$ for $\zeta > 1$ and so the positive term has a smaller magnitude than the negative term. So therefore, overall, p_1 is also negative.

KEY POINT

For an overdamped second order system the poles are real and negative.

A pole with a large magnitude is far away from the imaginary axis. It is poles that are close to the imaginary axis that take the longest time to decay.

Consider the unit step response given in Eqn. [10.13]. As both p_1 and p_2 are negative, the response terms associated with these system poles will decay with time to zero. They are therefore transient terms. The rate of decay of each of these transient terms depends on the magnitude of its associated pole. The larger the magnitude of the pole the faster the transient term decays.

Figure 10.2 shows the three components that make up the unit step response of an overdamped second order system. The figure also shows the associated pole–zero plot and the total unit step response.

If we examine Eqn. [10.11] then we see that B' is negative. For this reason the transient term associated with p_1 starts with a negative value as we can see by examining Figure 10.2.

The situation is more complicated for C' as can be seen from examining Eqn. [10.10]. There is a negative term and a positive term. However, we note that $\zeta > \sqrt{\zeta^2 - 1}$ and so $\zeta/\sqrt{\zeta^2 - 1} > 1$. Therefore the positive term has a larger magnitude than the negative term and so overall C' is slightly positive. Therefore the transient associated with p_2 starts with a positive value as we can see by examining Figure 10.2.

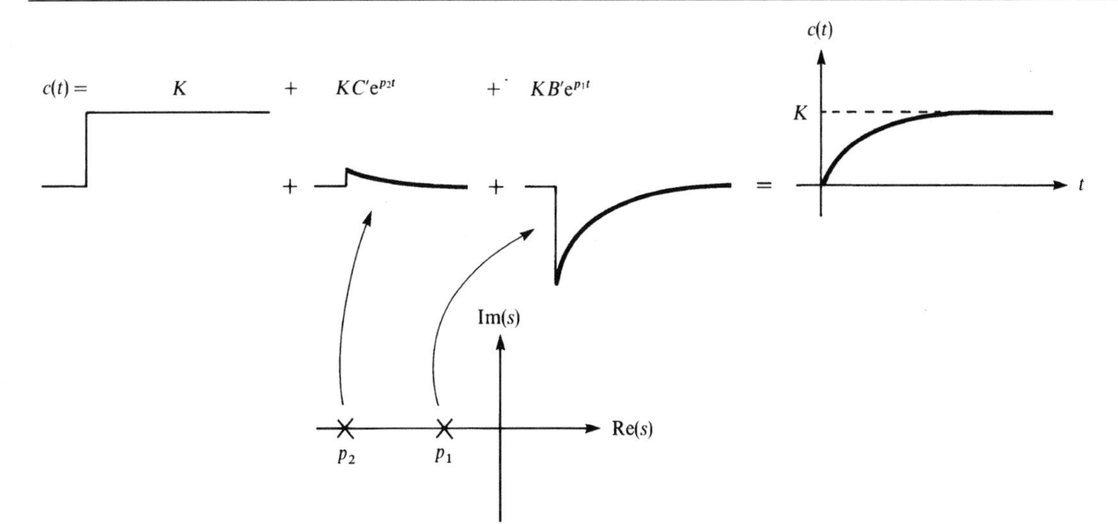

Figure 10.2 Unit step response of an overdamped second order system

Examining the step response for an overdamped system we can confirm that it does not overshoot its final value. It also takes longer to reach the final value than is the case for a critically damped system, although this cannot be confirmed until we examine the critically damped second order system in the next section.

KEY POINT

> The step response of an overdamped second order system does not overshoot its final value.

Examining Figure 10.2 we see that the step response looks rather like that of a first order system. This is especially true if p_2 is much further away from the imaginary axis than p_1. The response is then dominated by p_1 and p_1 is said to be a **dominant pole**. This concept will be discussed further in Chapter 11 when we deal with pole–zero plots more generally.

If we wish to gain a feel for the time scale with which an overdamped second order system responds then the pole nearest the imaginary axis is the important one. It is the time constant of this pole that determines the settling time. Recall, from the discussion of first order systems, that at four time constants the response is within 2% of its final value and it is considered to have settled.

Example

10.3 Calculate the unit step response of the overdamped second order system with transfer function

$$G(s) = \frac{6}{s^2 + 7s + 6}$$

Sketch the unit step response of the system. Determine the time constants of the poles and hence calculate the settling time of the system.

Solution The characteristic equation is

$$s^2 + 7s + 6 = 0$$

This factorises to $(s+1)(s+6)=0$ and hence the system poles are

$$p_1 = -1, \quad p_2 = -6$$

We can evaluate the d.c. gain of the system by using the cover-up rule. So,

$$K = \left.\frac{6}{s^2 + 7s + 6}\right|_{s=0} = \frac{6}{6} = 1$$

So using Eqn. [10.13] we have

$$c(t) = 1(1 + B'e^{-t} + C'e^{-6t})$$
$$c(t) = 1 + B'e^{-t} + C'e^{-6t}$$

In order to evaluate B' and C' we need to determine the damping ratio ζ. Comparing the transfer function with standard form we obtain

$$\omega_n^2 = 6$$
$$2\zeta\omega_n = 7$$
$$\zeta = \frac{7}{2\sqrt{6}} = 1.429$$

Note that $\zeta > 1$ confirming that the system is overdamped. Using Eqns [10.11] and [10.12] we have

$$B' = -\frac{1}{2} - \frac{1.429}{2\sqrt{1.429^2 - 1}} = -1.2$$

and

$$C' = -\frac{1}{2} + \frac{1.429}{2\sqrt{1.429^2 - 1}} = 0.2$$

So the unit step response of the system is

$$c(t) = 1 - 1.2e^{-t} + 0.2e^{-6t}$$

The unit step response for the system is shown in Figure 10.3.

Recall that the time constant of a pole is the reciprocal of the magnitude of the pole.

The time constant of p_1 is $\tau_1 = 1/1 = 1$ second. The time constant of p_2 is $\tau_2 = 1/6$ second. Therefore the settling time of the system is dominated by p_1. So the settling time of the system, t_s, is given by

$$t_s = 4\tau_1$$
$$t_s = 4 \times 1$$
$$t_s = 4 \text{ s}$$

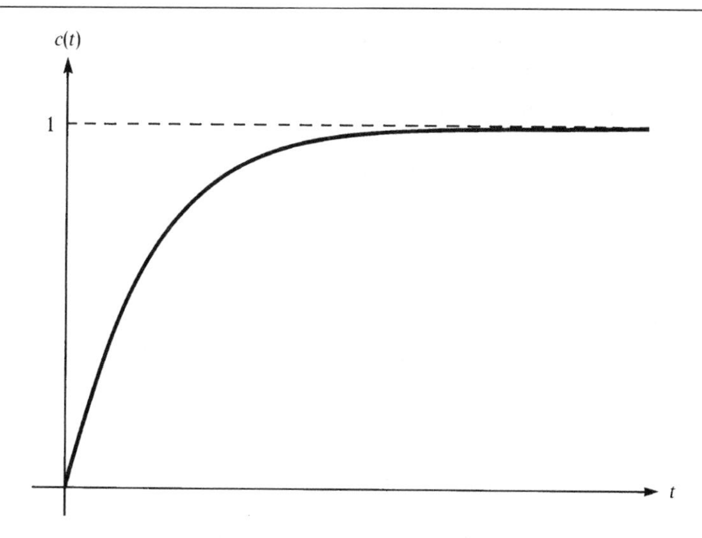

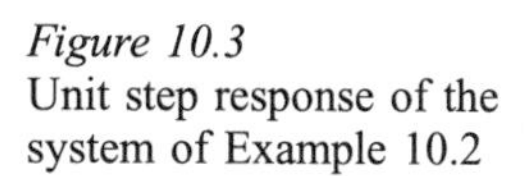
Figure 10.3
Unit step response of the system of Example 10.2

Self-assessment questions 10.3

1. What type of poles does a second order overdamped system have?
2. Describe the components that make up the step response of a second order overdamped system.
3. What is meant by a dominant pole?

Exercises 10.3

Calculate the unit step response of the following overdamped second order systems. Determine the time constants of the poles and hence calculate the settling time of the system.

(a) $\dfrac{12}{s^2 + 5s + 6}$ (b) $\dfrac{9}{s^2 + 10s + 9}$ (c) $\dfrac{32}{4s^2 + 16s + 12}$

10.4 Critically damped second order system

When $\zeta = 1$, the system is critically damped. Let us consider the unit step response of such a system. Recalling Eqn. [10.2] and noting that $R(s) = 1/s$ we have

$$C(s) = \frac{K\omega_n^2}{s(s^2 + 2\zeta\omega_n s + \omega_n^2)}$$

For $\zeta = 1$ we can factorise the denominator to give

$$C(s) = \frac{K\omega_n^2}{s(s + \omega_n)^2}$$

Note that this expression has three poles. There is a pole due to the input signal at $s=0$, and there are two system poles at $s=-\omega_n$, that is, there is a double pole at $s=-\omega_n$.

We can split this expression into partial fractions (see Appendix 2). As there is a linear repeated factor we have

$$C(s)=\frac{K\omega_n^2}{s(s+\omega_n)^2}=\frac{A}{s}+\frac{B}{(s+\omega_n)^2}+\frac{C}{s+\omega_n} \qquad \text{Eqn. [10.14]}$$

Placing the partial fractions over a common denominator gives

$$\frac{K\omega_n^2}{s(s+\omega_n)^2}=\frac{A(s+\omega_n)^2+Bs+Cs(s+\omega_n)}{s(s+\omega_n)^2}$$

Equating numerators gives

$$K\omega_n^2=A(s+\omega_n)^2+Bs+Cs(s+\omega_n) \qquad \text{Eqn. [10.15]}$$

Putting $s=0$ into Eqn. [10.15] we have

$$K\omega_n^2=A(0+\omega_n)^2+0+0$$
$$A=K$$

Putting $s=-\omega_n$ into Eqn. [10.15] we have

$$K\omega_n^2=0-B\omega_n+0$$
$$B=-K\omega_n$$

Comparing s^2 coefficients in Eqn. [10.15] we have

$$0=A+C$$
$$C=-A$$

So knowing $A=K$ we have

$$C=-K$$

Therefore the unit step response is, using Eqn. [10.14] and the values for A, B and C,

$$C(s)=\frac{A}{s}+\frac{B}{(s+\omega_n)^2}+\frac{C}{s+\omega_n}$$
$$C(s)=\frac{K}{s}-\frac{K\omega_n}{(s+\omega_n)^2}-\frac{K}{s+\omega_n}$$
$$C(s)=K\left[\frac{1}{s}-\frac{\omega_n}{(s+\omega_n)^2}-\frac{1}{s+\omega_n}\right]$$

Inverting this expression using Table 8.1 gives

$$c(t)=K(1-\omega_n t\mathrm{e}^{-\omega_n t}-\mathrm{e}^{-\omega_n t})$$
$$c(t)=K\left[1-(1+\omega_n t)\mathrm{e}^{-\omega_n t}\right] \qquad \text{Eqn. [10.16]}$$

The main feature of this response is the exponentially decaying transient term. This is modified by the factor $1+\omega_n t$ but essentially looks like an exponential decay. Figure 10.4 shows the components that make up the unit step response together with the total response and the position of the system poles. The system is critically damped. This corresponds to the fastest the system can rise without overshooting the final value.

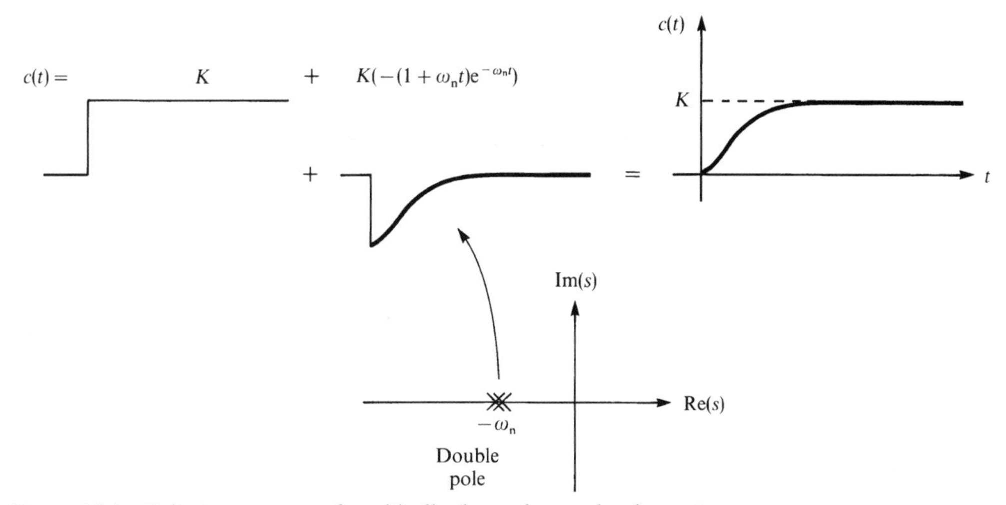

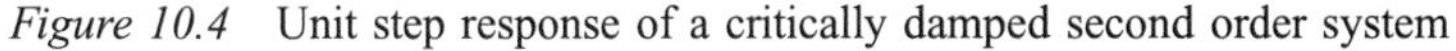
Figure 10.4 Unit step response of a critically damped second order system

The poles are real and equal for this system and this is shown as a double cross on the pole–zero plot to make it clear that there are two poles at the same point.

KEY POINT

For a critically damped second order system the poles are real and equal.

If we examine Eqn. [10.16] we see that the transient term is $-K(1+\omega_n t)e^{-\omega_n t}$. This term is dominated by the exponential component which has a time constant of $1/\omega_n$. So this is the effective time constant of the system.

KEY POINT

The effective time constant of a critically damped system is $1/\omega_n$ where ω_n is the undamped natural frequency of the system.

Example

10.4 Calculate the unit step response of the critically damped second order system with transfer function

$$G(s) = \frac{4}{s^2 + 4s + 4}$$

Sketch the step response of system. Calculate the settling time of the system.

Solution We have $\omega_n^2 = 4$ and so $\omega_n = 2$. Using the cover-up rule the d.c. gain is $G(0) = 4/(0+0+4) = 1$ and so $K = 1$. Substituting these values into Eqn. [10.16] we obtain

$$c(t) = 1 - (1 + 2t)e^{-2t}$$

The unit step response for the system is shown in Figure 10.5.

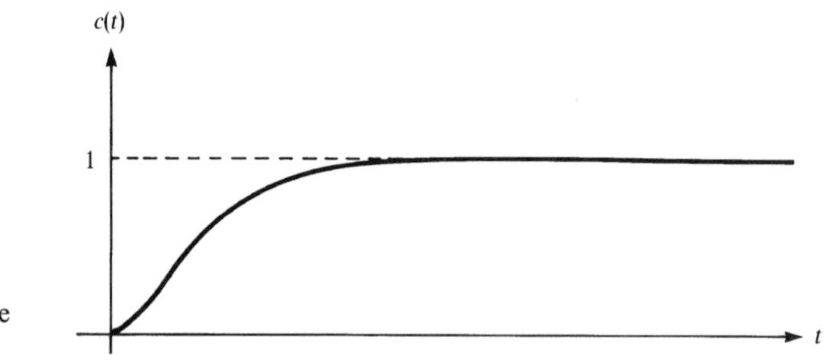

Figure 10.5
Unit step response of the system of Example 10.3

The effective time constant of the poles is $1/\omega_n = 1/2$ s. So the settling time of the system, t_s, is

$$t_s = 4 \times \frac{1}{2}$$

$$t_s = 2 \text{ s}$$

Self-assessment questions 10.4

1. What type of poles does a second order critically damped system have?
2. Describe the components that make up the step response of a critically damped second order system.
3. What is the effective time constant of a critically damped second order system?

Exercises 10.4

1. Calculate the step response of the following critically damped second order systems. Calculate the settling time of each system.

(a) $\dfrac{9}{s^2 + 6s + 9}$ (b) $\dfrac{64}{s^2 + 8s + 16}$ (c) $\dfrac{4}{4s^2 + 4s + 1}$

10.5 Underdamped second order system

When $\zeta < 1$, the system is underdamped. Let us consider the unit step response of such a system. Recalling Eqn. [10.2] and noting that $R(s) = 1/s$ we have

$$C(s) = \frac{K\omega_n^2}{s(s^2 + 2\zeta\omega_n s + \omega_n^2)}$$

We need to split this expression into partial fractions in order to allow the expression to be inverted using the standard forms of Table 8.1. We know that the quadratic factor has complex roots because $\zeta < 1$. For such a case it is common practice to retain the quadratic factor and not split it down to simple factors. So we have

$$C(s) = \frac{K\omega_n^2}{s(s^2 + 2\zeta\omega_n s + \omega_n^2)} = \frac{A}{s} + \frac{Bs + C}{s^2 + 2\zeta\omega_n s + \omega_n^2} \qquad \text{Eqn. [10.17]}$$

where A, B and C are coefficients to be determined.

If we put the partial fractions over a common denominator we obtain

$$\frac{K\omega_n^2}{s(s^2 + 2\zeta\omega_n s + \omega_n^2)} = \frac{A(s^2 + 2\zeta\omega_n s + \omega_n^2) + (Bs + C)s}{s(s^2 + 2\zeta\omega_n s + \omega_n^2)}$$

Equating the numerators of this expression gives

$$K\omega_n^2 = A(s^2 + 2\zeta\omega_n s + \omega_n^2) + (Bs + C)s$$

Comparing constant terms gives

$$K\omega_n^2 = A\omega_n^2$$
$$A = K$$

Comparing coefficients of s gives

$$0 = 2\zeta\omega_n A + C$$

So,

$$C = -2\zeta\omega_n K$$

Comparing coefficients of s^2 gives

$$0 = A + B$$
$$B = -K$$

Substituting values for A, B and C into Eqn. [10.17] gives

$$C(s) = \frac{A}{s} + \frac{Bs + C}{s^2 + 2\zeta\omega_n s + \omega_n^2}$$

$$C(s) = \frac{K}{s} - \frac{Ks + 2\zeta\omega_n K}{s^2 + 2\zeta\omega_n s + \omega_n^2}$$

$$C(s) = K\left(\frac{1}{s} - \frac{s + 2\zeta\omega_n}{s^2 + 2\zeta\omega_n s + \omega_n^2}\right)$$

If we examine Table 8.1 we see that there are some expressions with quadratic denominators, namely, the damped sine and the damped cosine. We therefore need to work towards these. We see that the denominators are of the form $(s+a)^2+b^2$ and so we first need to convert the denominator of the right-hand term of $C(s)$ into this form. We note that

$$(s+a)^2+b^2=s^2+2as+a^2+b^2$$

Comparing coefficients we see that $a=\zeta\omega_n$ and $a^2+b^2=\omega_n^2$. So,

$$\begin{aligned} b^2 &= \omega_n^2-a^2 \\ &= \omega_n^2-(\zeta\omega_n)^2 \\ &= \omega_n^2-\zeta^2\omega_n^2 \\ &= (1-\zeta^2)\omega_n^2 \end{aligned}$$

Therefore,

$$b=\sqrt{1-\zeta^2}\omega_n$$

Substituting in the values for a and b we can write

$$C(s)=K\left[\frac{1}{s}-\frac{s+2\zeta\omega_n}{(s+\zeta\omega_n)^2+(\sqrt{1-\zeta^2}\omega_n)^2}\right]$$

We now need to adjust the numerators. First we concentrate on obtaining the correct numerator for the damped cosine. The table entry is

$$\frac{s+a}{(s+a)^2+b^2}$$

and so we note that the numerator should be $s+\zeta\omega_n$. Therefore we have

$$C(s)=K\left[\frac{1}{s}-\frac{s+\zeta\omega_n}{(s+\zeta\omega_n)^2+(\sqrt{1-\zeta^2}\omega_n)^2}-\frac{\zeta\omega_n}{(s+\zeta\omega_n)^2+(\sqrt{1-\zeta^2}\omega_n)^2}\right]$$

We now concentrate on obtaining the correct numerator for the damped sine. The table entry is

$$\frac{b}{(s+a)^2+b^2}$$

and so we need to obtain a numerator of $\sqrt{1-\zeta^2}\omega_n$. Adjusting the numerator to obtain this term gives

$$C(s)=K\left[\frac{1}{s}-\frac{s+\zeta\omega_n}{(s+\zeta\omega_n)^2+(\sqrt{1-\zeta^2}\omega_n)^2}-\frac{\zeta}{\sqrt{1-\zeta^2}}\frac{\sqrt{1-\zeta^2}\omega_n}{(s+\zeta\omega_n)^2+(\sqrt{1-\zeta^2}\omega_n)^2}\right]$$

We can now invert this expression using Table 8.1. We use the values of a and b obtained earlier and note that the damped sine term is multiplied by the factor $\zeta/\sqrt{1-\zeta^2}$. So we have

$$c(t) = K\left[1 - e^{-\zeta\omega_n t}\cos\left(\sqrt{1-\zeta^2}\,\omega_n t\right) - \frac{\zeta}{\sqrt{1-\zeta^2}}e^{-\zeta\omega_n t}\sin\left(\sqrt{1-\zeta^2}\,\omega_n t\right)\right]$$

This can be written as

$$c(t) = K\left\{1 - \frac{e^{-\zeta\omega_n t}}{\sqrt{1-\zeta^2}}\left[\sqrt{1-\zeta^2}\cos\left(\sqrt{1-\zeta^2}\,\omega_n t\right) + \zeta\sin\left(\sqrt{1-\zeta^2}\,\omega_n t\right)\right]\right\}$$

This is also a very difficult derivation but again it is the result that is important to engineers. The derivation is only included for instruction and completeness.

Finally, we can combine the damped cosine and damped sine terms using the standard trigonometric identity

$$\sin(\theta + \phi) = \sin\theta\cos\phi + \cos\theta\sin\phi$$

We put $\theta = \sqrt{1-\zeta^2}\,\omega_n t$ and $\cos\phi = \zeta$. Since $\cos\phi = \zeta$ then $\sqrt{1-\zeta^2} = \sin\phi$. So finally we have

$$c(t) = K\left[1 - \frac{e^{-\zeta\omega_n t}}{\sqrt{1-\zeta^2}}\sin(\sqrt{1-\zeta^2}\,\omega_n t + \phi)\right] \qquad \text{Eqn. [10.18]}$$

where $\phi = \cos^{-1}\zeta$.

At first sight this expression looks extremely complicated but it is possible to understand its form with some thought. Again we see that the steady state term is a step as was the case for the critically damped and overdamped systems. The transient term is essentially an exponential factor multiplied by a sinusoidal factor. Note that, as ζ is a constant, the factor $1/\sqrt{1-\zeta^2}$ is purely a number. The exponential term has a decay time constant of $1/\zeta\omega_n$ and in fact this is the effective time constant for the system. The sinusoid is constrained within a **decaying exponential envelope** and we say it is a **trapped sinusoid**. To see why this is so examine Figure 10.6 which shows a typical transient term for an underdamped second order system. The decaying exponential envelope is indicated by broken lines. It consists of a plot of the $e^{-\zeta\omega_n t}/\sqrt{1-\zeta^2}$ term and its mirror image reflected in the time axis. Recall that the maximum value of the sine function is 1 and the minimum value is -1. When the sine term of Eqn. [10.18] has a value of -1 then this corresponds to touching the bottom of the exponential envelope. For other values of the sine function the transient term lies within the exponential envelope. The angular frequency of the sinusoid is $\omega_n\sqrt{1-\zeta^2}$ and is known as the **damped natural frequency**, ω_d. So, $\omega_d = \omega_n\sqrt{1-\zeta^2}$. Note also that the sinusoid has a phase shift ϕ.

To summarise, the transient term is a decaying sinusoid of angular frequency ω_d trapped in a decaying exponential envelope whose decay rate is governed by the time constant $1/\zeta\omega_n$.

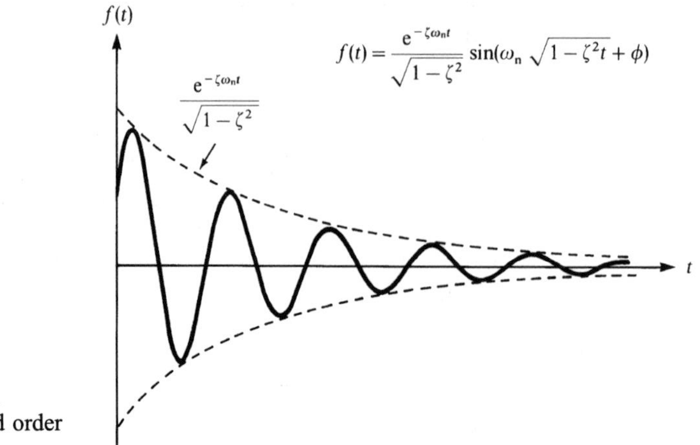

Figure 10.6
Transient term of an underdamped second order system

KEY POINT

The damped natural frequency ω_d is the angular frequency of the transient term. It is related to the undamped natural frequency ω_n by the formula

$$\omega_d = \omega_n\sqrt{1-\zeta^2}$$

If we examine the pole–zero plot for an underdamped second order system, shown in Figure 10.7, this provides valuable information concerning the relationships between the quantities we have discussed.

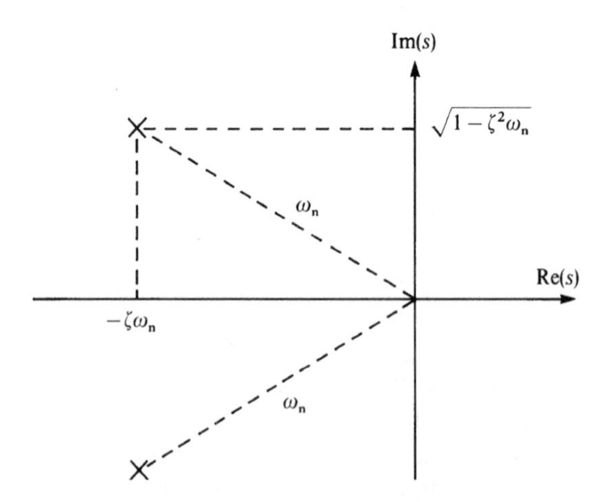

Figure 10.7
Pole–zero plot for an underdamped second order system

Recalling Eqn. [10.3] which gives the values of the second order system poles, we have

$$s = -\zeta\omega_n \pm \omega_n\sqrt{\zeta^2 - 1}$$

Now we know that $\zeta < 1$ so we can write

$$s = -\zeta\omega_n \pm \sqrt{1-\zeta^2}\omega_n j$$

Note that the poles have the same real component but the imaginary components have opposite signs. They are said to be a **complex conjugate pair**. We see that the real component is $-\zeta\omega_n$ which determines the decay rate of the exponential envelope. The imaginary component, $\sqrt{1-\zeta^2}\omega_n$, is the damped natural frequency. These are the two important quantities for determining the shape of the transient response and they can be read directly from the pole–zero plot. A further useful piece of information, which can help in remembering the pole–zero plot, is that the length from the origin to the pole is ω_n. This is easily shown by using Pythagoras's theorem. So,

$$\begin{aligned}
|p_1| &= \sqrt{(-\zeta\omega_n)^2 + (\sqrt{1-\zeta^2}\omega_n)^2} \\
&= \sqrt{\zeta^2\omega_n^2 + (1-\zeta^2)\omega_n^2} \\
&= \sqrt{\omega_n^2} \\
&= \omega_n
\end{aligned}$$

Similarly,

$$|p_2| = \omega_n$$

The unit step response of an underdamped second order system together with its various components is shown in Figure 10.8. Note that the response overshoots its final value.

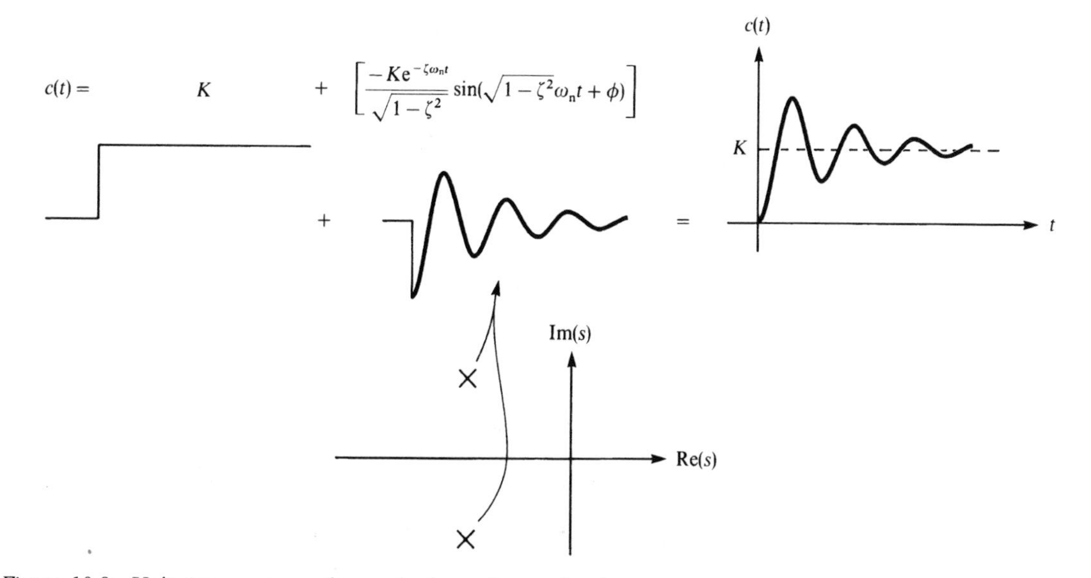

Figure 10.8 Unit step response of an underdamped second order system

KEY POINT

For an underdamped second order system the poles form a complex conjugate pair.

Let us now calculate the unit step response of an underdamped system.

Example

10.5 Calculate the unit step response of the underdamped system with transfer function

$$G(s) = \frac{32}{s^2 + s + 16}$$

Sketch the step response of the system. Determine the settling time of the system.

Solution We have $\omega_n^2 = 16$ and so $\omega_n = 4$. We also have

$$2\zeta\omega_n = 1$$

$$\zeta = \frac{1}{2\omega_n} = \frac{1}{2 \times 4} = 0.125$$

Using the cover-up rule we have

$$K = G(s)\Big|_{s=0} = \frac{32}{16} = 2$$

Also,

$$\omega_d = \sqrt{1 - \zeta^2}\omega_n = \sqrt{1 - 0.125^2} \times 4 = 3.969$$

We have

$$\frac{1}{\sqrt{1 - \zeta^2}} = \frac{1}{\sqrt{1 - 0.125^2}} = 1.008$$

Also,

$$2\zeta\omega_n = 1$$

$$\zeta\omega_n = \frac{1}{2} = 0.5$$

Finally,

$$\phi = \cos^{-1}\zeta = 1.445 \text{ radian}$$

Recalling Eqn. [10.18] and substituting in the correct values we obtain

$$c(t) = 2[1 - 1.008e^{-0.5t} \sin(3.969t + 1.445)]$$

This response is shown in Figure 10.9.

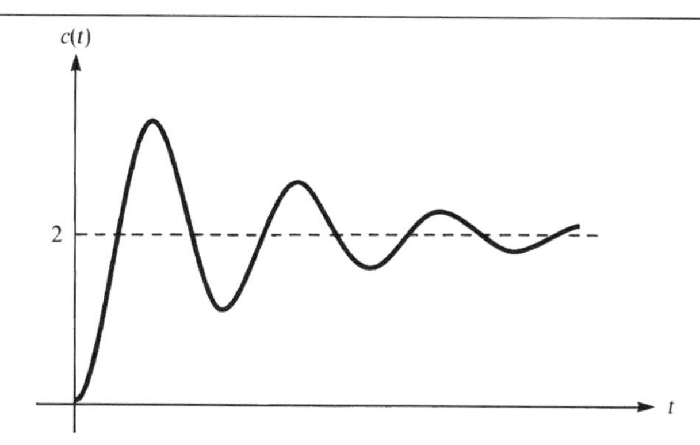

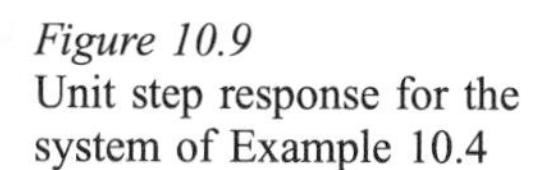
Figure 10.9
Unit step response for the system of Example 10.4

The effective time constant of the system is

$$\frac{1}{\zeta\omega_n} = \frac{1}{0.5} = 2$$

So the settling time of the system is $4 \times 2 = 8$ s.

Self-assessment questions 10.5

1. What type of poles does a second order underdamped system have?
2. Describe the components that make up the step response of an underdamped second order system.

Exercises 10.5

Calculate the unit step response of the following underdamped second order systems:

(a) $\dfrac{9}{s^2+s+9}$ (b) $\dfrac{50}{s^2+4s+25}$ (c) $\dfrac{16}{2s^2+2s+8}$

Calculate the settling time of each system.

10.6 Some practical second order systems

Having examined the three types of second order systems we are now in a position to look at some practical second order systems.

Example

10.6 Consider again the spring–mass–damper system of Example 2.3. Carry out the following:

(a) Derive expressions for the d.c. gain, the damping ratio and the undamped natural frequency of the system in terms of the constants K, M and B.

(b) Given that $M = 1$ kg and $K = 1$ N m^{-1}, calculate values of B for the system to be:

(i) overdamped with $\zeta = 3$
(ii) critically damped with $\zeta = 1$
(iii) underdamped with $\zeta = 0.2$

(c) Calculate the unit step response for the three cases of part (b). Plot these responses on a single graph.

Solution (a) Recall the system differential equation for this system, which is

$$f = M\frac{\mathrm{d}^2x}{\mathrm{d}t^2} + B\frac{\mathrm{d}x}{\mathrm{d}t} + Kx$$

where f is the system input and x is the system output.

We first form a transfer function for this system by taking Laplace transforms of the system differential equation, assuming zero initial conditions. We have

$$F(s) = Ms^2X(s) + BsX(s) + KX(s)$$

Factorising gives

$$F(s) = (Ms^2 + Bs + K)X(s)$$

Finally the transfer function $G(s)$ is given by

$$G(s) = \frac{X(s)}{F(s)} = \frac{1}{Ms^2 + Bs + K}$$

Dividing the numerator and the denominator by M to obtain the transfer function in standard form we have

$$G(s) = \frac{1/M}{s^2 + (B/M)s + K/M}$$

We can now compare this transfer function with the general transfer function for a second order system. In order to avoid confusing the d.c. gain with the spring constant we shall denote d.c. gain by K'. We have

$$\frac{1/M}{s^2 + (B/M)s + K/M} = \frac{K'\omega_n^2}{s^2 + 2\zeta\omega_n s + \omega_n^2} \qquad \text{Eqn. [10.19]}$$

Comparing the constant terms in the denominator polynomials we have

$$\omega_n^2 = \frac{K}{M}$$

and so the undamped natural frequency is given by

$$\omega_n = \sqrt{\frac{K}{M}} \qquad \text{Eqn. [10.20]}$$

Comparing the coefficients of s in the denominator polynomials we have

$$2\zeta\omega_n = \frac{B}{M}$$

and so

$$\zeta = \frac{B}{2M\omega_n}$$

Substituting in the value of ω_n in Eqn. [10.20] we have

$$\zeta = \frac{B}{2M}\frac{1}{\omega_n}$$
$$= \frac{B}{2M}\frac{1}{\sqrt{K/M}}$$
$$= \frac{B}{2M}\sqrt{\frac{M}{K}}$$

Cancelling a factor of $\sqrt{M}$ we have

$$\zeta = \frac{B}{2\sqrt{M}}\frac{1}{\sqrt{K}}$$

So the damping ratio is given by

$$\zeta = \frac{B}{2\sqrt{MK}} \qquad \text{Eqn. [10.21]}$$

Comparing the numerator coefficients we have

$$K'\omega_n^2 = \frac{1}{M}$$
$$K' = \frac{1}{M\omega_n^2}$$

So we have

$$K' = \frac{1}{MK/M}$$

So the d.c. gain of the system is given by

$$K' = \frac{1}{K} \qquad \text{Eqn. [10.22]}$$

(b) Recalling Eqn. [10.21] and rearranging to make B the subject of the equation we have

$$B = 2\zeta\sqrt{MK}$$

We are given that $M = 1$ and $K = 1$ and so

$$B = 2\zeta \qquad \text{Eqn. [10.23]}$$

(i) For $\zeta = 3$ we have $B = 2 \times 3 = 6$ N s m^{-1}.
(ii) For $\zeta = 1$ we have $B = 2 \times 1 = 2$ N s m^{-1}.
(iii) For $\zeta = 0.2$ we have $B = 2 \times 0.2 = 0.4$ N s m^{-1}.

(c) (i) Substituting $M = 1$, $K = 1$ and $B = 6$ in the transfer function we have

$$\frac{X(s)}{F(s)} = \frac{1}{Ms^2 + Bs + K} = \frac{1}{s^2 + 6s + 1}$$

and so

$$X(s) = \frac{F(s)}{s^2 + 6s + 1}$$

For a unit step input $F(s) = 1/s$ and so

$$X(s) = \frac{1}{s(s^2 + 6s + 1)} \qquad \text{Eqn. [10.24]}$$

There is a signal pole at $s = 0$. The system poles are obtained by solving the quadratic equation

$$s^2 + 6s + 1 = 0$$

Using the formula for solving a quadratic equation we have

$$\begin{aligned} s &= \frac{-6 \pm \sqrt{36 - 4}}{2} \\ &= \frac{-6 \pm \sqrt{32}}{2} \end{aligned}$$

Taking a factor of 4 out of the square root gives

$$s = \frac{-6 \pm 2\sqrt{8}}{2}$$

Cancelling a factor of 2 we have

$$\begin{aligned} s &= -3 \pm 2.828 \\ &= -5.828, \; -0.172 \end{aligned}$$

We can now split $X(s)$ into partial fractions in order to be able to invert it. We have

$$X(s) = \frac{1}{s(s^2 + 6s + 1)} = \frac{A}{s} + \frac{B}{s + 5.828} + \frac{C}{s + 0.172}$$

Placing the partial fraction over a common denominator gives

$$\frac{1}{s(s^2 + 6s + 1)} = \frac{A(s^2 + 6s + 1) + Bs(s + 0.172) + Cs(s + 5.828)}{s(s^2 + 6s + 1)}$$

Comparing numerators we have

$$1 = A(s^2 + 6s + 1) + Bs(s + 0.172) + Cs(s + 5.828)$$

Gathering terms gives

$$1 = s^2(A + B + C) + s(6A + 0.172B + 5.828C) + A \qquad \text{Eqn. [10.25]}$$

Comparing the constant terms of Eqn. [10.25] we have

$$A = 1$$

Comparing the coefficients of s in Eqn. [10.25] we have

$$6A + 0.172B + 5.828C = 0$$

Using $A = 1$ this reduces to

$$6 + 0.172B + 5.828C = 0 \qquad \text{Eqn. [10.26]}$$

Comparing the coefficients of s^2 in Eqn. [10.25] we have

$$A + B + C = 0$$

Using $A = 1$ this reduces to

$$1 + B + C = 0 \qquad \text{Eqn. [10.27]}$$

Substituting Eqn. [10.27] into Eqn. [10.26] to eliminate C we have

$$6 + 0.172B + 5.828(-1 - B) = 0$$
$$6 + 0.172B - 5.828 - 5.828B = 0$$
$$0.172 - 5.656B = 0$$

So,

$$B = \frac{0.172}{5.656} = 0.0304$$

Finally, substituting the values of A and B into Eqn. [10.27] to obtain C gives

$$1 + 0.0304 + C = 0$$
$$C = -1.0304$$

We can now write

$$X(s) = \frac{1}{s} + \frac{0.0304}{s + 5.828} - \frac{1.0304}{s + 0.172}$$

Inverting this expression using Table 8.1 we obtain

$$x(t) = 1 + 0.0304\mathrm{e}^{-0.5828t} - 1.0304e^{-0.172t}$$

This expression is plotted in Figure 10.10.

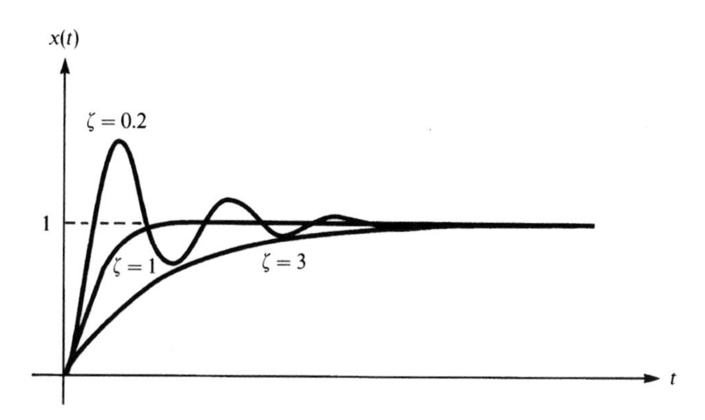

Figure 10.10
Unit step response of a spring–mass–damper system for three different values of damping ratio

(ii) Substituting $M=1$, $K=1$ and $B=2$ in the transfer function we have

$$\frac{X(s)}{F(s)}=\frac{1}{Ms^2+Bs+K}=\frac{1}{s^2+2s+1}$$

and so

$$X(s)=\frac{F(s)}{s^2+2s+1}$$

For a unit step input $F(s)=1/s$ and so

$$X(s)=\frac{1}{s(s^2+2s+1)}$$

Factorising the denominator we have

$$X(s)=\frac{1}{s(s+1)^2}$$

Splitting this expression into partial fractions we have

$$X(s)=\frac{1}{s^2+2s+1}=\frac{A}{s}+\frac{B}{(s+1)^2}+\frac{C}{s+1}$$

where A, B and C are constants to be determined. Placing the partial fractions over a common denominator gives

$$\frac{1}{s(s^2+2s+1)}=\frac{A(s^2+2s+1)+Bs+Cs(s+1)}{s(s^2+2s+1)}$$

Comparing numerators we have

$$1=A(s^2+2s+1)+Bs+Cs(s+1)$$

Collecting terms gives

$$1=s^2(A+C)+s(2A+B+C)+A \qquad \text{Eqn. [10.28]}$$

Comparing constant terms in Eqn. [10.28] we have

$$A=1$$

Comparing coefficients of s^2 in Eqn. [10.28] we have

$$A+C=0$$

and so

$$C=-A=-1$$

Finally, comparing coefficients of s in Eqn. [10.28] we have

$$2A+B+C=0$$
$$B=-2A-C$$

Using the values of A and C already obtained gives

$$B=-2\times 1-(-1)=-2+1=-1$$

We can now write

$$X(s) = \frac{1}{s} - \frac{1}{(s+1)^2} - \frac{1}{s+1}$$

Using Table 8.1 to invert this expression we have

$$x(t) = 1 - te^{-t} - e^{-t}$$
$$x(t) = 1 - e^{-t}(t+1)$$

This expression is also plotted in Figure 10.10.

(iii) Substituting $M=1$, $K=1$ and $B=0.4$ in the transfer function we have

$$\frac{X(s)}{F(s)} = \frac{1}{Ms^2 + Bs + K} = \frac{1}{s^2 + 0.4s + 1}$$

and so

$$X(s) = \frac{F(s)}{s^2 + 0.4s + 1}$$

For a unit step input $F(s) = 1/s$ and so

$$X(s) = \frac{1}{s(s^2 + 0.4s + 1)}$$

In this case we avoid splitting the quadratic factor because the denominator polynomial has complex roots. This is a consequence of the system being underdamped.

Splitting this expression into partial fractions we have

$$X(s) = \frac{1}{s(s^2 + 0.4s + 1)} = \frac{A}{s} + \frac{Bs + c}{s^2 + 0.4s + 1}$$

Placing the partial fractions over a common denominator gives

$$\frac{1}{s(s^2 + 0.4s + 1)} = \frac{A(s^2 + 0.4s + 1) + (Bs + C)s}{s(s^2 + 0.4s + 1)}$$

Comparing numerators we have

$$1 = A(s^2 + 0.4s + 1) + (Bs + C)s$$

Collecting terms gives

$$1 = (A + B)s^2 + (0.4A + C)s + A \qquad \text{Eqn. [10.29]}$$

Comparing the constant terms in Eqn. [10.29] we have

$$A = 1$$

Comparing coefficients of s^2 in Eqn. [10.29] we have

$$A + B = 0$$

and so

$$B = -1$$

Finally, comparing coefficients of s in Eqn. [10.29] we have

$$0.4A + C = 0$$
$$C = -0.4 \times 1$$
$$= -0.4$$

We can now write

$$X(s) = \frac{1}{s} - \left[\frac{s + 0.4}{s^2 + 0.4s + 1}\right]$$

In order to be able to invert this expression we need to convert the terms of the expression into the standard forms of Table 8.1.

First we complete the square for the denominator quadratic. This gives

$$X(s) = \frac{1}{s} - \left[\frac{s + 0.4}{(s + 0.2)^2 + 0.96}\right]$$
$$= \frac{1}{s} - \left[\frac{s + 0.4}{(s + 0.2)^2 + 0.98^2}\right]$$

Examining the entry in Table 8.1 for the damped cosine we see that we need a numerator of $s + 0.2$. So we write

$$X(s) = \frac{1}{s} - \left[\frac{s + 0.2}{(s + 0.2)^2 + 0.98^2} + \frac{0.2}{(s + 0.2)^2 + 0.98^2}\right]$$

Examining the entry in Table 8.1 for the damped sine we see that we need a numerator of 0.98. Now $0.2 = 0.204 \times 0.98$ and so we write

$$X(s) = \frac{1}{s} - \left[\frac{s + 0.2}{(s + 0.2)^2 + 0.98^2} + 0.204 \times \frac{0.98}{(s + 0.2)^2 + 0.98^2}\right]$$

We can now invert $X(s)$. This gives

$$x(t) = 1 - e^{-0.2t}\cos(0.98t) - 0.204\, e^{-0.2t}\sin(0.98t)$$

This response is also plotted in Figure 10.10.

Example

10.7 Consider the *RLC* circuit in Figure 10.11. Derive a system differential equation for this circuit with input signal v_i and output signal v_o. Hence obtain expressions for the undamped natural frequency of the circuit and the damping ratio.

Solution Using Kirchhoff's voltage law we have

$$v_i = v_R + v_L + v_o \qquad \text{Eqn. [10.30]}$$

For the resistor we have

$$v_R = iR \qquad \text{Eqn. [10.31]}$$

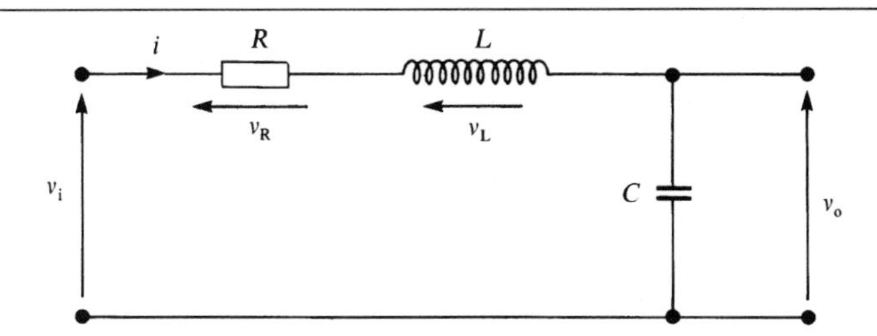

Figure 10.11
RLC circuit for Example 10.7

For the inductor we have

$$v_L = L\frac{di}{dt} \qquad \text{Eqn. [10.32]}$$

For the capacitor we have

$$i = C\frac{dv_o}{dt} \qquad \text{Eqn. [10.33]}$$

Substituting Eqns [10.31] and [10.32] into Eqn. [10.30] to eliminate v_R and v_L gives

$$v_i = iR + L\frac{di}{dt} + v_o \qquad \text{Eqn. [10.34]}$$

We now need to eliminate i from Eqns [10.33] and [10.34]. Unfortunately i occurs in derivative form in Eqn. [10.34], so differentiating Eqn. [10.33] we have

$$\frac{di}{dt} = C\frac{d^2v_o}{dt^2} \qquad \text{Eqn. [10.33a]}$$

We can now substitute Eqns [10.33] and [10.33a] into Eqn. [10.34] to eliminate i. This gives

$$v_i = RC\frac{dv_o}{dt} + LC\frac{d^2v_o}{dt^2} + v_o$$

Rearranging into standard form gives the system differential equation. So,

$$v_i = LC\frac{d^2v_o}{dt^2} + RC\frac{dv_o}{dt} + v_o$$

Taking Laplace transforms and assuming zero initial conditions to form a transfer function we have

$$V_i(s) = LCs^2V_o(s) + RCsV_o(s) + V_o(s)$$

Rearranging we have

$$\frac{V_o(s)}{V_i(s)} = \frac{1}{LCs^2 + RCs + 1}$$

Dividing through by LC to put the transfer function into standard form we have

$$\frac{V_o(s)}{V_i(s)} = \frac{1/LC}{s^2 + (R/L)s + 1/LC}$$

Comparing this transfer function with the general transfer function for a second order system we have

$$\frac{K\omega_n^2}{s^2 + 2\zeta\omega_n s + \omega_n^2} = \frac{1/LC}{s^2 + (R/L)s + 1/LC} \qquad \text{Eqn. [10.35]}$$

Comparing the constant terms of the denominators of Eqn. [10.35] we have

$$\omega_n^2 = \frac{1}{LC}$$

So the undamped natural frequency of the system is given by

$$\omega_n = \frac{1}{\sqrt{LC}} \qquad \text{Eqn. [10.36]}$$

Note that this expression does not involve R. In this circuit it is the resistor that causes the damping. If $R = 0$ then the oscillations in the system do not decay with time. Their frequency is determined by the values of L and C.

Comparing the coefficients of s in the denominators of Eqn. [10.35] we have

$$2\zeta\omega_n = \frac{R}{L}$$

$$\zeta = \frac{R}{2L\omega_n}$$

Substituting in the value of ω_n from the previous section gives

$$\zeta = \frac{R}{2L/\sqrt{LC}} = \frac{R\sqrt{LC}}{2L}$$

Cancelling a common factor of $\sqrt{L}$ we have

$$\zeta = \frac{R\sqrt{C}}{2L}$$

This is the damping ratio of the system.

Self-assessment questions 10.6

1. Describe how varying the damping coefficient of a spring–mass–damper system affects the shape of the step response.
2. What determines the natural frequency of the *RLC* circuit of Figure 10.11?

Exercises 10.6

1. Consider again the torsional pendulum system of Example 3.2. Carry out the following:

   (a) Derive expressions for the d.c. gain, the damping ratio and the undamped natural frequency of the system in terms of constants B, J and K.

(b) Given that $J = 2$ kg m^2 and $K = 8$ N m rad^{-1}, calculate values of B for the system to be:

(i) overdamped with $\zeta = 2$
(ii) critically damped with $\zeta = 1$
(iii) underdamped with $\zeta = 0.5$

(c) Calculate the unit step responses for the three case of part (b).

2. Consider Figure 4.10 which shows an armature-controlled d.c. motor. It is desired to control the angular speed of the motor, ω. The motor has an armature inductance L_a and the load is seated in bearings with a damping coefficient B. Derive a system differential equation relating the system input, which is the armature voltage v_a, to the system output ω. Derive expressions for the d.c. gain of the system, the damping ratio and the undamped natural frequency.

Test and assignment exercises 10

1. Calculate K, ζ and ω_n from the transfer functions of the following second order systems:

   (a) $G(s) = \dfrac{10}{s^2 + 4s + 1}$ (b) $G(s) = \dfrac{4}{s^2 + 4s + 4}$ (c) $G(s) = \dfrac{15}{s^2 + s + 32}$

   State whether the systems are overdamped, critically damped or underdamped.
2. Calculate the unit step response of the second order system of Test and assignment exercise 10.1.
3. Consider the coupled tank system of Example 5.2. Assume that $A_1 = A_2$ and $R_1 = R_2$. Show that the system will always be overdamped. Derive expressions for the damping ratio and the undamped natural frequency of this system.
4. Consider the mercury in glass thermometer of Example 6.7. Derive expressions for the d.c. gain, the damping ratio and the undamped natural frequency of this system.

11 Higher order systems

Objectives

This chapter:

- describes the characteristics of higher order systems
- relates the characteristics of higher order systems to the position of the system poles and zeros
- defines the terms used to characterise the step response of a general system
- explains how system identification can be used to obtain a mathematical model of a system

11.1 Introduction

In Chapters 9 and 10 we spent a great deal of time analysing the response of first and second order systems. The reason for this is that higher order systems can be broken down into a mixture of first and second order systems. It is therefore only necessary to understand their behaviour in detail in order to understand the response of higher order systems. We shall explore this concept in this chapter. Often it is not possible to obtain a mathematical model of a system directly. For example, it may not be possible to dismantle a system to determine the various components. An alternative is to carry out tests on the system and from these tests deduce a mathematical model for the system. This activity is known as system identification.

11.2 Step response of a third order system

The presence of zeros in a transfer function does

There are several possible pole–zero configurations for a third order system. We shall not consider the case when there are system zeros in order to avoid

not change the form of the system response. The main effect of zeros is to change the initial magnitude of the exponential transient terms.

too much complexity. For convenience, assume the system has a transfer function $G(s)$, an input signal $R(s)$ and an output signal $C(s)$. The block diagram for this system is shown in Figure 9.12. The various possible pole configurations for a third order system are shown in Figure 11.1.

Note that we are only considering stable systems. This corresponds to all the system poles being on the left-hand side of the s plane and the transient terms decaying to zero with increasing time. We demonstrate that this is the case later in the chapter.

Examining Figure 11.1 we see that cases (a) and (b) have coincident poles. We shall not discuss these further as they are relatively rare: in practice the poles will have slightly different positions for most systems. Case (c) is where there are three real different poles. Case (d) is where there is a real pole and a complex conjugate pair of poles. We discuss this case later in this section.

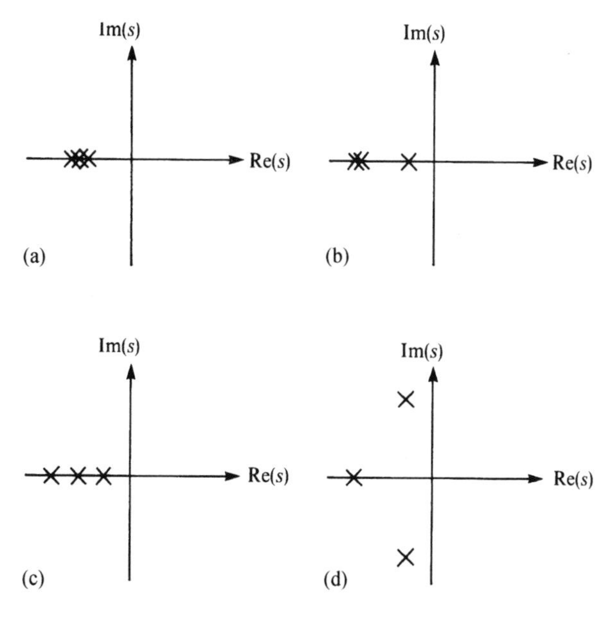

Figure 11.1
Possible pole configurations for a third order system

11.2.1 Third order systems with three real poles

We first consider case (c). For this case we can write the transfer function as

$$G(s) = \frac{K}{(1 + \tau_1 s)(1 + \tau_2 s)(1 + \tau_3 s)}$$

where τ_1, τ_2 and τ_3 are the time constants of the three poles and K is the d.c. gain of the system. Removing the factor τ_1 from the first bracketed term of the denominator gives

$$G(s) = \frac{K}{\tau_1(1/\tau_1 + s)(1 + \tau_2 s)(1 + \tau_3 s)}$$

Similarly, removing the factor τ_2 from the second bracketed term and the factor τ_3 from the third bracketed term we have

$$G(s) = \frac{K}{\tau_1\tau_2\tau_3(1/\tau_1 + s)(1/\tau_2 + s)(1/\tau_3 + s)}$$

which can be written

$$G(s) = \frac{K/\tau_1\tau_2\tau_3}{(s + 1/\tau_1)(s + 1/\tau_2)(s + 1/\tau_3)}$$

If we apply a unit step input to this system then $R(s) = 1/s$ and the system response is given by

$$C(s) = G(s)R(s) = \frac{K/\tau_1\tau_2\tau_3}{s(s + 1/\tau_1)(s + 1/\tau_2)(s + 1/\tau_3)}$$

Separating this expression into partial fractions gives

$$C(s) = \frac{A}{s} + \frac{B}{s + 1/\tau_1} + \frac{C}{s + 1/\tau_2} + \frac{D}{s + 1/\tau_3}$$

where A, B, C and D are constants which can be calculated.

If we invert this expression we obtain

$$c(t) = A + Be^{-t/\tau_1} + Ce^{-t/\tau_2} + De^{-t/\tau_3}$$

We see that the step response contains a steady state term due to the signal pole at $s = 0$. It also contains three transients terms, one for each of the system poles, $-1/\tau_1$, $-1/\tau_2$ and $-1/\tau_3$. These transient terms are exactly the same as those of a first order system. The only difference between a first order system and this type of system is that a first order system has one steady state term and one transient term whereas this third order system has one steady state term and three transient terms.

KEY POINT

A third order system with three real poles can be thought of as a combination of three first order systems.

Figure 11.2 shows the relationship between the pole positions and the terms of the step response for $\tau_1 = 1$, $\tau_2 = 2$, $\tau_3 = 3$ and $K = 10$. Note for this case that $A = 10$, $B = -5$, $C = 40$ and $D = -45$.

It is clear that with a knowledge of the step response of a first order system it is straightforward to understand the step response of this particular third order system.

Example

11.1 Figure 11.3 shows three tanks in series. The tanks have outlet valves with resistance R. Assume these are linear elements for simplicity. The cross-sectional areas of the tanks are A_1, A_2 and A_3 and the height of the liquid in

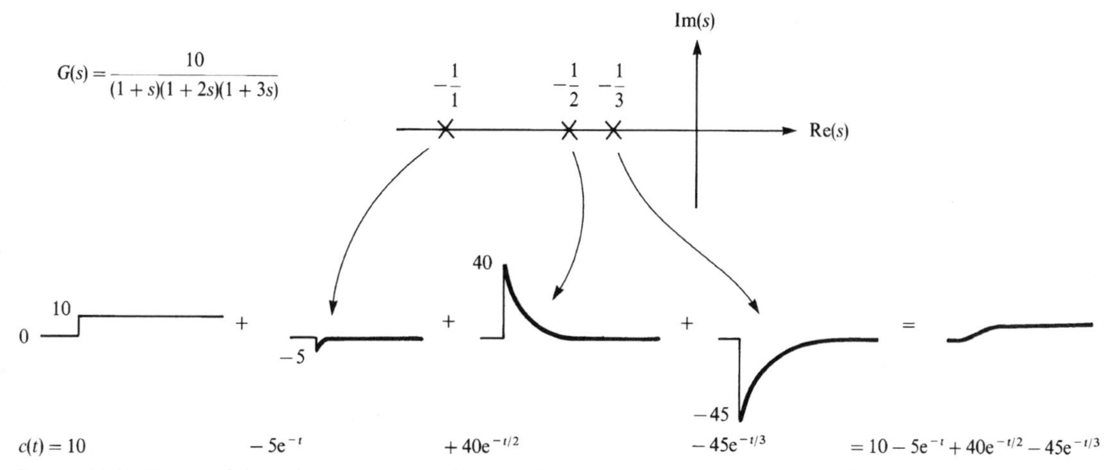

Figure 11.2 Terms of the unit step response for a third order system with three real poles

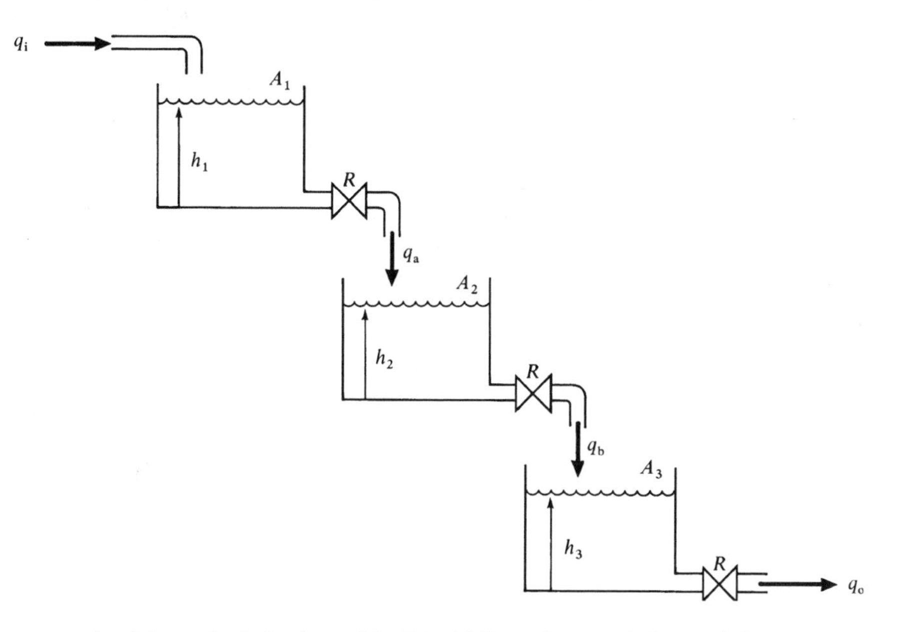

Figure 11.3
Three tanks in series

each of the tanks is h_1, h_2 and h_3. Liquid flows from tank 1 to tank 2 at a rate q_a and liquid flows from tank 2 to tank 3 at a rate q_b. The density of the liquid is ρ. Derive a transfer function relating the flow out of tank 3, q_o, to the flow into tank 1, q_i. Hence calculate the response of the system to a unit step input for the case $R = 5$ N s m^{-5}, $A_1 = 2$ m^2, $A_2 = 4$ m^2, $A_3 = 10$ m^2, $\rho = 1$ kg m^{-3} and $g = 10$ m s^{-2} for convenience.

Solution For tank 1 we have

$$q_i - q_a = A_1 \frac{dh_1}{dt} \qquad \text{Eqn. [11.1]}$$

and

$$\rho g h_1 = q_a R \qquad \text{Eqn. [11.2]}$$

For tank 2 we have

$$q_a - q_b = A_2 \frac{dh_2}{dt} \qquad \text{Eqn. [11.3]}$$

and

$$\rho g h_2 = q_b R \qquad \text{Eqn. [11.4]}$$

For tank 3 we have

$$q_b - q_o = A_3 \frac{dh_3}{dt} \qquad \text{Eqn. [11.5]}$$

and

$$\rho g h_3 = q_o R \qquad \text{Eqn. [11.6]}$$

We need to eliminate five intermediate variables, namely, q_a, q_b, h_1, h_2 and h_3. We have six equations and so this is possible. However, with a little thought it is quite easy to form an overall transfer function for this system and avoid the tedious process of forming a system differential equation prior to obtaining an overall transfer function. Let us first examine Eqns [11.1] and [11.2]. We note that the variable h_1 is confined to these two equations. Examining Figure 11.3 we see that this is because tank 1 only interacts with the rest of the system by means of its output flow q_a. The height of liquid in tank 1, h_1, does not directly interact with the rest of the system. As we only require a transfer function for the system we can immediately take the Laplace transform of Eqns [11.1] and [11.2] and assume zero initial conditions. This gives

$$Q_i(s) - Q_a(s) = A_1 s H_1(s) \qquad \text{Eqn. [11.7]}$$

and

$$\rho g H_1(s) = Q_a(s) R \qquad \text{Eqn. [11.8]}$$

Eliminating $H_1(s)$ from Eqns [11.7] and [11.8] gives

$$Q_i(s) - Q_a(s) = \frac{A_1 R}{\rho g} s Q_a(s)$$

$$Q_i(s) = Q_a(s)\left(1 + \frac{A_1 R}{\rho g} s\right)$$

$$\frac{Q_a(s)}{Q_i(s)} = \frac{1}{1 + (A_1 R/\rho g)s} = \frac{1}{1 + C_1 R s} \qquad \text{Eqn. [11.9]}$$

where C_1 is the capacitance of tank 1 and is equal to $A_1/\rho g$.

By a similar process we can combine the equations for tank 2, namely Eqns [11.3] and [11.4], to obtain

$$\frac{Q_b(s)}{Q_a(s)} = \frac{1}{1 + C_2 R s} \qquad \text{Eqn. [11.10]}$$

where C_2 is the capacitance of tank 2 and is equal to $A_2/\rho g$.

For tank 3, combining Eqns [11.5] and [11.6] gives

$$\frac{Q_o(s)}{Q_b(s)} = \frac{1}{1 + C_3 R s} \qquad \text{Eqn. [11.11]}$$

where C_3 is the capacitance of tank 3 and is equal to $A_3/\rho g$.

The overall transfer function for the system is obtained by combining Eqns [11.9], [11.10] and [11.11] to give

$$\frac{Q_o(s)}{Q_i(s)} = \frac{Q_a(s)}{Q_i(s)}\frac{Q_b(s)}{Q_a(s)}\frac{Q_o(s)}{Q_b(s)} = \frac{1}{(1 + C_1Rs)(1 + C_2Rs)(1 + C_3Rs)}$$

A block diagram for the system is shown in Figure 11.4.

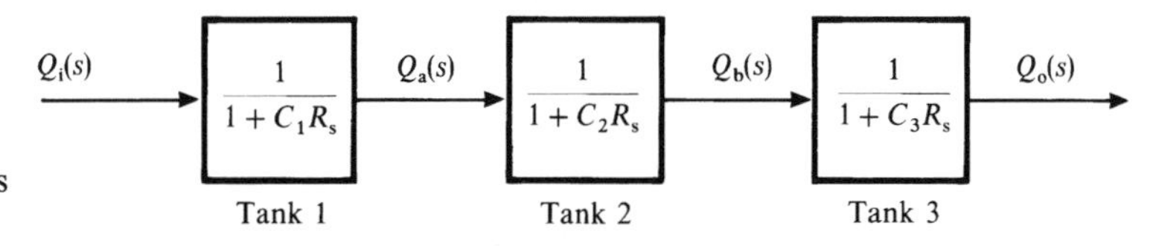

Figure 11.4
Block diagram for the tanks in series

Using the values given, the overall transfer function becomes

$$\frac{Q_o(s)}{Q_i(s)} = \frac{1}{(1 + (2 \times 5/(1 \times 10))s)(1 + (4 \times 5/(1 \times 10))s)(1 + (10 \times 5/(1 \times 10))s)} = \frac{1}{(1 + s)(1 + 2s)(1 + 5s)}$$

For a unit step input we have

$$Q_i(s) = \frac{1}{s}$$

and so

$$Q_o(s) = \frac{1}{s(1 + s)(1 + 2s)(1 + 5s)}$$

We can write this in standard form as

$$Q_o(s) = \frac{1}{10s(s + 1)(s + 0.5)(s + 0.2)} = \frac{0.1}{s(s + 1)(s + 0.5)(s + 0.2)}$$

In order to invert this expression we need to split it into partial fractions. This gives

$$Q_o(s) = \frac{0.1}{s(s + 1)(s + 0.5)(s + 0.2)} = \frac{A}{s} + \frac{B}{s + 1} + \frac{C}{s + 0.5} + \frac{D}{s + 0.2}$$

where A, B, C and D are constants to be determined.

Placing the partial fractions over a common denominator gives

$$\frac{0.1}{s(s + 1)(s + 0.5)(s + 0.2)} = \frac{A(s + 1)(s + 0.5)(5 + 0.2) + Bs(s + 0.5)(s + 0.2) + Cs(s + 1)(s + 0.2) + Ds(s + 1)(s + 0.5)}{s(s + 1)(s + 0.5)(s + 0.2)}$$

Equating the numerators of this expression gives

$$0.1 = A(s+1)(s+0.5)(s+0.2) + Bs(s+0.5)(s+0.2) + Cs(s+1)(s+0.2) + Ds(s+1)(s+0.5)$$

Putting $s = 0$ gives

$$0.1 = A(1)(0.5)(0.2)$$

$$A = 1$$

Putting $s = -1$ gives

$$0.1 = B(-1)(-0.5)(-0.8)$$

$$B = \frac{-0.1}{0.5 \times 0.8} = -0.25$$

Putting $s = -0.5$ gives

$$0.1 = C(-0.5)(0.5)(-0.3)$$

$$C = \frac{0.1}{0.5 \times 0.5 \times 0.3} = 1.333$$

Putting $s = -0.2$ gives

$$0.1 = D(-0.2)(0.8)(0.3)$$

$$D = \frac{-0.1}{0.2 \times 0.8 \times 0.3} = -2.083$$

So we have

$$Q_o(s) = \frac{1}{s} - \frac{0.25}{s+1} + \frac{1.333}{s+0.5} - \frac{2.083}{s+0.2}$$

Taking inverse Laplace transforms gives

$$q_o(t) = 1 - 0.25e^{-t} + 1.333e^{-0.5t} - 2.083e^{-0.2t}$$

The flow rate out of the system, $q_o(t)$, has units of $m^3\ s^{-1}$.

11.2.2 Third order system with one real pole and two complex poles

Let us now examine case (d) when the system has one real pole and two complex poles which form a complex conjugate pair. Because this system has two complex poles it is convenient to retain a quadratic factor in the denominator of the transfer function. We examined this case when we looked at an underdamped second order system in Section 10.5. There is also a linear factor generated by the single real pole. So we can write the transfer function for this system as

$$G(s) = \frac{K\omega_n^2}{(1+\tau s)(s^2 + 2\zeta\omega_n s + \omega_n^2)}$$

where τ is the time constant of the real pole, K is the d.c. gain of the system and ζ and ω_n are the damping ratio and the undamped natural frequency of the complex poles. Removing a factor τ from the first bracketed term of the denominator to obtain $G(s)$ in standard form gives

$$G(s) = \frac{K\omega_n^2}{\tau(1/\tau + s)(s^2 + 2\zeta\omega_n s + \omega_n^2)}$$

Rearranging we have

$$G(s) = \frac{K\omega_n^2/\tau}{(s + 1/\tau)(s^2 + 2\zeta\omega_n s + \omega_n^2)}$$

If we apply a unit step input to this system then $R(s) = 1/s$ and

$$C(s) = G(s)R(s)$$

$$C(s) = \frac{K\omega_n^2/\tau}{s(s + 1/\tau)(s^2 + 2\zeta\omega_n s + \omega_n^2)}$$

Splitting this expression into partial fractions gives

$$C(s) = \frac{A}{s} + \frac{B}{s + 1/\tau} + \frac{Cs + D}{s^2 + 2\zeta\omega_n s + \omega_n^2}$$

where A, B, C and D are constants to be determined.

We shall not invert this expression for reasons of space. However, it is not necessary to do so in order to understand what it consists of. The term A/s is a steady state term which arises as a result of the signal pole at $s = 0$. The term $B/(s + 1/\tau)$ is a transient term equivalent to that of a first order system. It has a time constant τ. The term $(Cs + D)/(s^2 + 2\zeta\omega_n s + \omega_n^2)$ is a transient term equivalent to that of a second order underdamped system with a system zero at $s = -D/C$. Its form is essentially the same as that of a standard second order underdamped system.

Figure 11.5 shows the relationship between the pole positions and the terms of the unit step response for $\tau = 1$, $\zeta = 0.5$, $\omega_n = 2$ and $K = 10$. Note that for this case $A = 10$, $B = -13.33$, $C = 3.333$ and $D = -6.667$. We therefore see that the time response of this third order system can be broken down into a mixture of first and second order responses.

KEY POINT

> A third order system with one real pole and two complex poles can be regarded as a combination of a first order system and a second order underdamped system.

Example

11.2 Figure 11.6 shows an electrical system consisting of two stages. The first stage is an *RLC* circuit and the second stage is an operational amplifier circuit. These are separated by a broken line. For convenience the voltages and currents have been marked on the figure. In order to prevent the second stage of the system loading the first stage of the system, the two stages have been separated by a buffer amplifier.

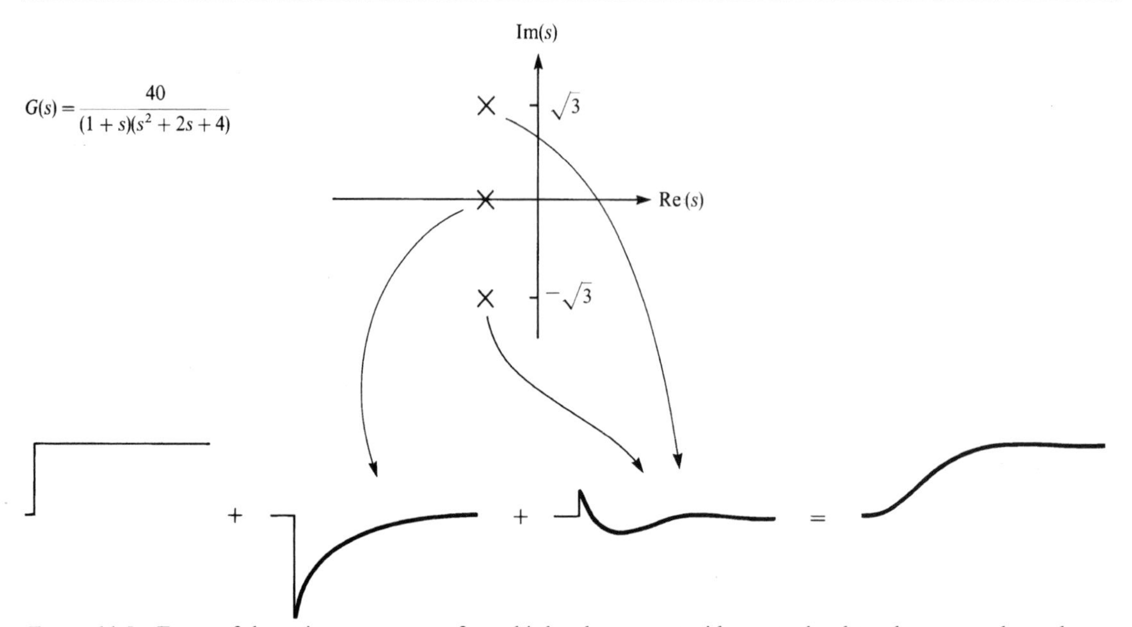

Figure 11.5 Terms of the unit step response for a third order system with one real pole and two complex poles

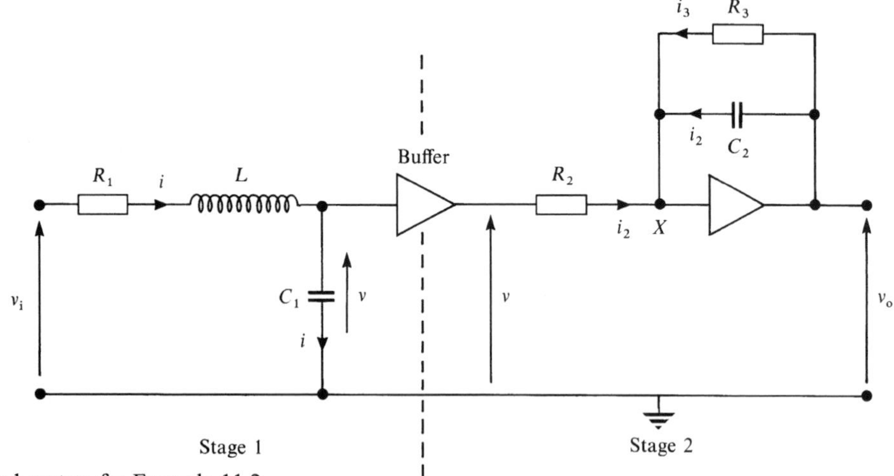

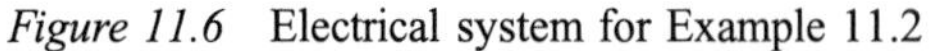

Figure 11.6 Electrical system for Example 11.2

A **buffer** amplifier has a gain of 1 and so the input voltage is the same as the output voltage. It draws negligible current from any circuit that it is connected to and so provides a convenient way of isolating stages of an electronic circuit. A buffer amplifier is depicted by a triangle.

Assume the operational amplifier stage draws so little current that it can be assumed to be equal to zero. Derive an overall transfer function for the system by combining the individual transfer functions for each of the two stages. Calculate the unit step response of the system when $R_1 = 1\ \Omega$, $L = 0.5$ H, $C_1 = 0.25$ F, $R_2 = 10\ \text{k}\Omega$, $R_3 = 10\ \text{k}\Omega$ and $C_2 = 100\ \mu$F.

Solution First we examine stage 1. Because the second stage does not load the first stage we can analyse each stage separately. Assuming a current i flows in stage

1 we have

$$v_{\mathrm{i}} = iR_1 + L\frac{\mathrm{d}i}{\mathrm{d}t} + v \qquad \text{Eqn. [11.12]}$$

and

$$i = C_1\frac{\mathrm{d}v}{\mathrm{d}t} \qquad \text{Eqn. [11.13]}$$

Combining Eqns [11.12] and [11.13] to eliminate i gives

$$v_i = R_1C_1\frac{\mathrm{d}v}{\mathrm{d}t} + LC_1\frac{\mathrm{d}^2v}{\mathrm{d}t^2} + v \qquad \text{Eqn. [11.14]}$$

Taking the Laplace transform of Eqn. [11.14] and assuming zero initial conditions gives

$$V_{\mathrm{i}}(s) = R_1CsV(s) + LC_1s^2V(s) + V(s)$$

Factorising we have

$$V_{\mathrm{i}}(s) = V(s)(LC_1s^2 + R_1C_1s + 1)$$

Rearranging we obtain

$$\frac{V(s)}{V_{\mathrm{i}}(s)} = \frac{1}{LC_1s^2 + R_1C_1s + 1} \qquad \text{Eqn. [11.15]}$$

Eqn. [11.15] is the transfer function for the *RLC* stage of the system.

Let us now examine stage 2 of the circuit. We have assumed that the operational amplifier draws negligible current. Also, point X is a virtual earth point and so can be assumed to have a voltage of zero. We discussed this concept in Section 7.3. Using Kirchhoff's current law we can write

$$i_1 + i_2 + i_3 = 0 \qquad \text{Eqn. [11.16]}$$

Also for R_2 we have

$$v = i_1R_2 \qquad \text{Eqn. [11.17]}$$

For C_2 we have

$$i_2 = C_2\frac{\mathrm{d}v_{\mathrm{o}}}{\mathrm{d}t} \qquad \text{Eqn. [11.18]}$$

and for R_3

$$v_{\mathrm{o}} = i_3R_3 \qquad \text{Eqn. [11.19]}$$

Substituting Eqns [11.17], [11.18] and [11.19] into Eqn. [11.16] to eliminate i_1, i_2 and i_3 gives

$$\frac{v}{R_2} + C_2\frac{\mathrm{d}v_{\mathrm{o}}}{\mathrm{d}t} + \frac{v_{\mathrm{o}}}{R_3} = 0 \qquad \text{Eqn. [11.20]}$$

Taking the Laplace transform of Eqn. [11.20] and assuming zero initial conditions gives

$$\frac{V(s)}{R_2} + C_2sV_{\mathrm{o}}(s) + \frac{V_{\mathrm{o}}(s)}{R_3} = 0$$

Rearranging gives

$$\frac{V(s)}{R_2} = -\frac{V_o(s)}{R_3} - C_2 s V_o(s) = -V_o(s)\left(\frac{1}{R_3} + C_2 s\right)$$

So

$$\frac{V(s)}{R_2} = -V_o(s)\frac{1 + R_3 C_2 s}{R_3}$$

$$\frac{V_o(s)}{V(s)} = -\frac{R_3}{R_2(R_3 C_2 s + 1)} \qquad \text{Eqn. [11.21]}$$

Eqn. [11.21] is the transfer function for the operational amplifier stage of the system. Note the presence of the negative sign because the operational amplifier stage inverts the input voltage.

Combining Eqns [11.15] and [11.21] to eliminate $V(s)$ gives the overall transfer function for the system. We have

$$\frac{V_o(s)}{V_i(s)} = \frac{V_o(s)}{V(s)}\frac{V(s)}{V_i(s)} = -\frac{R_3}{R_2(R_3 C_2 s + 1)(LC_1 s^2 + R_1 C_1 s + 1)}$$

Substituting in the component values given earlier we have

$$\frac{V_o(s)}{V_i(s)} = -\frac{10^4}{10^4(10^4 \times 10^{-4} \times s + 1)(0.5 \times 0.25 \times s^2 + 1 \times 0.25 \times s + 1)}$$

Simplifying we have

$$\frac{V_o(s)}{V_i(s)} = -\frac{1}{(s+1)(0.125s^2 + 0.25s + 1)}$$

Multiplying the numerator and denominator by 8 gives

$$\frac{V_o(s)}{V_i(s)} = -\frac{8}{(s+1)(s^2 + 2s + 8)}$$

Calculate the complex poles yourself, using the formula for solving a quadratic equation.

This system has one real pole at $s = -1$ and two complex poles at $s = -1 \pm 2.646\text{j}$.

For a unit step $V_i(s) = 1/s$ and so

$$V_o(s) = -\frac{8}{s(s+1)(s^2 + 2s + 8)}$$

Splitting into partial fractions gives

$$V_o(s) = \frac{A}{s} + \frac{B}{s+1} + \frac{Cs + D}{s^2 + 2s + 8}$$

where A, B, C and D are constants to be determined.

Putting the partial fractions over a common denominator gives

$$\frac{-8}{s(s+1)(s^2 + 2s + 8)} = \frac{A(s+1)(s^2 + 2s + 8) + Bs(s^2 + 2s + 8) + (Cs + D)s(s+1)}{s(s+1)(s^2 + 2s + 8)}$$

Equating the numerators of this expression we have

$$-8 = A(s+1)(s^2 + 2s + 8) + Bs(s^2 + 2s + 8) + (Cs + D)s(s+1)$$

Expanding brackets gives

$$-8 = A(s^3 + 2s^2 + 8s + s^2 + 2s + 8) + B(s^3 + 2s^2 + 8s) + Cs^3 + Cs^2 + Ds^2 + Ds$$

Gathering terms we have

$$-8 = s^3(A + B + C) + s^2(3A + 2B + C + D) + s(10A + 8B + D) + 8A$$

Putting $s = 0$ gives

$$-8 = 8A$$
$$A = -1$$

Putting $s = -1$ gives

$$-8 = B(-1)((-1)^2 - 2 + 8)$$
$$-8 = -B(7)$$
$$B = \frac{8}{7}$$

Equating the s^3 terms gives

$$0 = A + B + C$$
$$C = -(A + B)$$
$$= -\left(-1 + \frac{8}{7}\right)$$
$$C = -\frac{1}{7}$$

Equating the s terms gives

$$0 = 10A + 8B + D$$
$$D = -(10A + 8B)$$
$$= -\left(-10 \times 1 + 8 \times \frac{8}{7}\right)$$
$$= -\left(\frac{64}{7} - 10\right) = \frac{6}{7}$$

So finally we can write

$$V_o(s) = -\frac{1}{s} + \frac{8}{7(s+1)} + \frac{6 - s}{7(s^2 + 2s + 8)}$$

Before inverting this expression we need to split up the term $(6 - s)/[7(s^2 + 2s + 8)]$ to match the standard forms available in Table 8.1. First we complete the square for the denominator:

$$\frac{6 - s}{7(s^2 + 2s + 8)} = -\frac{1}{7}\left[\frac{s - 6}{(s + 1)^2 + 7}\right]$$

Rearranging we have

$$\frac{6 - s}{7(s^2 + 2s + 8)} = -\frac{1}{7}\left[\frac{s - 6}{(s + 1)^2 + (\sqrt{7})^2}\right]$$

Splitting the expression in order to match the exponential sine and exponential cosine entries in Table 8.1 gives

$$\frac{6-s}{7(s^2+2s+8)} = -\frac{1}{7}\left[\frac{s+1}{(s+1)^2+(\sqrt{7})^2} - \frac{7}{(s+1)^2+(\sqrt{7})^2}\right]$$

$$= -\frac{1}{7}\left[\frac{s+1}{(s+1)^2+(\sqrt{7})^2} - \frac{7}{\sqrt{7}}\frac{\sqrt{7}}{(s+1)^2+(\sqrt{7})^2}\right]$$

These two terms appear in Table 8.1. So we have

$$V_o(s) = -\frac{1}{s} + \frac{8}{7(s+1)} - \frac{1}{7}\frac{s+1}{(s+1)^2+(\sqrt{7})^2} + \frac{1}{\sqrt{7}}\frac{\sqrt{7}}{(s+1)^2+(\sqrt{7})^2}$$

Inverting this expression gives

$$v_o(t) = -1 + \frac{8}{7}e^{-t} - \frac{1}{7}e^{-t}\cos(\sqrt{7}t) + \frac{1}{\sqrt{7}}e^{-t}\sin(\sqrt{7}t)$$

This is the unit step response of the system. Note the decaying sinusoid terms arise as a result of the complex conjugate pair of system poles. The decaying exponential transient term arises as a result of the real system pole. The steady state term arises as a result of the signal pole.

Self-assessment questions 11.2

1. State the various possible pole-zero configurations for a third order system.
2. Sketch the response of a third order system with three negative real poles none of which are coincident.
3. Sketch the response of a third order system with one negative real pole and a complex conjugate pair of poles with negative real parts.

11.3 Step response of an *n*th order system

We now consider the step response of an nth order system. We assume the system input is $R(s)$, the system output is $C(s)$ and the transfer function of the system is $G(s)$. Consider a general nth order system with a transfer function $G(s)$ given by

For example

$$G(s) = \frac{3s^2+2s+4}{s^3+s^2+4s+1}$$

Here $m=2$ and $n=3$ and the coefficient of s^2 is 3.

$$G(s) = \frac{P(s)}{Q(s)}$$

where $P(s)$ and $Q(s)$ are polynomials in s and $P(s)$ is of order m and $Q(s)$ is of order n. For most physical systems $m < n$ and so the order of the numerator polynomial is less than the order of the denominator polynomial. Note that the coefficient of s^m in $P(s)$ is not necessarily unity.

If we apply a unit step to this system we have $R(s) = 1/s$ and the system output is given by

$$C(s) = G(s)R(s) = \frac{P(s)}{sQ(s)}$$

As $Q(s)$ is of order n it can be factorised to give n factors. So we can write

Recall that the poles are the roots of the characteristic equation. They are the values of s that cause the transfer function to tend towards infinity.

$$C(s) = \frac{P(s)}{s(s-p_1)(s-p_2)\cdots(s-p_n)}$$

where $(s-p_1)$, $(s-p_2)$, ..., $(s-p_n)$ are the factors of $Q(s)$ and $p_1, p_2, \ldots,$ p_n are the poles of the nth order system. Note that some of these poles may be complex.

We can separate this expression into partial fractions. This gives

$$C(s) = K\left(\frac{1}{s} + \frac{A_1}{s-p_1} + \frac{A_2}{s-p_2} + \cdots + \frac{A_n}{s-p_n}\right)$$

where K is the d.c. gain of the system and $A_1, A_2, \ldots, A_n$ are constants to be determined. Inverting this expression we obtain

$$c(t) = K(1 + A_1 e^{p_1 t} + A_2 e^{p_2 t} + \cdots + A_n e^{p_n t})$$

If the poles have negative real parts then the exponential terms will eventually decay to zero. These terms are then transient terms and there is a transient term for each of the system poles. This is the requirement for a **stable system.**

KEY POINT

A stable system is one in which the output signal does not keep growing with time in response to a bounded input signal.

The use of the word bounded reflects the fact that if the input signal keeps increasing then this may lead to an output signal that continues to increase.

For this case, the only term that remains with time is the term that arises from the signal pole at $s = 0$. This gives a constant term of magnitude K. This is a **steady state** term.

If any of the poles have positive real parts then the associated transient term will increase with time. This corresponds to an **unstable system**.

KEY POINT

For an unstable system, the output signal continues to grow with increasing time in response to a bounded input.

A final possibility is that all the poles have negative real parts or zero real parts, corresponding to them being on the imaginary axis in the s plane. For this case the system is said to be a **marginally stable system.**

KEY POINT

To summarise, each system pole gives rise to a transient term in the step response. For stability all of these poles must have negative real parts. In other words, they must all lie in the left-hand side of the s plane. If any system pole lies in the right-hand side of the s plane then the system is unstable.

We see that the position of a system pole in the s plane governs the form of the associated transient term. In fact the further a pole is to the left of the imaginary axis, the faster its transient term decays. Also, the larger the imaginary component of a complex pair of system poles, the higher the frequency of the decaying sinusoid. This information, together with the shape of transient terms associated with system poles on the imaginary axis and on the right-hand side of the s plane is summarised in Figure 11.7.

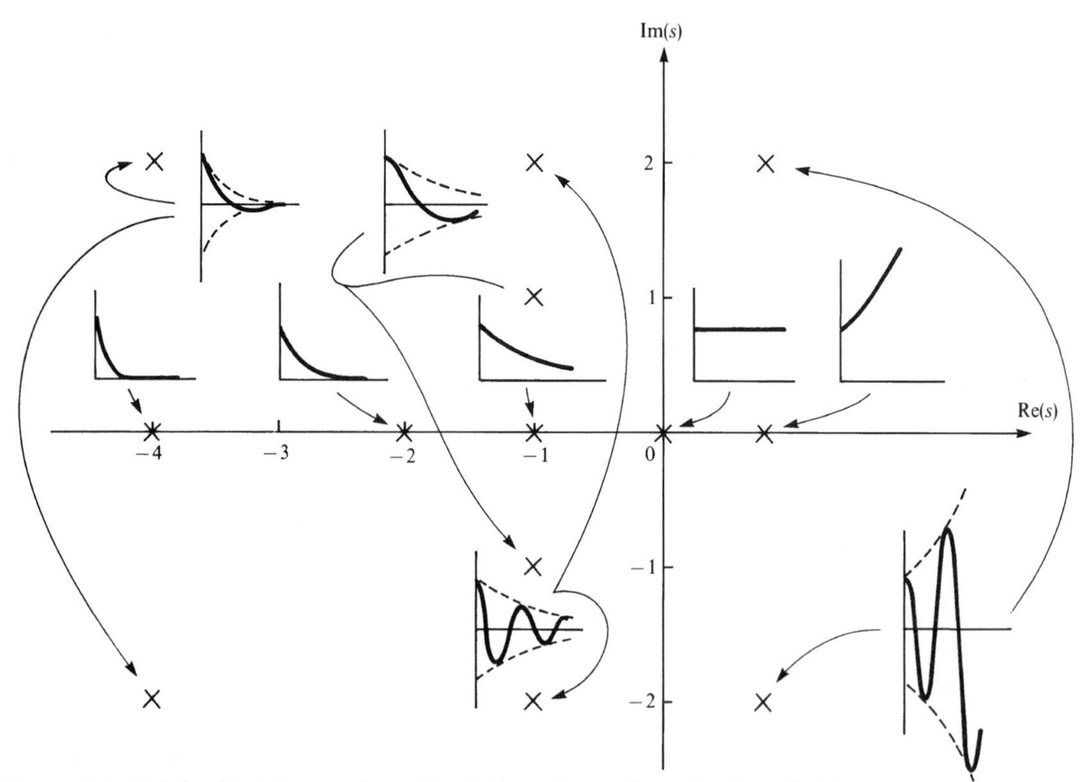

Figure 11.7 Relationship between pole position in the s plane and associated transient term

Self-assessment questions 11.3

1. State the requirement on the system poles for a system to be stable and explain why this is the case.

2. If a system has some poles with positive real parts then will the system be stable or unstable?
3. A negative real system pole is moved further away from the imaginary axis. Describe the effect on its associated transient term.
4. A complex conjugate pair of system poles with negative real parts have their imaginary component doubled. Describe the effect on the associated transient term.

11.4 Performance characteristics of a system

It is possible to characterise a first order system completely by specifying its time constant and d.c. gain. With a second order system all that is required is the undamped natural frequency, the damping ratio and the d.c. gain. However, with higher order systems no such simple parameters exist to characterise the performance of the system. For this reason engineers have to resort to specifying certain performance characteristics which are deemed to be important for a system. Whenever a system is being designed it is these characteristics that are used as a basis for the design. Figure 11.8 shows some of the main characteristics for a step response. For convenience a unit step input is assumed and the desired output is also assumed to be unity.

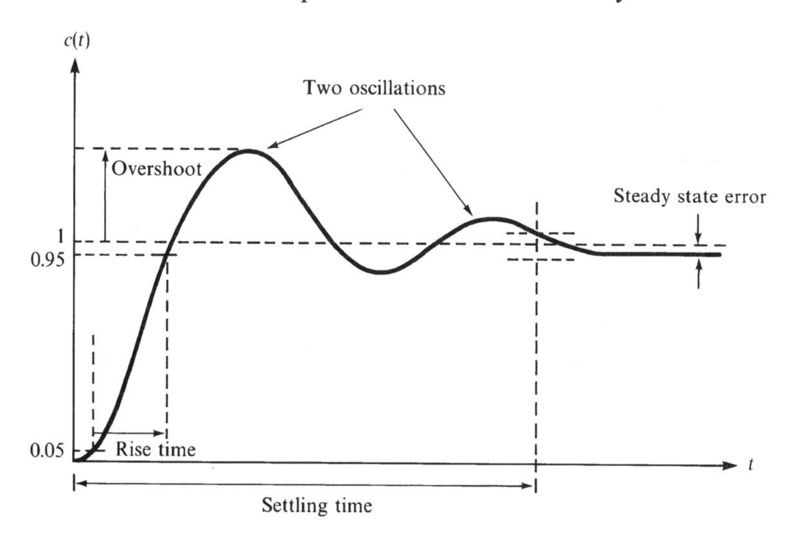

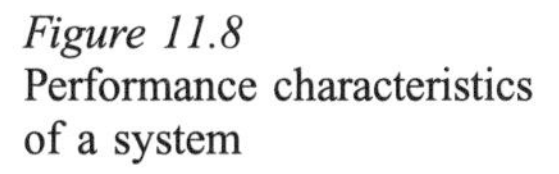
Figure 11.8
Performance characteristics of a system

The amount by which a system overshoots its final value can be very important. In fact many engineering systems have to be designed to have no overshoot. **Overshoot** is usually expressed as a percentage of the desired final value.

When a response finally comes within a certain percentage band of the desired value and stays there it is considered to have settled. Typical values are

$\pm$**2%** of the desired value. For example, for a first order system the system is within 2% of its final value after four time constants. The time required to reach this band is known as the **settling time** of the system.

The **steady state error** is the amount by which the final value differs from the desired value when the system settles to a constant value. This is often expressed as a percentage of the desired value.

Another quantity of interest is the **rise time** of the system. Definitions of rise time vary but a common one is the time to go from 5% to 95% of the desired value. This is a useful definition because it avoids determining the exact start time for systems containing noise and it avoids determining the finish time for overdamped systems which can be difficult to decide.

Finally, another commonly used characteristic is the number of times the system oscillates before it settles, that is, before it reaches the settling band and stays within it. This is most easily found by counting the number of peaks in the oscillatory response.

Example 11.3

Calculate the following from the step response shown in Figure 11.9:

(a) maximum percentage overshoot
(b) settling time assuming a 10% settling band
(c) rise time
(d) number of oscillations before the output signal remains within the settling band

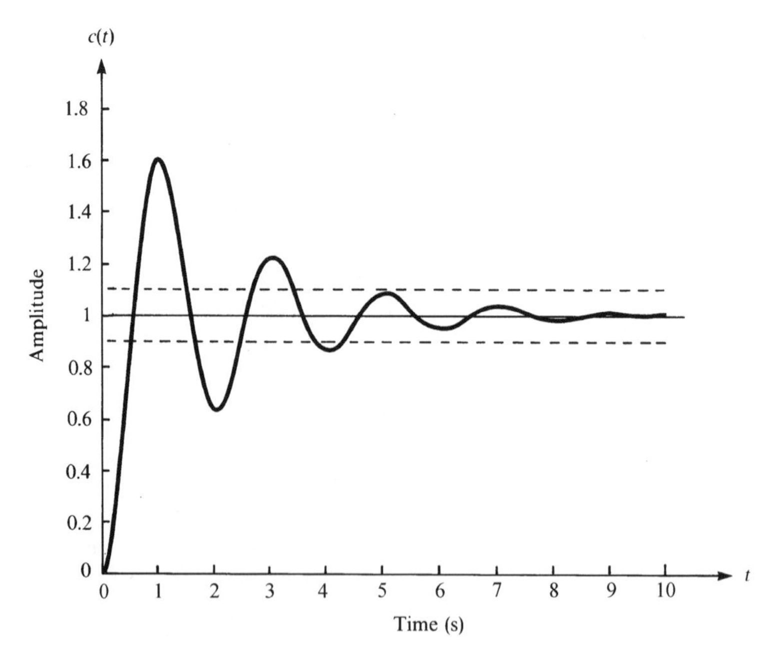

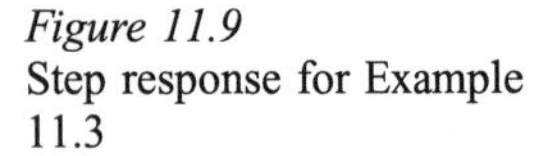
Figure 11.9
Step response for Example 11.3

(e) steady state percentage error.

Solution

(a) The maximum percentage overshoot is 60%.

(b) The settling time is 4.2 s.
(c) This is difficult to estimate given the compressed time scale. It is approximately 0.4 s.
(d) The number of oscillations before the system settles is 2.
(e) There is no steady state error.

Self-assessment questions 11.4

1. Explain why it is necessary to define general performance characteristics for systems of order 3 and above.
2. Define the settling time of a system.
3. Define the maximum overshoot of a system.
4. Define the rise time of a system.
5. Suggest reasons why a rise time from 0% to 100% may not be so useful for characterising a system.
6. Define the steady state error of a system.

Exercises 11.4

1. Determine the following from the unit step response shown in Figure 11.10:
   (a) maximum percentage overshoot
   (b) settling time assuming a 10% settling band
   (c) rise time
   (d) number of oscillations before the output signal remains within the settling band
   (e) steady state error.

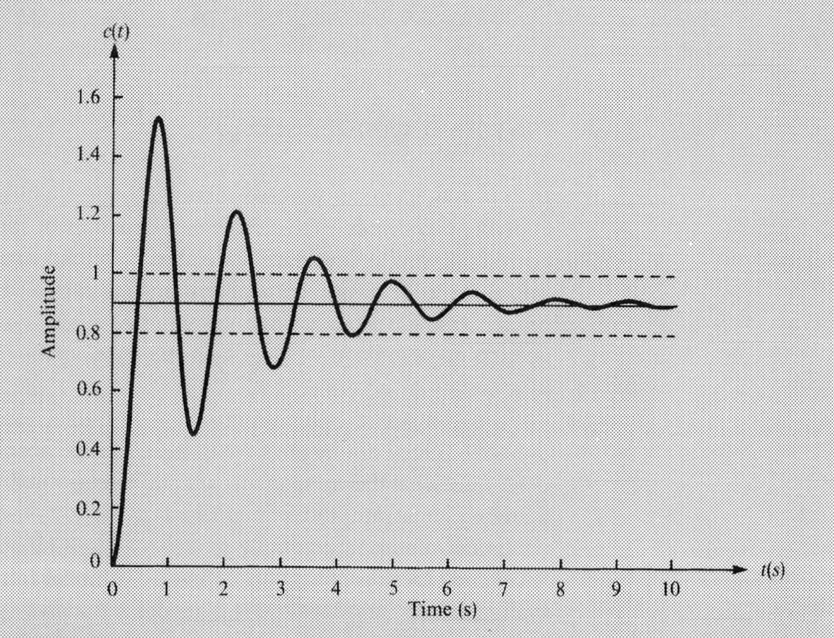

Figure 11.10
Step response for Exercises 11.4.2

2. Figure 11.11 shows the unit step response of an underdamped system. Calculate the following:
   (a) maximum percentage overshoot
   (b) settling time assuming a 5% settling band

(c) rise time
(d) number of oscillations before the output signal remains within the settling band
(e) steady state error.

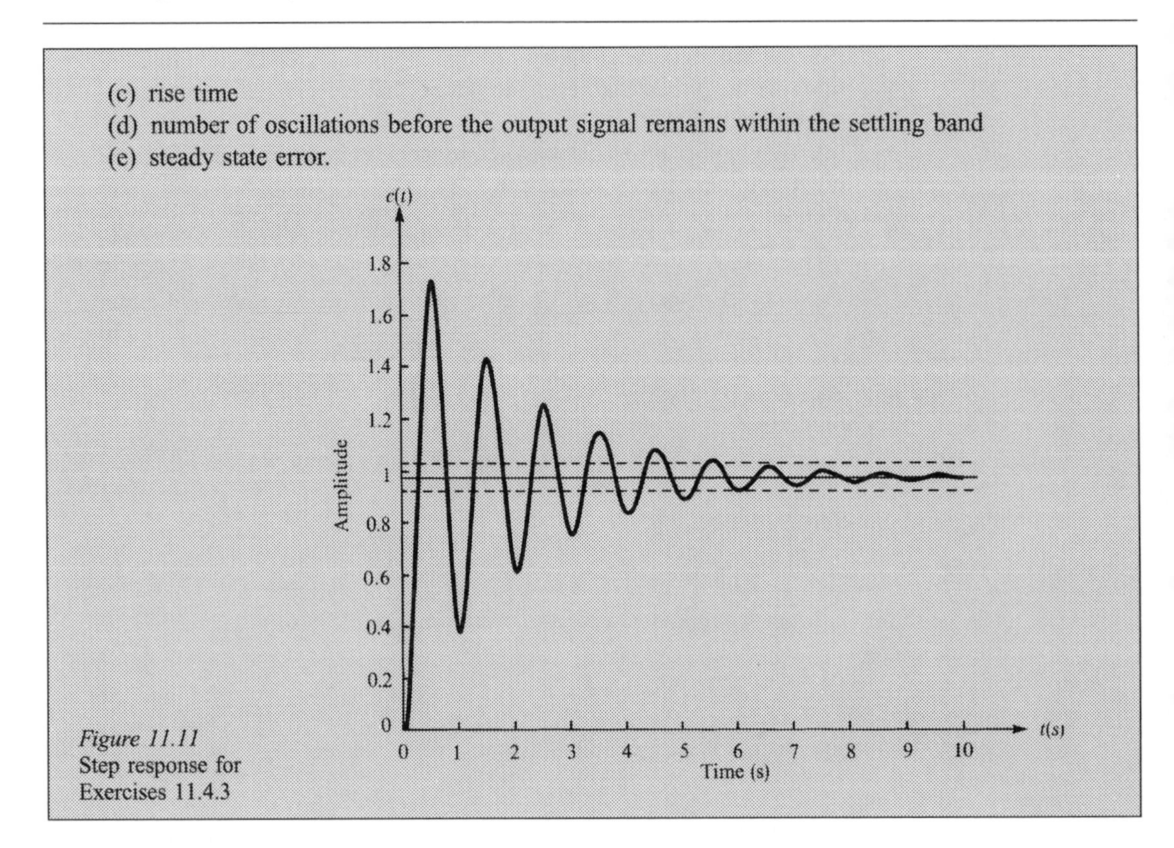

Figure 11.11
Step response for Exercises 11.4.3

11.5 System identification

In previous chapters we have obtained mathematical models of systems by applying physical laws to simple combinations of idealised components. This is a good way to introduce the subject of mathematical modelling but it is important to be aware that real systems are rarely so simple. Even so, simple models can often predict the behaviour of a real system adequately for engineering purposes and it is a golden rule in modelling to use as simple a model of a system as possible.

Problems can occur when trying to model some real systems found in industry. It may not be possible to gain internal access to a system in order to determine what the components are and obtain suitable models for them. Also, many systems are too complicated to break down into simple idealised components and yet a simple model may still be adequate for analysis. The way round these problems is to develop a model from tests carried out on the system rather than developing a model by analysing a set of interconnected components. The process of obtaining a mathematical model from system testing is known as **system identification**.

We have not the space here to describe all the various methods of system identification and so we will confine ourselves to a brief description of some of

observe how the system responds. The results are then used to develop a mathematical model for the system.

Example

11.4 A unit step input is applied to a system. Figure 11.12 shows the response of the system to the unit step input. Deduce a mathematical model for the system.

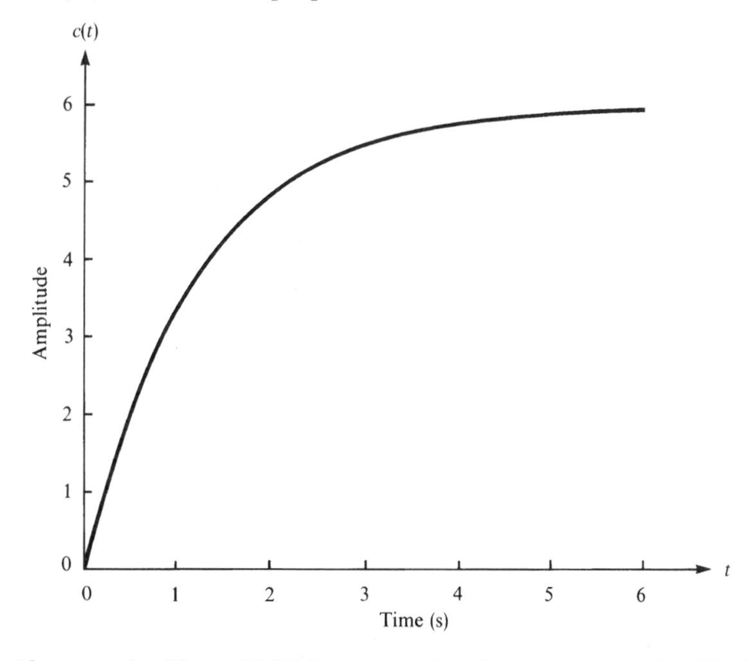

Figure 11.12
Unit step response for the system of Example 11.4

Solution If we examine Figure 11.12 then we see that the step response is indicative of a first order system. The system may be more complicated than this but we shall assume it is a linear first order system.

A unit step input is applied to the system and the final value of the output is 6. A unit step has a height of 1; therefore the d.c. gain of the system, K, is 6.

One way to calculate the time constant of the system is to calculate the time the response takes to reach 0.63 of its final height. In this case this corresponds to the response reaching a height of $0.63 \times 6 = 3.78$. Examining Figure 11.12 we see that this roughly corresponds to a value of $t = 1.2$ s. We deduce $\tau = 1.2$ for this system.

Finally, the transfer function for the system, $G(s)$, is

$$G(s) = \frac{K}{1 + \tau s} = \frac{6}{1 + 1.2s}$$

This is a mathematical model for the system written as a transfer function.

In Example 11.4 we assumed the system was linear and first order from our knowledge of first order systems. One way of confirming that a system is linear is to carry out several tests on the system with different strength input signals. If the shape of the response remains the same and the ratio of the final output value over the step input height remains the same then this is a strong indication that the system is linear. For example, in Example 11.4 if when we applied a step input of height 10 the shape of the response remained the same and its final value was 60 then this would suggest a linear system.

Identifying a second order system is more difficult. If the system is overdamped then it may be that one pole dominates the response because it has a much longer time constant and so the transient term due to the other pole decays very quickly. If so, the response will look very similar to that of a first order system and the dominant time constant can be calculated. Calculating the d.c. gain is straightforward but identifying the other time constant is more difficult. Methods do exist but they are beyond the scope of this book.

It is possible to identify an underdamped second order system merely by carrying out a step test. A typical unit step response for an underdamped second order system is shown in Figure 11.13.

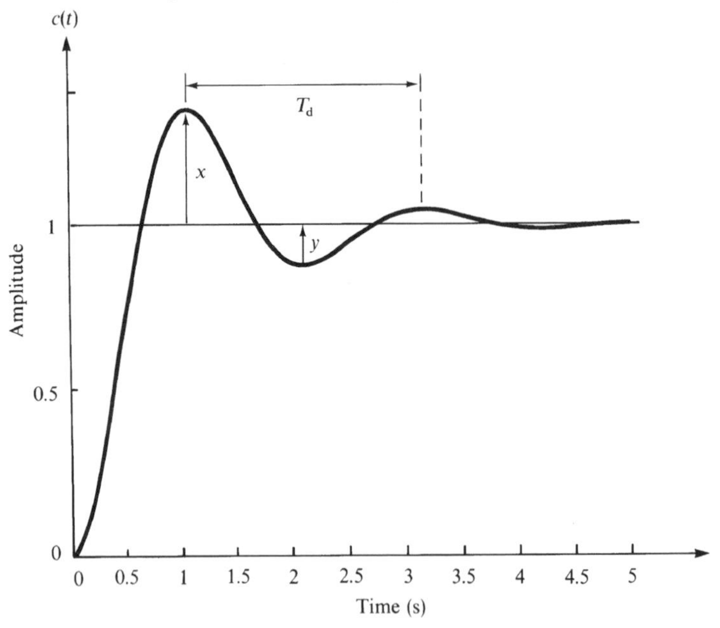

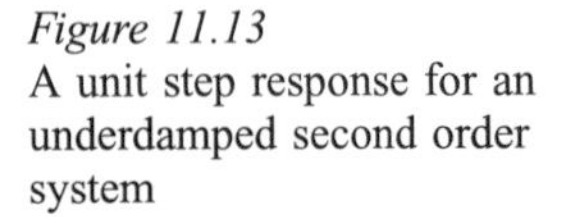
Figure 11.13
A unit step response for an underdamped second order system

The damping ratio ζ can be calculated by measuring the size of the first overshoot, x and the first undershoot y. The formula to calculate ζ is

This formula is stated without proof.

$$\log_e\left(\frac{x}{y}\right) = \frac{\pi\zeta}{\sqrt{1-\zeta^2}}$$

The damped natural frequency ω_d can be calculated by measuring the time between successive peaks, T_d, using the formula

$$\omega_d = \frac{2\pi}{T_d}$$

Having calculated ζ and ω_d it is straightforward to calculate the undamped natural frequency ω_n using the relationship

$$\omega_d = \omega_n\sqrt{1-\zeta^2}$$

Example

11.5 A unit step input is applied to a system and the response is shown in Figure 11.14. Assuming the system can be approximated by a linear second order model, derive a mathematical model for the system.

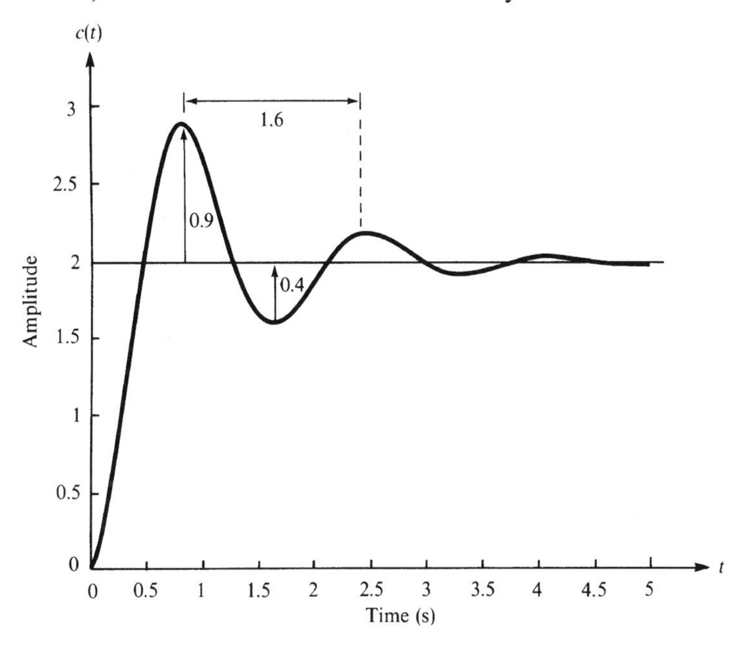

Figure 11.14
A unit step response for the system of Example 11.5

Solution The first overshoot has a value of $x = 0.9$. The first undershoot has a value of $y = 0.4$. So we have

$$\log_e\left(\frac{0.9}{0.4}\right) = \frac{\pi\zeta}{\sqrt{1-\zeta^2}}$$

Simplifying gives

$$0.8109 = \frac{\pi\zeta}{\sqrt{1-\zeta^2}}$$

Squaring both sides gives

$$0.6576 = \frac{\pi^2\zeta^2}{1-\zeta^2}$$

Hence,

$$0.6576(1-\zeta^2) = 9.870\zeta^2$$

Removing brackets gives

$$0.6576 - 0.6576\zeta^2 = 9.876\zeta^2$$

Collecting terms we have

$$0.6576 = 10.534\zeta^2$$

Simplifying we have

$$\zeta^2 = 0.06243$$

Finally we have

$$\zeta = 0.250$$

Note that this is only an approximate value for the damping ratio. In order to obtain an accurate value the values of x and y would need to be measured using a more accurate scale.

It is possible to obtain an estimate of the damped natural frequency by measuring the time difference between successive peaks of the response. We have

$$T_{\mathrm{d}} = 1.6$$

Hence,

$$\omega_{\mathrm{d}} = \frac{2\pi}{T_{\mathrm{d}}} = \frac{2\pi}{1.6} = 3.93 \quad \mathrm{rad\ s^{-1}}$$

Now the undamped natural frequency is related to the damped natural frequency by the formula

$$\omega_{\mathrm{d}} = \omega_{\mathrm{n}}\sqrt{1-\zeta^2}$$

Hence we have

$$\omega_{\mathrm{n}} = \frac{\omega_{\mathrm{d}}}{\sqrt{1-\zeta^2}} = \frac{3.93}{\sqrt{1-0.250^2}} = 4.06 \quad \mathrm{rad\ s^{-1}}$$

Finally, the d.c. gain of the system is 2. Therefore a suitable mathematical model for the system is

$$\frac{K\omega_{\mathrm{n}}^2}{s^2 + 2\zeta\omega_{\mathrm{n}}s + \omega_{\mathrm{n}}^2} = \frac{2 \times 4.06^2}{s^2 + 2 \times 0.250 \times 4 \times s + 4.06^2}$$

$$= \frac{33.0}{s^2 + 2.0s + 16.5}$$

So far we have only considered the use of the step signal to identify a mathematical model of a system. It is also possible to use impulse and ramp signals to extend the range of systems that can be identified. Collectively these are known as **time domain methods**. It is also possible to apply a sinusoidal signal to the system and see how the system responds as the frequency of the input signal is varied. These methods are known as **frequency domain methods**. Finally, it is possible to apply random signals to a system. This is a powerful method of identifying systems and has the added advantage of not disturbing a system unduly, thus making it suitable for testing systems that are in continuous use.

Self-assessment questions 11.5

1. Explain why it is sometimes necessary to obtain a mathematical model for a system by using test signals.
2. State the three main types of system identification techniques.
3. What needs to be done to identify a first order system from a step test?
4. What needs to be done to identify a second order underdamped system from a step test?

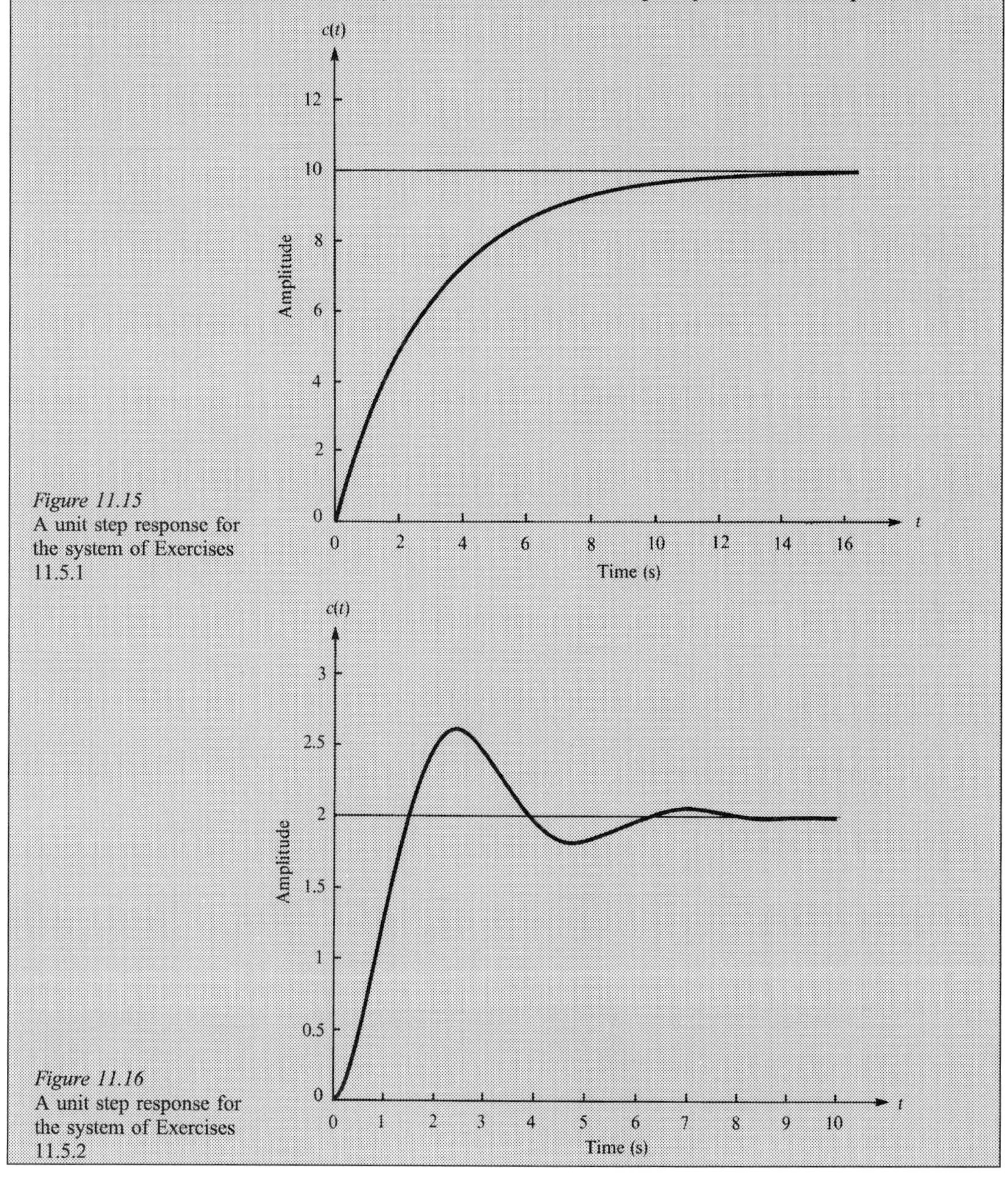

Figure 11.15
A unit step response for the system of Exercises 11.5.1

Figure 11.16
A unit step response for the system of Exercises 11.5.2

Exercises 11.5

1. A unit step is applied to a system and Figure 11.15 shows the system response. Identify a mathematical model for the system.

2. A unit step is applied to a system and Figure 11.16 shows the system response. Identify a mathematical model for the system.

Test and assignment exercises 11

1. Consider the three tanks in series in Example 11.1. Calculate the response of this system to a unit step input when $R = 10$ N s m^{-5}, $A_1 = 4$ m^2, $A_2 = 3$ m^2, $A_3 = 2$ m^2, $\rho = 1$ kg m^{-3} and $g = 10$ m s^{-2} for convenience.
2. Consider the electrical system of Example 11.2. Calculate the response of this system to a unit step input when $R_1 = 2\ \Omega$, $L = 0.2$ H, $C_1 = 0.5$ F, $R_2 = 5$ kΩ, $R_3 = 20$ kΩ and $C_2 = 200\ \mu$F.
3. Consider the position control system of Example 4.8. Calculate the response of this system to a unit step input when $R_1 = 2\ \Omega$, $L_a = 1$ H, $B = 2$ N m s rad^{-1}, $K_T = 0.5$ N m rad^{-1} and $K_e = 1$ V s rad^{-1}.

powerful method of identifying systems and has the added advantage of not disturbing a system unduly, thus making it suitable for testing systems that are in continuous use.

12 Feedback and Control

Objectives

This chapter:

- explains the difference between open loop and closed loop control
- describes how the performance of a system can be changed by the use of feedback
- simplifies block diagrams using standard rules
- explains the importance of stability when designing a control system
- explains what is meant by proportional control
- designs control systems containing a proportional controller
- introduces root locus design

12.1 Introduction

So far we have mainly concerned ourselves with the analysis of a system, without seeking to change its performance. In this chapter we look at how to control the performance of a system in order to improve it in some way. We first examine the difference between open and closed loop control. We then look at how the performance of a system can be improved by the use of feedback. We also examine the potential problems that feedback control introduces. Finally, the topic of control system design is introduced through the technique of root locus design. Control engineering is an extremely important area of engineering as control systems are found in systems as diverse as robotics and flight control of aircraft. Indeed the human body itself contains many control systems; such is the ubiquitous nature of feedback control.

12.2 Open and closed loop control

We have used the s domain symbols for these signals but clearly they have time domain equivalents. For example, $f(t)$ is the time domain equivalent of $R(s)$.

Although we concentrate on closed loop control, it is worth briefly discussing another form of control which is widely used, namely, **open loop control**. Consider the system shown in Figure 12.1. The aim of this form of control is to adjust the input signal in order to obtain a desired output signal. The input signal is called the **reference input** and given the symbol $R(s)$. The output signal $C(s)$ is called the **controlled output** to reflect the fact that the aim of the control system is to control this value. The system under control is referred to as the **plant**. It has a transfer function $G_p(s)$. A simple example of an open loop control system is shown in Figure 12.2.

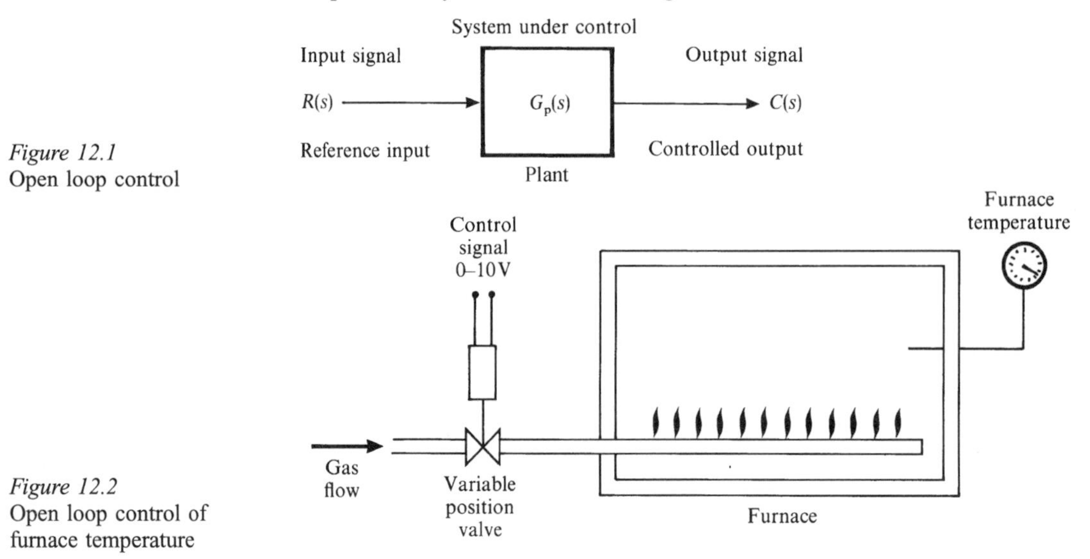

Figure 12.1
Open loop control

Figure 12.2
Open loop control of furnace temperature

The system consists of a furnace which is heated by gas. The flow rate of gas to the furnace can be varied by adjusting the position of a valve situated in the gas inlet pipe. Usually for such a system the valve is turned by a motor and is known as a motorised valve. The motor–valve unit is self-contained and all that is required is a voltage signal between say 0 and 10 V to adjust the valve from fully closed to fully open. The aim of the control system is to maintain the furnace temperature at a particular value. So in this case the furnace temperature is the controlled output. The reference input is the control signal to the valve which can be varied between 0 V corresponding to it being completely closed and 10 V corresponding to it being completely open. A block diagram for the system is shown in Figure 12.3.

An advantage of this system is its simplicity. However, there are several disadvantages. If the system is truly open loop then a schedule of valve settings must be used to obtain a particular furnace temperature, based on

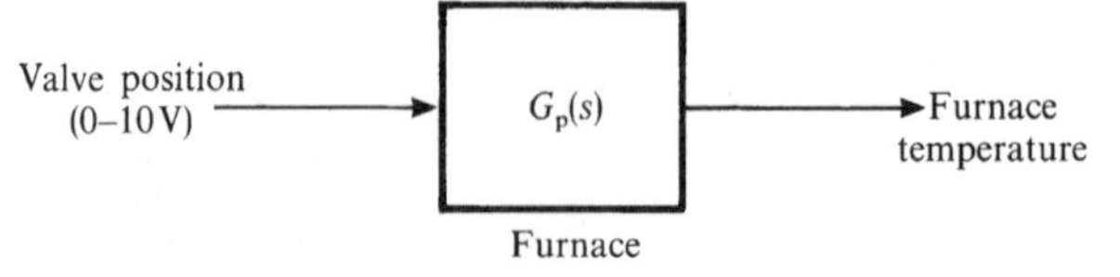

Figure 12.3
Block diagram of open loop control of furnace temperature

experience. So, for example, if it is known that a control signal of 5 V, corresponding to the valve position being 50% open, gives a furnace temperature of 330°C, then an operator could adjust the valve position to obtain the particular furnace temperature desired. The problem with this procedure is that it takes no account of what the actual furnace temperature is. So, for example, if it was a particularly cold day then the heat losses out of the furnace would be greater, thus requiring a higher gas flow rate to obtain a particular furnace temperature. An open loop control system cannot deal with this effectively. A further possibility is that the gas pressure may change, thus changing the gas flow rate for a fixed valve position, or the furnace lining may be damaged, thus increasing the heat losses. None of these would be taken into account by an open loop control system.

KEY POINT

> An open loop control system is simple to implement but does not take account of the actual value of the controlled variable when deciding on the value of reference input. As such it is vulnerable to changes in conditions with the effect that the controlled output can fluctuate with varying conditions.

The other main form of control is **closed loop control** or **feedback control**. These terms are used interchangeably by engineers. With this form of control account is taken of the actual value of the controlled output and if this deviates from the desired value then action is taken to restore it to the correct value. The general block diagram for closed loop control is shown in Figure 12.4.

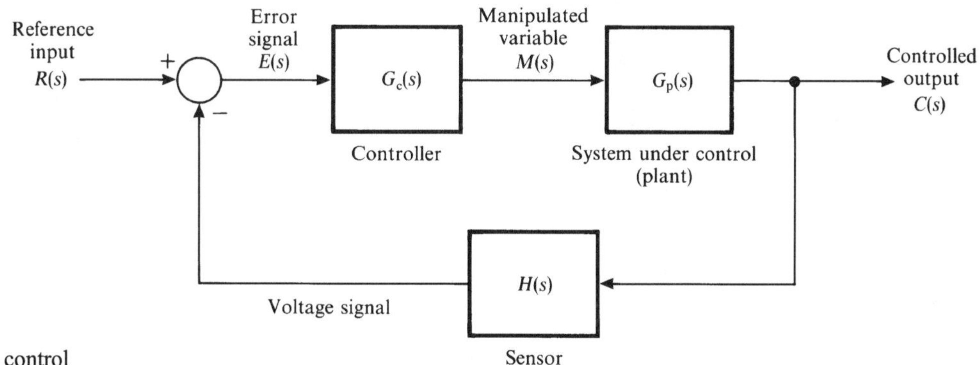

Figure 12.4
Closed loop control

The controlled output is measured by means of a sensor. The sensor is usually connected to some electronic circuitry which produces a voltage proportional to the controlled output. Although this is depicted by a transfer function $H(s)$, in practice it is usually a straight gain block. The output signal from the sensor is compared with the reference input and the difference between the two is known as the **error signal**. This is a measure of how far away the controlled output is from the desired value. This error signal is sent to a controller with transfer function $G_c(s)$. This produces an output to **drive** the system under control, which has a transfer function $G_p(s)$. This output is known as the **manipulated variable**. It is now clear why this type of system is known as feedback control. It is because the output signal from the system is fed back and influences the system. It is common practice to refer to this type of feedback control as **negative feedback** because the signal is introduced into the summing junction using a negative sign.

KEY POINT

> The advantage of closed loop control is that it is possible to take account of changes in conditions and make adjustments accordingly. Therefore closed loop control systems tend to be much more accurate than open loop control systems. The disadvantage is that there is an increased complexity due to the need to use a sensor and a certain amount of electronics. Another disadvantage is that a closed loop control system can become unstable if care is not taken in its design.

To reinforce this discussion of closed loop control let us examine a closed loop control system for control of furnace temperature. The block diagram for this is shown in Figure 12.5. We see that a temperature sensor measures the value of the furnace temperature and produces a voltage signal proportional to this temperature – recall that $H(s)$ is usually a simple gain. The reference input and voltage signal are compared and the difference between the two is the error signal which is sent to the controller. The controller then adjusts the valve position thus changing the gas flow rate to the furnace. This in turn changes the temperature of the furnace thus reducing the error signal.

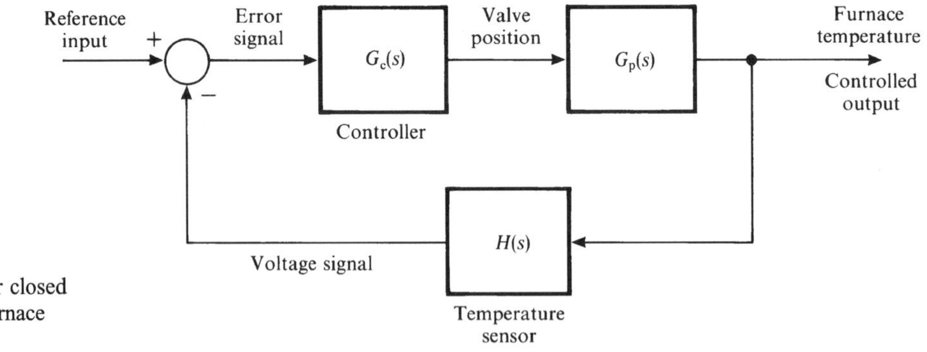

Figure 12.5
Block diagram for closed loop control of furnace temperature

Note throughout the discussions of this section that we have represented the control systems with blocks containing transfer functions. Behind this is the assumption that the systems are linear. We shall only be dealing with linear systems but it is important to note that nonlinearities are present in many practical systems.

Self-assessment questions 12.2

1. Explain the advantages and disadvantages of open loop control.
2. Explain the advantages and disadvantages of closed loop control.
3. What is meant by the terms reference input and controlled output?
4. Why is it necessary to have a sensor in a closed loop control system?

Exercises 12.2

1. Draw a block diagram for the closed loop control of the height of liquid in a chemical tank. This is illustrated in Figure 12.6. The flow rate of liquid going into the tank can be varied by applying a control voltage to a motorised valve. A level sensor is available which, with the use of appropriate electronics, provides a voltage proportional to the height of liquid in the tank.

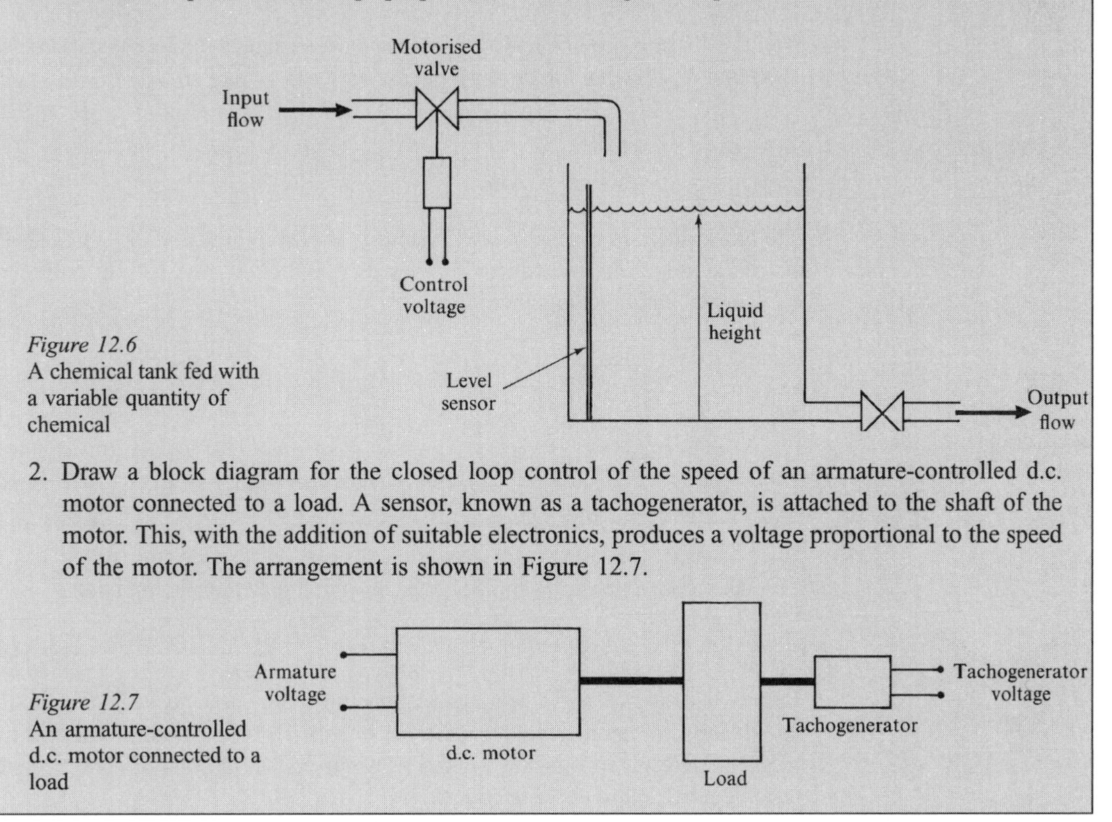

Figure 12.6
A chemical tank fed with a variable quantity of chemical

Figure 12.7
An armature-controlled d.c. motor connected to a load

2. Draw a block diagram for the closed loop control of the speed of an armature-controlled d.c. motor connected to a load. A sensor, known as a tachogenerator, is attached to the shaft of the motor. This, with the addition of suitable electronics, produces a voltage proportional to the speed of the motor. The arrangement is shown in Figure 12.7.

12.3 Block diagram reduction

When designing control systems it is a frequent requirement to simplify block diagrams involving several transfer functions. It is convenient to develop rules for this **block diagram reduction** for configurations that are commonly met. Using these rules can save a great deal of time compared with laboriously working them out each time. Many of these rules exist but for our purposes it is sufficient to consider just two. Consider the block diagram shown in Figure 12.8.

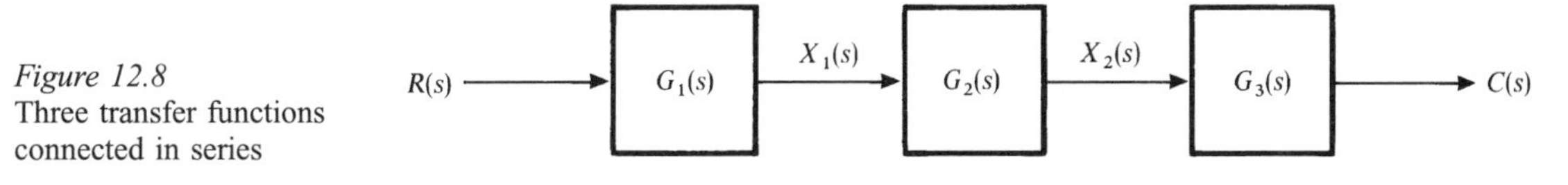

Figure 12.8
Three transfer functions connected in series

We wish to obtain a single transfer function for this system. In other words, we wish to obtain a direct relationship between $C(s)$ and $R(s)$ and so reduce the diagram to a single block. First we note that

$$X_1(s) = G_1(s)R(s)$$
$$X_2(s) = G_2(s)X_1(s)$$
$$C(s) = G_3(s)X_2(s)$$

We have three equations and two intermediate variables to eliminate, namely $X_1(s)$ and $X_2(s)$. Combining these equations gives

$$C(s) = G_1(s)G_2(s)G_3(s)R(s)$$
$$\frac{C(s)}{R(s)} = G_1(s)G_2(s)G_3(s)$$

The block diagram for the overall system is shown in Figure 12.9. We deduce that to combine blocks connected in series it is merely necessary to multiply the transfer functions together.

R(s) → $G_1(s)\,G_2(s)\,G_3(s)$ → C(s)

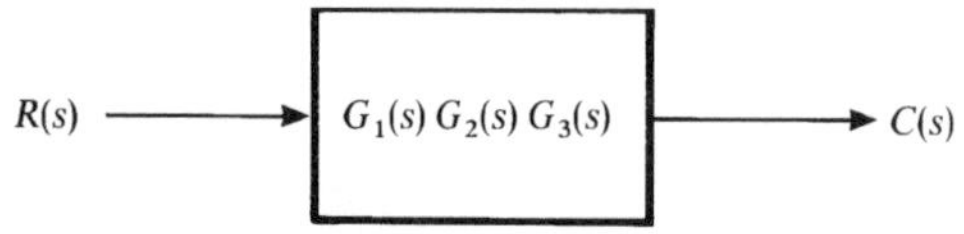

Figure 12.9
Block diagram for the overall transfer function

> Transfer functions connected in series can be reduced to a single transfer function by multiplying the individual transfer functions together.

This is because of the algebraic relationship that exists between the input signal and an output signal in the s domain. It is the main reason why engineers like to work in the s domain. It makes the analysis of complicated systems much more straightforward.

Consider Figure 12.10 which shows a generalised negative feedback loop. The term $G(s)$ is known as the **forward transfer function** and the term $H(s)$ is known as the **reverse transfer function**. This configuration appears so often that it is useful to deduce a rule for reducing it to a single block. We have

$$C(s) = G(s)E(s) \qquad \text{Eqn. [12.1]}$$
$$X(s) = H(s)C(s) \qquad \text{Eqn. [12.2]}$$
$$E(s) = R(s) - X(s) \qquad \text{Eqn. [12.3]}$$

We require a relationship between $C(s)$ and $R(s)$. We have three equations and two intermediate variables to eliminate, namely, $E(s)$ and $X(s)$. Combining Eqns [12.2] and [12.3] to eliminate $X(s)$ gives

$$E(s) = R(s) - H(s)C(s) \qquad \text{Eqn. [12.4]}$$

Combining Eqns [12.1] and [12.4] to eliminate $E(s)$ gives

$$C(s) = G(s)[R(s) - H(s)C(s)]$$

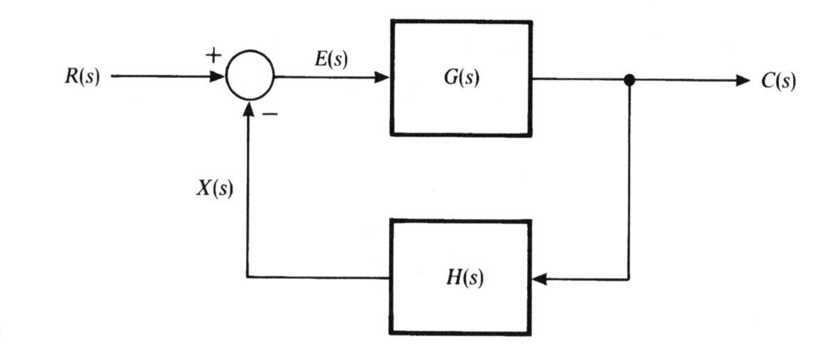

Figure 12.10
A negative feedback loop

Multiplying into the bracket we have

$$C(s) = G(s)R(s) - G(s)H(s)C(s)$$

Collecting terms involving $C(s)$ on the left-hand side gives

$$C(s) + G(s)H(s)C(s) = G(s)R(s)$$

Factorising gives

$$C(s)[1 + G(s)H(s)] = G(s)R(s)$$

Finally, dividing through we obtain

$$\frac{C(s)}{R(s)} = \frac{G(s)}{1 + G(s)H(s)}$$

KEY POINT

The formula for reducing a negative feedback loop is

$$\frac{C(s)}{R(s)} = \frac{G(s)}{1 + G(s)H(s)}$$

where $G(s)$ is the forward transfer function and $H(s)$ is the reverse transfer function.

This is a general rule for reducing a negative feedback loop to a single block.

Example

12.1 Reduce the system shown in Figure 12.11 to a single transfer function.

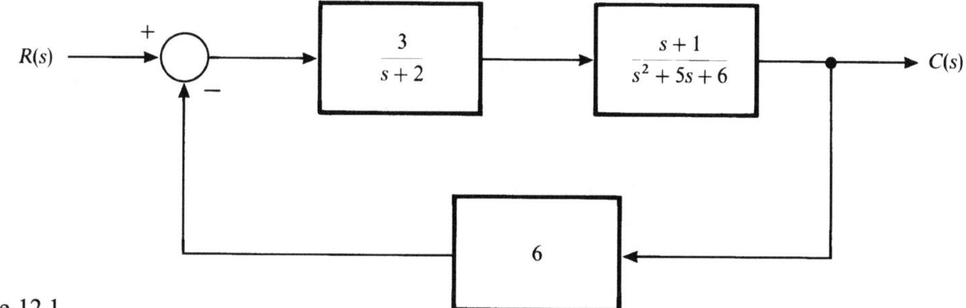

Figure 12.11
System for Example 12.1

Solution First we need to obtain the forward transfer function. This is straightforward as we simply multiply together the two transfer functions in the forward part of the loop. So,

$$G(s) = \frac{3}{s+2} \times \frac{s+1}{s^2+5s+6}$$
$$= \frac{3(s+1)}{(s+2)(s^2+5s+6)}$$

Next we note that the reverse transfer function is $H(s) = 6$. So finally we have

$$\frac{C(s)}{R(s)} = \frac{G(s)}{1+G(s)H(s)} = \frac{\dfrac{3(s+1)}{(s+2)(s^2+5s+6)}}{1+\dfrac{3(s+1)\times 6}{(s+2)(s^2+5s+6)}}$$

Putting the denominator term over a common denominator gives

$$\frac{C(s)}{R(s)} = \frac{\dfrac{3(s+1)}{(s+2)(s^2+5s+6)}}{\dfrac{(s+2)(s^2+5s+6)+18(s+1)}{(s+2)(s^2+5s+6)}}$$

Cancelling the denominators gives

$$\frac{C(s)}{R(s)} = \frac{3(s+1)}{(s+2)(s^2+5s+6)+18s+18}$$

Expanding the denominator gives

$$\frac{C(s)}{R(s)} = \frac{3(s+1)}{s^3+5s^2+6s+2s^2+10s+12+18s+18}$$

Finally, tidying up the denominator we have

$$\frac{C(s)}{R(s)} = \frac{3(s+1)}{s^3+7s^2+34s+30}$$

This is the overall transfer function for the system. We note that the system is third order.

Self-assessment questions 12.3

1. State the rule for combining two transfer functions connected in series.
2. State the rule for reducing a negative feedback loop to a single block.

Exercises 12.3

1. Reduce the systems shown in Figure 12.12 to a single transfer function.

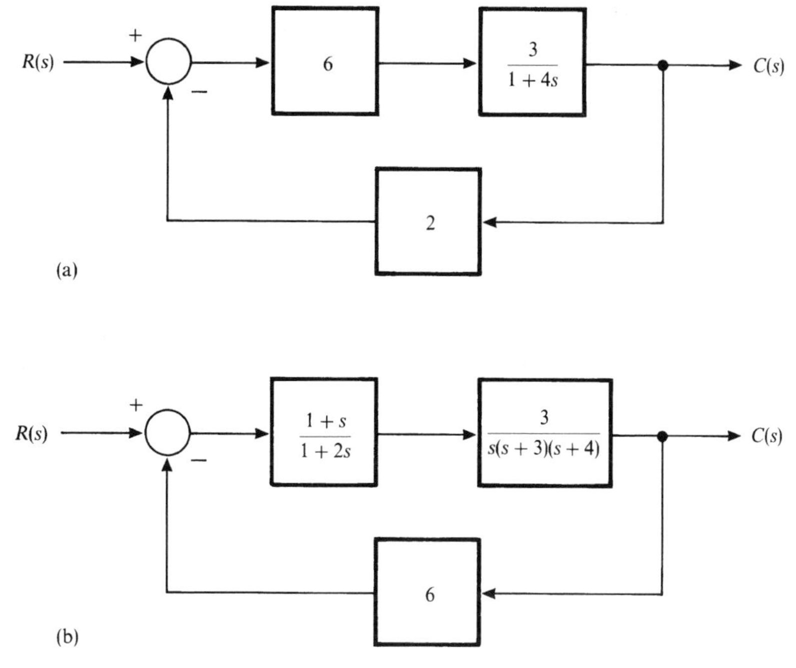

Figure 12.12
Systems for Exercise 12.3.1

12.4 Proportional control

One of the simplest and most common forms of control is proportional control. Consider the general closed loop control system shown in Figure 12.4. This shows a controller with transfer function, $G_c(s)$, a plant with transfer function $G_p(s)$ and a sensor with transfer function $H(s)$.

One of the simplest types of controller is a **proportional controller**. This has a transfer function $G_c(s) = K$ where K is a constant known as the controller **gain**. The controller usually consists of an amplifier which multiplies the error signal by a gain to produce the output from the controller, known as the manipulated variable.

KEY POINT

> For proportional control, the controller has a transfer function K which is a constant, known as the controller gain.

The value of the controller gain can be adjusted to obtain suitable characteristics for the closed loop system. We shall explore the use of proportional control by means of an example.

Example

12.2 Consider the proportional control system shown in Figure 12.13. The system consists of a d.c. motor connected into a negative feedback loop and controlled by a controller with gain K. The output from the system is the angular position

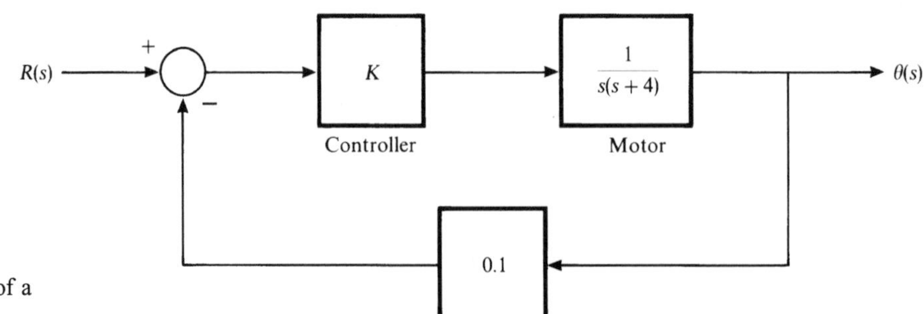

Figure 12.13
Proportional control of a d.c. motor

of the motor, $\theta(s)$. Calculate a value of K to achieve a critically damped system and hence determine the position of the closed loop system poles. Calculate the time constant of these poles and hence determine the settling time of the system when subjected to a step input.

Solution The first thing to do is to calculate the overall transfer function for the system. This involves reducing a negative feedback loop. The forward transfer function is

$$G(s) = \frac{K}{s(s+4)}$$

The reverse transfer function is $H(s) = 0.1$. So the overall transfer function is given by

$$\frac{\theta(s)}{R(s)} = \frac{G(s)}{1 + G(s)H(s)} = \frac{K/s(s+4)}{1 + 0.1K/s(s+4)}$$

Simplifying this expression gives

$$\frac{\theta(s)}{R(s)} = \frac{K/s(s+4)}{(s(s+4) + 0.1K)/s(s+4)} = \frac{K}{s^2 + 4s + 0.1K}$$

The characteristic equation of this system is

$$s^2 + 4s + 0.1K = 0$$

The roots of a standard quadratic equation $ax^2 + bx + c = 0$ are equal if $b^2 - 4ac = 0$. This is clear from examining the formula for determining the roots which is

$$x = \frac{-b \pm \sqrt{b^2 - 4ac}}{2a}$$

Recall from Section 10.4 that for a critically damped system we require the roots of the characteristic equation to be real and equal. So for this system we require

$$4^2 - 4 \times 1 \times 0.1K = 0$$

$$16 - 0.4K = 0$$

$$0.4K = 16$$

$$K = 40$$

Hence, when $K = 40$, the system is critically damped. The roots of the characteristic equation are given by

$$s = \frac{-4 \pm \sqrt{4^2 - 4 \times 1 \times 0.1 \times 40}}{2 \times 1}$$

$$= \frac{-4 \pm \sqrt{0}}{2} = -\frac{4}{2} = -2$$

So there is a double pole at $s = -2$.

Recall that for a real pole the time constant is the reciprocal of the magnitude of the pole.

Here we have $s = -2$ and so $2 = 1/\tau$. Therefore $\tau = 1/2 = 0.5$ second. It is useful to remember this reciprocal relationship between the pole position and the time constant as it is often needed. Remember that for a system the settling time is four times the time constant of the dominant pole(s). So in this case the settling time of the system is $4 \times 0.5 = 2$ seconds.

Example

12.3 Consider the control system of Figure 12.14. A furnace is heated by means of a steam jacket which surrounds the furnace. A steam valve controls the flow of steam to the steam jacket. A sensor is used to measure the temperature of the furnace. The system is controlled by means of a proportional controller with an initial gain setting of 5.

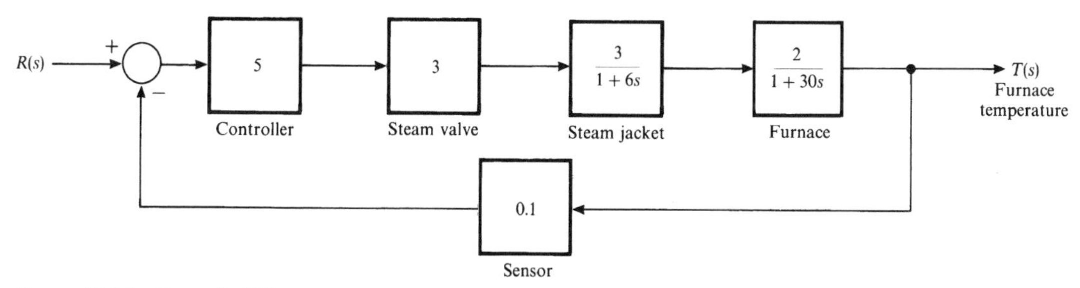

Figure 12.14 Control of furnace temperature

(a) Determine the damping ratio and sketch the position of the closed loop poles in the s plane.

(b) A redesign of the steam jacket allows its time constant to be halved. The gain of the controller is adjusted so that the damping ratio is unchanged. This ensures that the overshoot does not change and so like is being compared with like. Show how the modification moves the closed loop poles in the s plane. Discuss whether or not the characteristics of the closed loop system have been improved.

Solution (a) We need to calculate the overall transfer function of the system by reducing the negative feedback loop. The forward transfer function is given by

$$G(s) = 5 \times 3 \times \frac{3}{1+6s} \times \frac{2}{1+30s}$$

$$= \frac{90}{(1+6s)(1+30s)}$$

$$= \frac{90}{180s^2 + 36s + 1}$$

The reverse transfer function is $H(s) = 0.1$. So the overall transfer function is given by

$$\frac{T(s)}{R(s)} = \frac{G(s)}{1 + G(s)H(s)} = \frac{\dfrac{90}{180s^2 + 36s + 1}}{1 + \dfrac{90}{180s^2 + 36s + 1} \times 0.1}$$

Simplifying this expression we have

$$\frac{T(s)}{R(s)} = \frac{\dfrac{90}{180s^2 + 36s + 1}}{\dfrac{180s^2 + 36s + 1 + 9}{180s^2 + 36s + 1}} = \frac{90}{180s^2 + 36s + 10}$$

This is a second order system because the denominator of the transfer function is a quadratic polynomial. However, before we can compare terms with the standard form discussed in Chapter 10 we first need to obtain a unity coefficient for the s^2 term. This is achieved by dividing both numerator and denominator by 180. This gives

$$\frac{T(s)}{R(s)} = \frac{0.5}{s^2 + 0.2s + 0.5556} = \frac{K\omega_n^2}{s^2 + 2\zeta\omega_n s + \omega_n^2}$$

Comparing terms we see that

$$\omega_n^2 = 0.05556$$
$$\omega_n = 0.2357$$

Also,

$$2\zeta\omega_n = 0.2$$
$$\zeta = \frac{0.2}{2\omega_n} = \frac{0.2}{2 \times 0.2357} = 0.4243$$

We see that the system is underdamped because $\zeta < 1$. The closed loop poles are obtained by solving the characteristic equation $s^2 + 0.2s + 0.05556 = 0$. This gives

$$s = \frac{-0.2 \pm \sqrt{0.2^2 - 4 \times 0.05556}}{2}$$
$$= \frac{-0.2 \pm 0.4268\mathrm{j}}{2}$$
$$= -0.1 \pm 0.2134\mathrm{j}$$

These poles are plotted in Figure 12.15.

(b) The original steam jacket has a transfer function of $3/(1 + 6s)$. Comparing this transfer function with that for a standard first order system, $K/(1 + \tau s)$, we see that it has a time constant of 6 s. Therefore the new steam jacket has a time constant of 3 s. So the transfer function of the new steam jacket is $3/(1 + 3s)$. If we denote the proportional controller gain by M then the forward transfer function of the new control system is

$$G(s) = M \times 3 \times \frac{3}{1 + 3s} \times \frac{2}{1 + 30s} = \frac{18M}{(1 + 3s)(1 + 30s)}$$

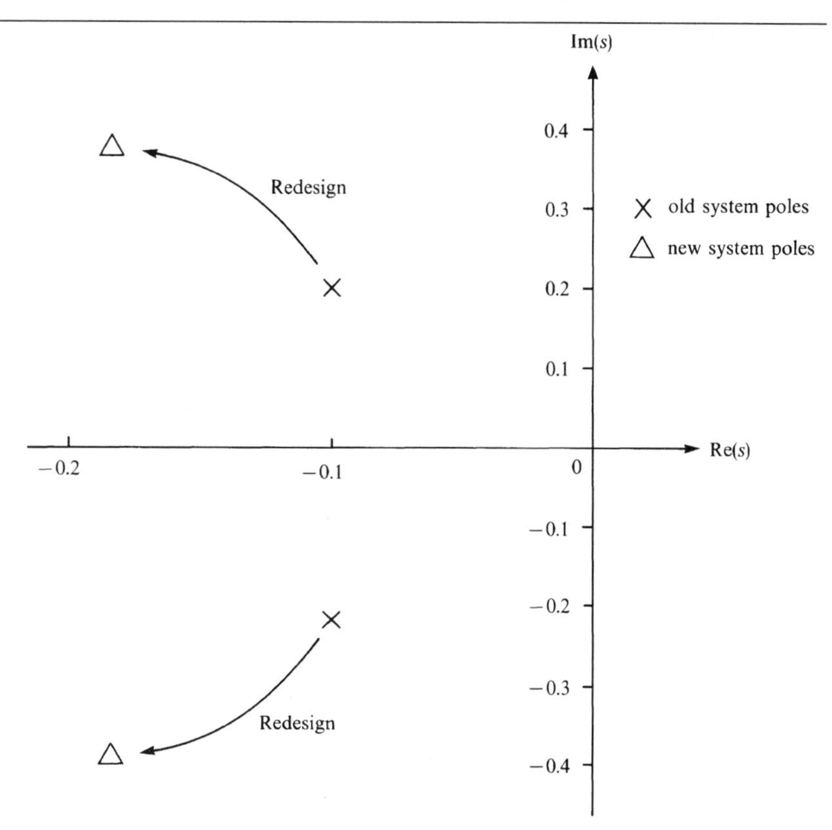

Figure 12.15 Pole–zero plot for Example 12.3

The reverse transfer function is still 0.1 and so the overall transfer function of the new system is

$$\frac{T(s)}{R(s)} = \frac{\dfrac{18M}{(1+3s)(1+30s)}}{1+\dfrac{18M \times 0.1}{(1+3s)(1+30s)}}$$

Simplifying this expression gives

$$\frac{T(s)}{R(s)} = \frac{18M}{(1+3s)(1+30s)+1.8M}$$

Expanding the denominator gives

$$\frac{T(s)}{R(s)} = \frac{18M}{90s^2+33s+1+1.8M}$$

Finally, dividing the numerator and denominator by 90 gives

$$\frac{T(s)}{R(s)} = \frac{0.2M}{s^2+0.3667s+(1+1.8M)/90}$$

Comparing this expression with the standard form for a second order system we have

$$\frac{0.2M}{s^2 + 0.3667s + (1 + 1.8M)/90} = \frac{K\omega_n^2}{s^2 + 2\zeta\omega_n s + \omega_n^2}$$

Comparing the coefficients of s in the denominator polynomials gives

$$2\zeta\omega_n = 0.3667$$

We require the damping coefficient to be the same as part (a), that is, $\zeta = 0.4243$; therefore,

$$\omega_n = \frac{0.3667}{2\zeta} = \frac{0.3667}{2 \times 0.4243} = 0.4321$$

Comparing the constant term of the denominator polynomial gives

$$\omega_n^2 = \frac{1 + 1.8M}{90}$$

$$1 + 1.8M = 90\omega_n^2$$

$$1.8M = 90\omega_n^2 - 1$$

$$M = \frac{90\omega_n^2 - 1}{1.8}$$

Substituting in the value of ω_n gives

$$M = \frac{90 \times 0.4321^2 - 1}{1.8} = 8.780$$

This is the required value of M to give the same damping coefficient as in part (a). Substituting this value into the characteristic equation gives

$$s^2 + 0.3667s + \frac{1 + 1.8M}{90} = 0$$

$$s^2 + 0.3667s + \frac{1 + 1.8 \times 8.780}{90} = 0$$

$$s^2 + 0.3667s + 0.1867 = 0$$

The closed loop poles of this new system are

$$s = \frac{-0.3667 \pm \sqrt{0.3667^2 - 4 \times 0.1867}}{2}$$

$$= \frac{-0.3667 \pm 0.7825\mathrm{j}}{2} = -0.183 \pm 0.3913\mathrm{j}$$

These poles are plotted in Figure 12.15. Note that the poles have moved further away from the imaginary axis. This means that the transient response terms due to the poles decay quicker. Therefore the system responds more quickly to the input signal and yet the overshoot remains the same because the damping ratio is the same in each case. Therefore the characteristics of the closed loop system have improved. This result could have been predicted. Reducing the time constants in a plant tends to make the plant respond more quickly when it forms part of a control system.

Self-assessment questions 12.4

1. Explain what is meant by the term 'proportional control'.
2. How can proportional control be used to alter the characteristics of a system?

Exercises 12.4

1. Figure 12.16 shows a proportional control system to control the height of liquid in a tank which forms part of a set of coupled tanks. The system can be assumed to be linear for convenience.

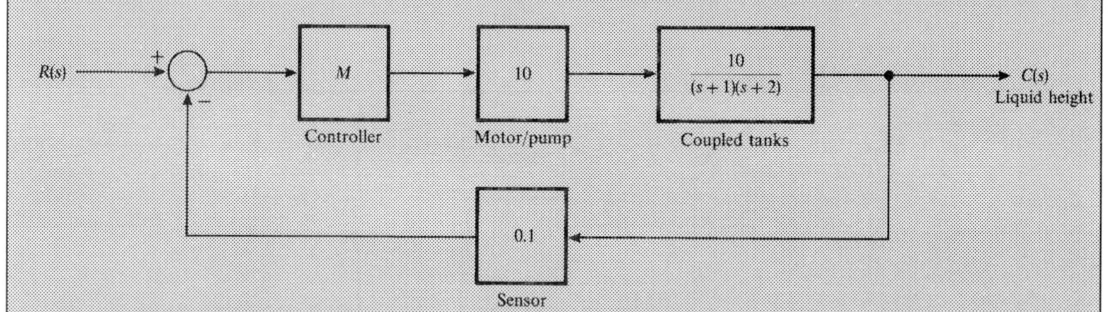

Figure 12.16 A liquid level control system

Calculate a value of controller gain M to make the control system critically damped. Hence calculate the values of the closed loop system poles.

2. Figure 12.17 shows a proportional control system to control the angular position of a motor. Calculate the closed loop system poles and the damping coefficient for the following values of controller gain M:

   (a) $M=1$
   (b) $M=2$
   (c) $M=5$

   In each case describe how the system would respond to a step input.

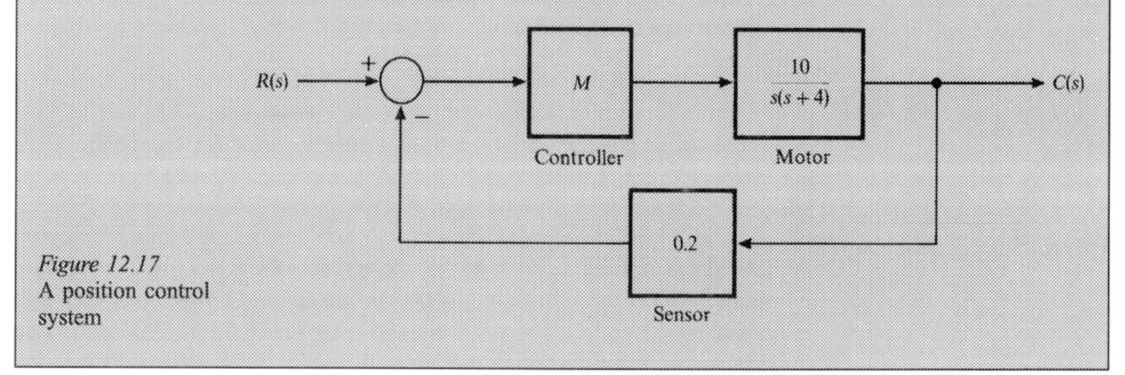

Figure 12.17 A position control system

12.5 Control system design

It is necessary to consult a specialised textbook on control systems engineering to obtain a full working knowledge of control systems design.

There are many different techniques for designing control systems. Several of them involve working in the frequency domain and so are beyond the scope of this book. We shall examine a technique that is based in the time domain, known as the **root locus** technique. Our treatment will be introductory owing to limitations of space.

Consider the control system shown in Figure 12.18. The system has a proportional controller of gain M. In order to calculate the poles of this system it is necessary to calculate an overall transfer function. This is achieved by reducing the negative feedback loop to a single transfer function. So, we have

$$\frac{C(s)}{R(s)} = \frac{MG_p(s)}{1 + MG_p(s)H(s)}$$

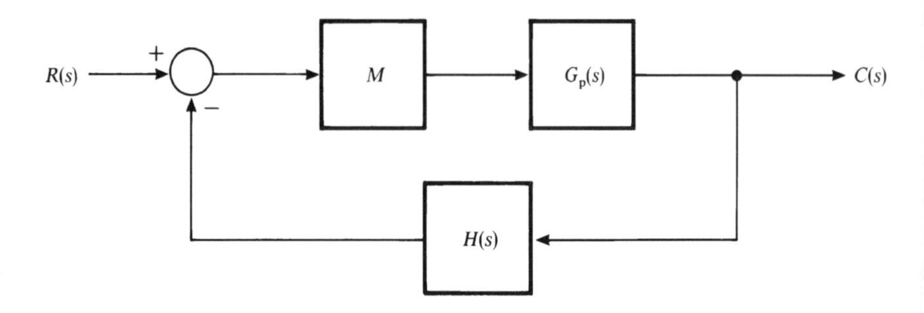

Figure 12.18
A proportional control system

The characteristic equation for this system is

KEY POINT

$$1 + MG_p(s)H(s) = 0$$

The roots of this equation correspond to the poles of the system. Note that they vary as M is changed. A plot of the system poles as M is varied is known as a **root locus plot**. By examining a root locus plot an engineer can decide on appropriate values of controller gain to obtain a suitable response of the control system to a particular input.

KEY POINT

A root locus plot allows an engineer to choose a suitable value of controller gain to satisfy a design specification for the control system.

A particularly important use of the root locus plot is to warn an engineer when a system is likely to be unstable. If the root locus plot crosses into the right-hand side of the s plane for any values of controller gain M, this means care must be taken when designing the controller.

KEY POINT

A root locus plot can be used to determine whether a system is likely to be unstable for certain values of controller gain.

Computer packages are available to carry out root locus plots for systems. However, it is also possible to develop rules to allow hand sketching of root locus plots. It is important to acquire the ability to do hand sketches as this greatly aids the understanding of control system design. We shall examine these rules shortly but first we develop a simple root locus plot by means of an example.

Example

12.4 The angular position of a d.c. motor is controlled by means of a closed loop proportional control system. The arrangement is shown in Figure 12.19. Draw a root locus plot for this system.

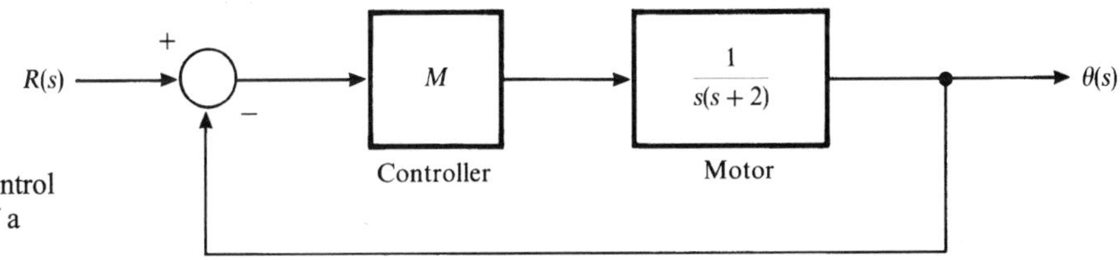

Figure 12.19
A control system to control the angular position of a motor

Solution Examining this control system we see that the forward transfer function is $G(s) = M/s(s+2)$ and the reverse transfer function is $H(s) = 1$. The overall transfer function for this system is given by

$$\frac{\theta(s)}{R(s)} = \frac{G(s)}{1 + G(s)H(s)}$$

$$= \frac{M/s(s+2)}{1 + M/s(s+2)}$$

$$= \frac{M}{s(s+2) + M}$$

$$= \frac{M}{s^2 + 2s + M}$$

So the characteristic equation of the system is

$$s^2 + 2s + M = 0$$

In order to obtain the root locus plot it is necessary to calculate the roots of the characteristic equation as M is increased. Consider the case when $M = 0$. The characteristic equation then reduces to $s^2 + 2s = 0$, that is, $s(s+2) = 0$. The roots of this equation are then $s = 0$ and $s = -2$. These points are marked as crosses on the root locus plot as they represent the start of the root locus. The reason why this is so will become clear later in this section. For $M > 0$ we have

$$s = \frac{-2 \pm \sqrt{4 - 4M}}{2} = \frac{-2 \pm 2\sqrt{1 - M}}{2} = -1 \pm \sqrt{1 - M}$$

For $0 < M < 1$ the roots of the characteristic equation are real and different. What is more, a little thought reveals that their values lie between $s = -2$ and $s = 0$. Therefore the section of the real axis for which $-2 < s < 0$ is part of the root locus. In this region the system is overdamped. When $M = 1$ then the roots of the characteristic equation are equal and have a value of $s = -1$. This is the point at which the root locus breaks away from the real axis and is known as a **breakaway point**. This corresponds to the point at which the system is critically damped. For $M > 1$ the roots of the characteristic equation are complex and form a complex conjugate pair. They are given by

$$s = \frac{-2 \pm \sqrt{4M - 4}\mathrm{j}}{2}$$

$$s = \frac{-2 \pm 2\sqrt{M - 1}\mathrm{j}}{2}$$

$$s = -1 \pm \sqrt{M - 1}\mathrm{j}$$

Note that the real part of the roots remains constant and has a value of -1. Also the complex parts of the roots are on opposite sides of the real axis of the s plane. This is the region in which the system is underdamped. The complete root locus is shown in Figure 12.20. Arrows are used to indicate the movement of the closed loop poles as M is increased.

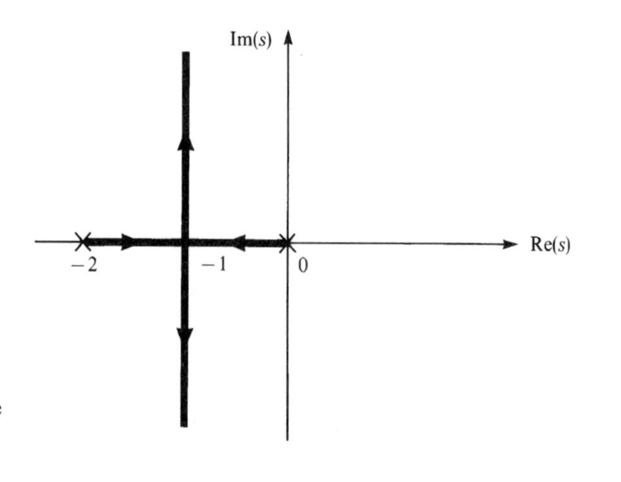

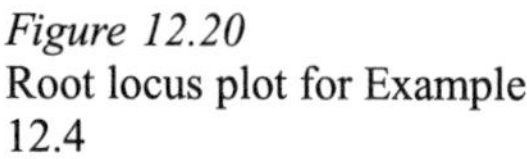
Figure 12.20
Root locus plot for Example 12.4

Example 12.4 illustrates many of the features associated with root locus plotting. However, before further progress can be made it is necessary to develop some rules for plotting a root locus. Consider again the general proportional control system of Figure 12.18. Recall that the characteristic equation of the system is,

$$1 + MG_{\mathrm{p}}(s)H(s) = 0$$

The term $MG_p(s)H(s)$ is known as the **loop gain function**. Its form depends on the type of system and feedback sensor. In addition, it also depends on the type of controller used. In this case we are using proportional control but it is possible to obtain root locus plots for other types of controller although the analysis of such controllers is beyond the scope of this book. By varying the value of M the poles of the closed loop system move in the s plane and hence trace out a root locus.

KEY POINT

The system characteristic equation for the majority of systems can be written as

$$1 + M\frac{N(s)}{D(s)} = 0 \qquad \text{Eqn. [12.5]}$$

where $N(s)$ is a polynomial in s of order m and $D(s)$ is a polynomial in s of order n.

For example, if $G_p(s) = 10(s+1)/(s^2+s+1)$ and $H(s) = 0.1$ then the characteristic equation is

$$1 + M \times 0.1 \times \frac{10(s+1)}{s^2+s+1} = 0$$

$$1 + \frac{M(s+1)}{s^2+s+1} = 0$$

Here $N(s) = s+1$ and so is of order 1, that is, $m = 1$. Also, $D(s) = s^2+s+1$ and so is of order 2, that is, $n = 2$.

The roots of the equation $N(s) = 0$ are called the **open loop zeros** of the system. The roots of the equation $D(s) = 0$ are called the **open loop poles** of the system. Consider again Eqn. [12.5]. Putting the left-hand side of the equation over a common denominator $D(s)$ gives

$$\frac{D(s) + MN(s)}{D(s)} = 0$$

Remember it is the roots of this characteristic equation that correspond to the poles of the closed loop system. Therefore we can drop the denominator of this expression to leave

KEY POINT

$$D(s) + MN(s) = 0 \qquad \text{Eqn. [12.6]}$$

Before developing the rules for root locus plotting it is convenient to introduce one further piece of notation. Recall from Example 12.4 that there were two parts to the root locus plot. The different parts of the root locus plot are known as the **branches** of the root locus.

Rule 1 The branches of the root locus plot start on the open loop poles of the system.

This is easily shown. Consider Eqn. [12.6]. The root locus plot starts when

$M = 0$. Putting $M = 0$ in Eqn. [12.6] gives

$$D(s) = 0$$

We see that the roots of the characteristic equation correspond to those of the open loop poles. The open loop poles are marked on the pole–zero plot by using small crosses as we saw in Example 12.4.

Rule 2 The number of branches of the root locus plot is equal to the number of open loop poles.

The number of open loop poles is equal to the number of roots of the equation $D(s) = 0$, that is, the number of open loop poles. Each of these open loop poles is the start of a branch of the root locus plot.

Rule 3 The branches of the root locus plot terminate on the open loop zeros of the system.

If M is made large in Eqn. [12.6] then the $D(s)$ term can be safely ignored because it is small in comparison. We then have $MN(s) \approx 0$ as the characteristic equation. This has roots corresponding to the roots of the open loop zeros and so the branches of the root locus terminate on the open loop zeros. The open loop zeros are marked on the pole–zero plot by using small circles.

Rule 4 Branches of the root locus that do not terminate on open loop zeros head towards infinity. If there are n open loop poles and m open loop zeros then the number of branches that head towards infinity is $n - m$.

For a physically realizable system the number of open loop zeros is less than or equal to the number of open loop poles. This is not proved. Therefore there may be fewer open loop poles than open loop zeros. For this case it is not possible for every branch to terminate on an open loop zero as there are not enough. Instead, some of the branches terminate on **zeros at infinity**. These can still be classified as zeros because allowing s to tend to infinity in the expression $N(s)/D(s)$ leads to this expression tending to zero. This is because the order of $N(s)$ is less than the order of $D(s)$, that is, the highest power in the numerator is of the form s^m and the highest power in the denominator is s^n. For example,

$$\lim_{s \to \infty} \frac{s + 3}{s^2 + 3s + 2} = 0$$

because the s^2 term in the denominator dominates the s term in the numerator.

Rule 5 Branches that head towards infinity approach asymptotes as they do so. The angle ϕ_{k+1} of these asymptotes is given by

$$\phi_{k+1} = \frac{(2k + 1)\pi}{n - m}$$

where $k = 0, 1, 2 \ldots, n - m - 1$.

This relationship is not proved here. An **asymptote** is a straight line which a curve approaches but never quite reaches. Asymptotes are usually drawn as broken lines on root locus plots.

Rule 6 The asymptotes intersect the real axis at a single point α given by

$$\alpha = \frac{(\Sigma \text{ open loop poles}) - (\Sigma \text{ open loop zeros})}{n - m}$$

This relationship is not proved here. Once α has been calculated and the angle of the asymptotes, using Rule 5, then it is possible to draw the asymptotes.

Rule 7 The branches of the root locus are symmetrical about the real axis.

This is true because the complex roots of the characteristic equation are complex conjugate pairs. It can be shown that this is always the case when a polynomial has real coefficients.

Rule 8 Sections of the real axis to the left of an odd total number of open loop singularities on this axis form part of the root locus plot.

The term singularity refers to either a pole or a zero. This relationship is not proved.

Rule 9 The breakaway points of the root locus from the real axis are obtained by solving the equation

$$\frac{\mathrm{d}}{\mathrm{d}s}\left[\frac{D(s)}{N(s)}\right] = 0$$

This result is not proved.

It is possible to use these rules to draw a range of root locus plots. More specialised rules do exist to improve these plots but the ones given are sufficient for most purposes. The important thing to remember is that often all a design engineer requires is a sketch. If an accurate plot is required then a computer can be used. It would be a useful exercise to look again at Example 12.4 and decide which rules are relevant to its construction. Let us now consider another example.

Example

12.5 A system has transfer function $1/s(s+1)(s+2)$. Obtain a root locus plot for this system when it is controlled using a closed loop proportional control system with unity feedback. Comment on the stability of the system. Calculate the value of controller gain corresponding to marginal stability.

Solution We are given that the system is unity feedback and so $H(s) = 1$. Assuming the controller has a gain M, the loop gain function is

$$\frac{MN(s)}{D(s)} = \frac{M}{s(s+1)(s+2)}$$

Examining the numerator polynomial we see that it is zero order and so $m = 0$. The denominator is third order and so $n = 3$. Therefore there are $n - m = 3 - 0 = 3$ zeros at infinity using Rule 4. The three open loop poles are $s = 0$, -1, -2 and these are marked as crosses on the pole–zero plot shown in Figure 12.21.

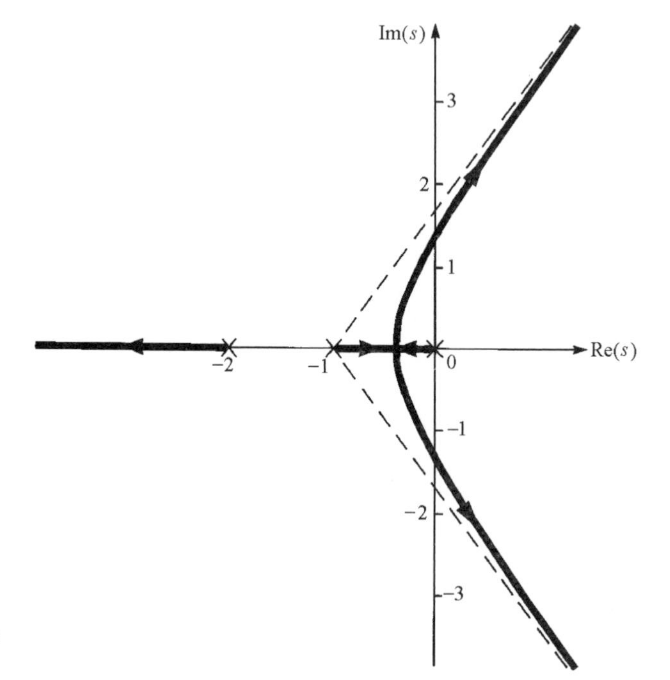

Figure 12.21
Root locus plot for Example 12.5

Examining the real axis we see that the first singularity is at $s = 0$. Therefore the section of the real axis to the left of this, up to the next singularity, is a part of the root locus by Rule 8, that is, $-1 \leq s \leq 0$ is part of the root locus. Also, using Rule 8, the section of the real axis to the left of the singularity at $s = -2$ is part of the root locus, that is, $s \leq -2$. The number of zeros at infinity is $n - m = 3 - 0 = 3$. Therefore, there are three asymptotes. The asymptote angles are given by, using Rule 5,

$$\phi_{k+1} = \frac{(2k+1)\pi}{3-0} = \frac{(2k+1)\pi}{3} \qquad k = 0, 1, 2$$

Substituting in the different values of k we obtain

$$\phi_1 = \frac{\pi}{3}$$

$$\phi_2 = \frac{3\pi}{3} = \pi$$

$$\phi_3 = \frac{5\pi}{3}$$

Using Rule 6, these asymptotes intersect the real axis at the point

$$\alpha = \frac{[0 + (-1) + (-2)] - 0}{3 - 0} = -\frac{3}{3} = -1$$

Recall that a control system becomes unstable if any of the poles are in the right-hand side of the s plane.

The complete root locus plot is shown in Figure 12.21. Examining this root locus we see that the system can go unstable if a large enough value of controller gain K is chosen. This causes the two dominant poles to cross into the right-hand half of the s plane.

We can calculate the value of M at which the root locus crosses the imaginary axis by noting that the equation describing the imaginary axis is $s = \mathrm{j}\omega$ where ω is a variable parameter. First we substitute $s = \mathrm{j}\omega$ in the characteristic equation to calculate the points at which the root locus cuts the imaginary axis. We have

$$1 + \frac{M}{s(s+1)(s+2)} = 0$$

and so

$$\frac{s(s+1)(s+2) + M}{s(s+1)(s+2)} = 0$$

Dropping the denominator of this expression because it is not needed we have

$$s(s+1)(s+2) + M = 0$$

Removing the brackets gives

$$s^3 + 3s^2 + 2s + M = 0$$

Substituting $s = \mathrm{j}\omega$ we have

$$(\mathrm{j}\omega)^3 + 3(\mathrm{j}\omega)^2 + 2\mathrm{j}\omega + M = 0$$

Removing the brackets gives

$$-\mathrm{j}\omega^3 - 3\omega^2 + 2\mathrm{j}\omega + M = 0$$

Gathering real and imaginary parts gives

$$(M - 3\omega^2) + \mathrm{j}(2\omega - \omega^3) = 0 \qquad \text{Eqn. [12.7]}$$

Comparing imaginary parts of Eqn. [12.7] we have

$$2\omega - \omega^3 = 0$$

$$\omega(2 - \omega^2) = 0$$

This has three solutions, namely, $\omega = 0$ which is the trivial solution and $\omega = \pm\sqrt{2}$. So the root locus cuts the imaginary axis at $s = \pm\sqrt{2}\mathrm{j}$. We can now calculate the value of M corresponding to marginal stability. Comparing real parts of Eqn. [12.7] we have

$$M - 3\omega^2 = 0$$

$$M = 3\omega^2$$

Using the value of ω already obtained we have

$$M = 3 \times 2 = 6$$

This is the value of controller gain at which the system is marginally stable.

We now consider a root locus plot in which there is an open loop zero.

Example

12.6 A system has transfer function $(s+3)/(s+1)(s+2)$. Obtain a root locus plot for this system when it is controlled using a closed loop proportional control system with unity feedback.

Solution The loop gain function for this system is

$$\frac{MN(s)}{D(s)} = \frac{M(s+3)}{(s+1)(s+2)}$$

There is one open loop zero at $s=-3$ and so $m=1$. There are two open loop poles at $s=-1$ and $s=-2$ and so $n=2$. These are plotted in Figure 12.22.

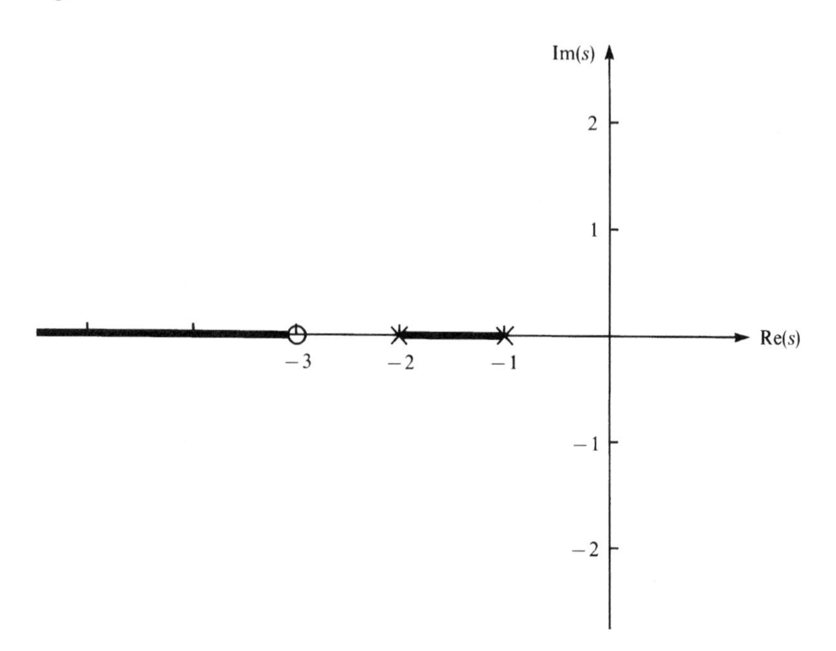

Figure 12.22
Partial root locus plot for Example 12.6

Using Rule 8, the sections of the real axis $-2 \leq s \leq -1$ and $s \leq -3$ are part of the root locus. There is one zero at infinity and so there is one asymptote. This has an angle given by, using Rule 5,

$$\phi_{k+1} = \frac{(2k+1)\pi}{2-1} \qquad k=0$$

Putting $k=0$ gives $\phi_1 = \pi$.

On the basis of the previous calculations it is possible to sketch a partial root locus plot. This is shown in Figure 12.22. Examining this figure we see that part of the root locus is missing. A root locus starts on the open loop poles and terminates on the open loop zeros. It is not possible to calculate the shape of the rest of the root locus using the rules given. In fact its shape is a circle. The circle must be symmetrical about the real axis by Rule 7 and so its centre can be located by calculating the breakaway points of the root locus from the real axis using Rule 9. We have

$$\frac{d}{ds}\left[\frac{N(s)}{D(s)}\right] = 0$$

$$\frac{d}{ds}\left[\frac{s+3}{(s+1)(s+2)}\right] = 0$$

$$\frac{d}{ds}\left(\frac{s+3}{s^2+3s+2}\right) = 0$$

Recall that

$$\frac{d}{dx}\left(\frac{u}{v}\right) = \frac{v\frac{du}{dx} - u\frac{dv}{dx}}{v^2}$$

This can be evaluated using the rules for differentiating a quotient.

Using $u = s+3$ and $v = s^2 + 3s + 2$ we have

$$\frac{du}{ds} = 1$$

$$\frac{dv}{ds} = 2s + 3$$

So finally we have

$$\frac{d}{ds}\left[\frac{N(s)}{D(s)}\right] = \frac{(s^2+3s+2)(1) - (s+3)(2s+3)}{(s^2+3s+2)^2} = 0$$

$$\frac{(s^2+3s+2)(1) - (2s^2+9s+9)}{(s^2+3s+2)^2} = 0$$

$$\frac{-s^2-6s-7}{(s^2+3s+2)^2} = 0$$

In order to find the roots of this equation it is only necessary to consider the numerator. Equating this to zero and multiplying by -1 gives

$$s^2 + 6s + 7 = 0$$

Solving this equation to obtain the breakaway points gives

$$s = \frac{-6 \pm \sqrt{36-28}}{2}$$

$$= \frac{-6 \pm \sqrt{8}}{2}$$

$$= -1.59, -4.41$$

It is now possible to complete the root locus plot for the system. This is shown in Figure 12.23.

We have now examined several root locus plots. By varying M it is possible to alter the position of the poles of the closed loop control system. Hence it is possible to change the way the control system responds to an input signal. Therefore root locus plots are used extensively by control engineers to choose suitable values for the controller gain.

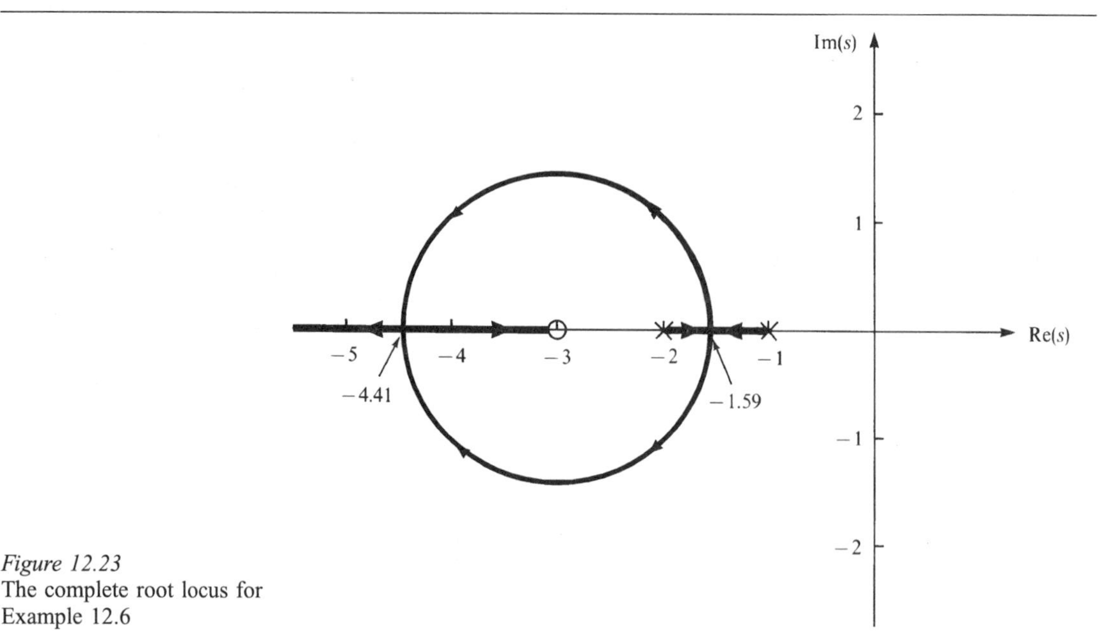

Figure 12.23
The complete root locus for Example 12.6

The root locus design technique can be used with a variety of different controllers. We have concentrated on proportional controllers but more sophisticated controllers are available and can be read about in more advanced books on control engineering.

Self-assessment questions 12.5

1. Describe what is meant by a root locus plot.
2. State the rules for drawing a root locus plot.

Exercises 12.5

1. Consider the general control system shown in Figure 12.18. Sketch root locus plots for the following plants assuming the control systems are unity feedback, that is, $H(s) = 1$:

   (a) $G_p(s) = \dfrac{1}{s+2}$

   (b) $G_p(s) = \dfrac{1}{(s+1)(s+3)}$

   (c) $G_p(s) = \dfrac{s+1}{(s+2)(s+3)}$

   (d) $G_p(s) = \dfrac{1}{(s+1)(s+2)(s+3)}$

2. Consider Figure 12.24 which shows a proportional control system to control the level of liquid in a tank which forms part of a coupled tank system. Sketch a root locus plot for this system.

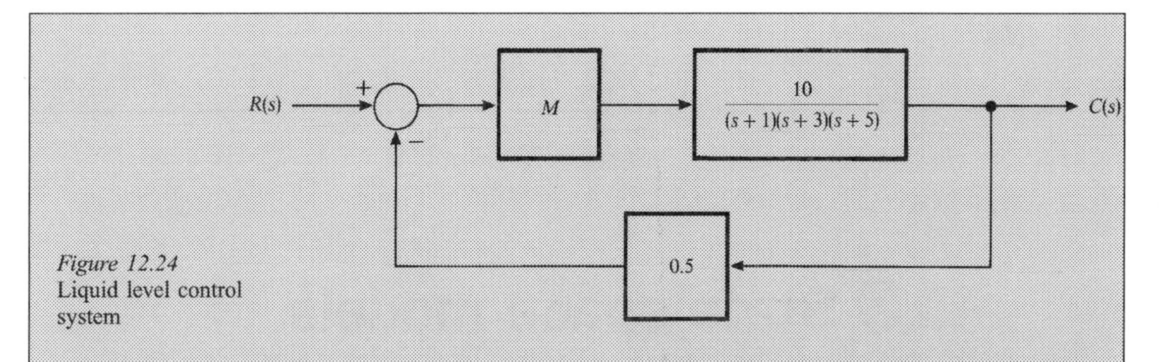

Figure 12.24
Liquid level control system

Test and assignment exercises 12

1. Draw a root locus diagram for the motor control system of Example 12.2.
2. Consider the room temperature control system of Figure 12.25. Carry out the following:
   (a) Calculate an overall transfer function for the system.
   (b) Calculate a value of controller gain to make the system critically damped.
   (c) Sketch a root locus plot for the system.

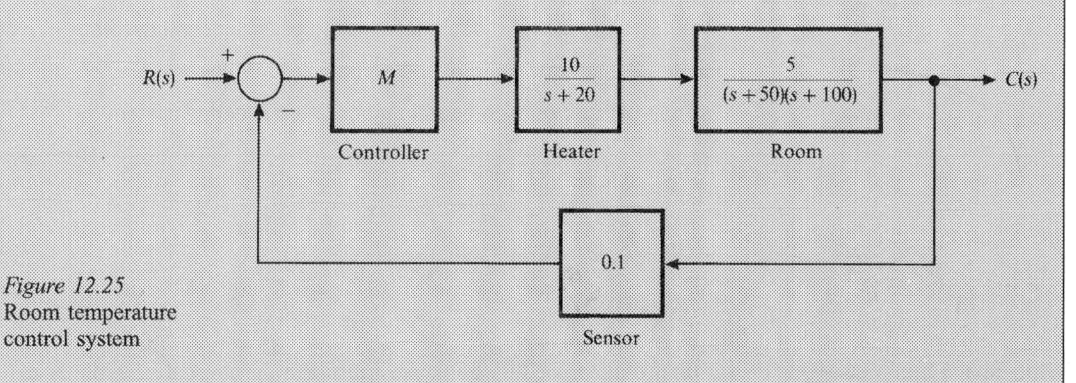

Figure 12.25
Room temperature control system

3. Draw a root locus plot for the furnace temperature control system of Example 12.3 before and after the steam jacket is redesigned.

13 State space models

Objectives

This chapter:

- explains the benefits of using a state space model
- describes how to choose appropriate state space variables for a system
- creates state space models for a range of engineering systems
- states the general state space equations for a linear system
- solves the state space equations for an engineering system

13.1 Introduction

State space methods are used extensively in the aerospace industries because of the complexity of the systems being analysed.

In the final chapter of this book we introduce an alternative type of mathematical model of an engineering system, namely, the state space model. This type of model is applicable to a larger class of systems than the models we have examined earlier in the book. It is particularly useful for analysing engineering systems that have several inputs and several outputs. State space modelling techniques have been used increasingly by engineers in recent years as engineering systems become more complicated. They lend themselves particularly well to computer solution. At first state space models appear to be more difficult to understand than transfer function models, but as familiarity is gained they become a more natural way of examining a system as they remain firmly in the time domain, rather than being in the less intuitive s domain.

13.2 Creating a state space model

A transfer function model of an engineering system is very useful but it does have its limitations. Essentially it relates the output of a system to its input and so relies on there only being one system input and one system output. Methods can be found to use the transfer function approach when a system has more than one input or output but the methods are clumsy. We also saw earlier that when forming a transfer function model it is necessary to assume that the initial conditions on a system are zero. Many systems have nonzero initial conditions and dealing with this when using a transfer function approach is again difficult.

An alternative is to use a **state space model** which happily deals with multiple inputs and outputs as well as taking care of nonzero initial conditions.

KEY POINT

A state space model models the internal state of a system rather than directly describing the relationship between the system inputs and outputs.

Coupled first order differential equations contain common variables.

At the heart of the approach is the concept of the **state variables** which are used to describe the current state of the system. These state variables are related to each other by means of a set of **state equations** which are a set of coupled first order differential equations.

KEY POINT

The number of state variables and state equations needed to model a system is equal to the order of the system.

We examined the concept of system order in earlier chapters. For convenience the state variables are gathered together to form a **state vector**. This allows the notation for writing a state model to be simplified as we shall see shortly. It is also necessary to form one or more **output equations** for the system in order to describe how the system outputs vary in response to variations in the state vector and any inputs. Before proceeding to a general formulation we first introduce the topic by way of an example.

Example

13.1 Figure 13.1 shows an RC circuit with input voltage v_i and output voltage v_o. Form a state space model for this system using the voltage v_C across the capacitor as the state variable.

Solution Let us first write the equations for the system. Using Kirchhoff's voltage law,

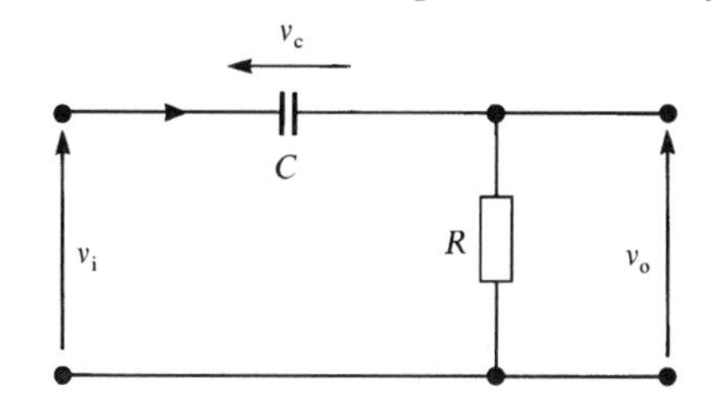

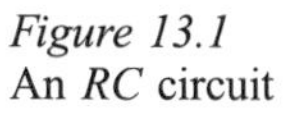

Figure 13.1
An RC circuit

we have

$$v_i = v_C + v_o \qquad \text{Eqn. [13.1]}$$

For the capacitor

$$i = C\frac{dv_C}{dt} \qquad \text{Eqn. [13.2]}$$

For the resistor

$$v_o = iR \qquad \text{Eqn. [13.3]}$$

We are told to use v_C as the state variable. Only one state variable is required as the system is first order. Therefore a state equation involving v_C is required. The equation also needs to avoid any reference to the system output so the state of the system can be calculated purely by a knowledge of the input signal to the system. Eqn. [13.2] is the basis of the state equation but we first need to eliminate i. So combining Eqn. [13.2] with Eqn. [13.3] to remove i gives

$$\frac{v_o}{R} = C\frac{dv_C}{dt}$$

$$v_o = RC\frac{dv_C}{dt} \qquad \text{Eqn. [13.4]}$$

We now need to eliminate v_o from Eqn. [13.4]. So combining with Eqn. [13.1] gives

$$v_i = v_C + RC\frac{dv_C}{dt}$$

This is the state equation for this system. It describes the variation of the state variable v_C purely in terms of the input signal to the system, v_i. For reasons that will be clear shortly we rearrange this equation to give

$$RC\frac{dv_C}{dt} = -v_C + v_i$$

Dividing by RC gives

$$\frac{dv_C}{dt} = -\frac{1}{RC}v_C + \frac{1}{RC}v_i \qquad \text{Eqn. [13.5]}$$

This is the standard form for such an equation. We also need to relate the output from the system, v_o, to the state variable v_C and the system input v_i. Eqn. [13.1], when rearranged, allows this. So we have

$$v_o = -v_C + v_i \qquad \text{Eqn. [13.6]}$$

This equation allows us to obtain the system output v_o given the state variable v_C and the system input v_i. Again the equation has been written in a standard form that will become clear later. Eqns [13.5] and [13.6] form the state space model for the RC circuit.

It is conventional to write the state equations in a standard form. This reduces the chances of errors being made and makes it easier to present the equations to a computer for solution. For a linear system we can write

KEY POINT

$\dot{\boldsymbol{x}} = A\boldsymbol{x} + B\boldsymbol{u}$	Eqn. [13.7]
$\boldsymbol{y} = C\boldsymbol{x} + D\boldsymbol{u}$	Eqn. [13.8]

where

At first sight these equations appear to be extremely complicated but with familiarity comes the realisation that they provide a very compact way of representing all linear engineering systems.

$\boldsymbol{x}$ is an n-component column vector known as the **state vector** representing the state of the nth order system
$\dot{\boldsymbol{x}}$ is the time derivative of the state vector
$\boldsymbol{u}$ is the **input vector** composed of the input signals to the system
$\boldsymbol{y}$ is the **output vector** composed of the output signals from the system
A is the **state matrix**
B is the **input matrix**
C is the **output matrix**
D is the **direct transmission matrix**

Note that it is not necessary to use any particular symbols when deriving the state equations and so whatever is convenient is the best rule. Let us now examine the state model derived in Example 13.1 and put it in standard form. We can write Eqn. [13.5] as

$$\dot{v}_C = -\frac{1}{RC}v_C + \frac{1}{RC}v_i$$

and so $A = -1/RC$ and $B = 1/RC$.

Note in this case that the matrices A and B are single constants. Recalling Eqn. [13.6] we have

$$v_o = -v_C + v_i$$

and so $C = -1$ and $D = 1$.

The matrices C and D are also single constants. This is the simplest possible configuration. We shall now examine a more complex case.

Example

13.2 Derive a state space model for the system shown in Figure 13.2. Use the displacement of the mass, x, and the velocity of the mass, v, as the state variables. Assume the system input is the force f and the system output is x.

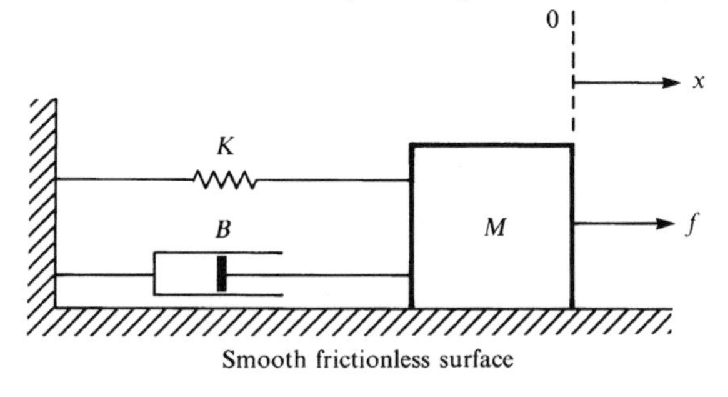

Figure 13.2
A spring–mass–damper system

Solution We have already examined this system in Example 2.3. The system differential equation is

Note in this case that the system output is also a state variable. This often occurs when modelling engineering systems.

$$f = M\frac{d^2x}{dt^2} + B\frac{dx}{dt} + Kx \qquad \text{Eqn. [13.9]}$$

The choice of state variables has already been made and is x and v. We can immediately write the first state equation. We note that $v = \frac{dx}{dt}$ and so

$$\dot{x} = v \qquad \text{Eqn. [13.10]}$$

The second state equation is obtained by noting that $\dot{v} = d^2x/dt^2$ and re-arranging Eqn. [13.9]. So,

$$M\frac{d^2x}{dt^2} = -Kx - B\frac{dx}{dt} + f$$

$$\frac{d^2x}{dt^2} = -\frac{K}{M}x - \frac{B}{M}\frac{dx}{dt} + \frac{f}{M}$$

$$\dot{v} = -\frac{K}{M}x - \frac{B}{M}v + \frac{f}{M} \qquad \text{Eqn. [13.11]}$$

The two state equations, Eqns [13.10] and [13.11], can now be written in standard form. This gives

$$\begin{pmatrix} \dot{x} \\ \dot{v} \end{pmatrix} = \begin{pmatrix} 0 & 1 \\ -\frac{K}{M} & -\frac{B}{M} \end{pmatrix} \begin{pmatrix} x \\ v \end{pmatrix} + \begin{pmatrix} 0 \\ \frac{1}{M} \end{pmatrix} f$$

Note that the state matrix is

$$A = \begin{pmatrix} 0 & 1 \\ -\frac{K}{M} & -\frac{B}{M} \end{pmatrix}$$

and the input matrix is

$$B = \begin{pmatrix} 0 \\ \frac{1}{M} \end{pmatrix}$$

Note, also, that the input vector is one dimensional as there is only one system input, namely, f.

All that remains is to obtain the output equation. This is straightforward as the output x is one of the state variables. So we can write

$$x = (1 \quad 0)\begin{pmatrix} x \\ v \end{pmatrix}$$

We see that the output matrix is

$$C = (1 \quad 0)$$

and the direct transmission matrix D is zero.

Let us now examine a system that has several inputs and outputs.

Example

13.3 Figure 13.3 shows two coupled tanks. The tanks have input flow rates q_{i1} and q_{i2} and output flow rates q_{o1} and q_{o2}. The tanks have vertical sides with cross-sectional areas A_1 and A_2 and liquid heights h_1 and h_2. Assume the outlet valves can all be modelled as linear elements with resistance R and that the liquid flow rate from tank 1 to tank 2 is q. Derive a state space model for this system using h_1 and h_2 as the state variables.

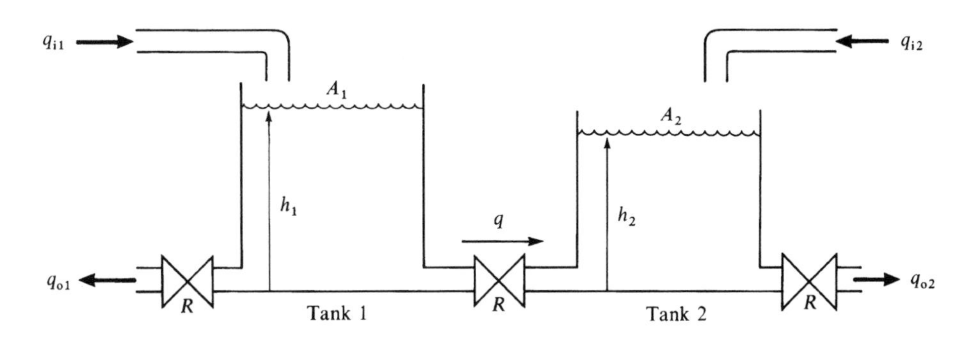

Figure 13.3 Two coupled tanks

Solution Before developing a state space model let us derive the system equations. For tank 1, using the conservation of mass, we have

$$q_{i1} - q_{o1} - q = A_1 \frac{dh_1}{dt} \qquad \text{Eqn. [13.12]}$$

For the left side valve we have

$$\rho g h_1 = q_{o1} R \qquad \text{Eqn. [13.13]}$$

For the right side valve we have

$$\rho g h_1 - \rho g h_2 = qR \qquad \text{Eqn. [13.14]}$$

For tank 2, using the conservation of mass, we have

$$q_{i2} + q - q_{o2} = A_2 \frac{dh_2}{dt} \qquad \text{Eqn. [13.15]}$$

For the right side valve we have

$$\rho g h_2 = q_{o2} R \qquad \text{Eqn. [13.16]}$$

Note that the only time that this is not true is when one of the state variables happens also to be an output variable.

By examining these equations we can see that Eqns [13.12] and [13.15] form the basis of the state equations as Eqn. [13.12] has a dh_1/dt term and Eqn. [13.15] has a dh_2/dt term. However, we need to remove any terms involving the output variables.

Starting with Eqn. [13.12] we can use Eqn. [13.13] to remove the q_{o1}

term and Eqn. [13.14] to remove the q term. This gives

$$q_{i1} - \frac{\rho g}{R} h_1 - \frac{\rho g h_1 - \rho g h_2}{R} = A_1 \frac{dh_1}{dt}$$

$$A_1 \frac{dh_1}{dt} = -\frac{2\rho g}{R} h_1 + \frac{\rho g}{R} h_2 + q_{i1}$$

$$\frac{dh_1}{dt} = -\frac{2\rho g}{RA_1} h_1 + \frac{\rho g}{RA_1} h_2 + \frac{q_{i1}}{A_1} \qquad \text{Eqn. [13.17]}$$

This is the first state equation.

Turning to Eqn. [13.15] we can remove the q term using Eqn. [13.14] and the q_{o2} term using Eqn. [13.16]. This gives

$$q_{i2} + \frac{\rho g h_1 - \rho g h_2}{R} - \frac{\rho g}{R} h_2 = A_2 \frac{dh_2}{dt}$$

$$A_2 \frac{dh_2}{dt} = \frac{\rho g}{R} h_1 - \frac{2\rho g}{R} h_2 + q_{i2}$$

$$\frac{dh_2}{dt} = \frac{\rho g}{RA_2} h_1 - \frac{2\rho g}{RA_2} h_2 + \frac{q_{i2}}{A_2} \qquad \text{Eqn. [13.18]}$$

This is the second state equation.

Putting Eqns [13.17] and [13.18] in standard form gives

$$\begin{pmatrix} \dot{h}_1 \\ \dot{h}_2 \end{pmatrix} = \begin{pmatrix} -\frac{2\rho g}{RA_1} & \frac{\rho g}{RA_1} \\ \frac{\rho g}{RA_2} & -\frac{2\rho g}{RA_2} \end{pmatrix} \begin{pmatrix} h_1 \\ h_2 \end{pmatrix} + \begin{pmatrix} \frac{1}{A_1} & 0 \\ 0 & \frac{1}{A_2} \end{pmatrix} \begin{pmatrix} q_{i1} \\ q_{i2} \end{pmatrix}$$

We can use Eqns [13.13] and [13.16] to obtain the output equations. These are

$$\begin{pmatrix} q_{o1} \\ q_{o2} \end{pmatrix} = \begin{pmatrix} \frac{\rho g}{R} & 0 \\ 0 & \frac{\rho g}{R} \end{pmatrix} \begin{pmatrix} h_1 \\ h_2 \end{pmatrix}$$

With practice it becomes easier to choose appropriate state variables to simplify the analysis that is being carried out.

Deciding on the appropriate state variables to model a system is to a large extent a matter of experience. Sometimes it is relatively easy as, for example, with the coupled tanks system discussed in Example 13.3. At other times a good choice is not so obvious. One of the main things to ensure is that the variables are linearly independent. For example, choosing the position of an object, x, and say $3x$ as state variables will lead to trouble as they are clearly linearly dependent. Another thing to ensure is that the number of state variables chosen is equal to the order of the system. If too many are used then this will make the model degenerate and if too few are used then they will be insufficient to capture the state of the system adequately.

Self-assessment questions 13.2

1. Explain how a state space model of a system differs from a transfer function model of a system.
2. Describe the advantages of using a state space model.
3. Explain the term state vector.

Exercises 13.2

1. Obtain a state space model for the torsional pendulum of Example 3.2. Use the angular displacement θ and the angular velocity ω of the pendulum as the state space variables. The system input is the applied torque T and the system output is θ.
2. Derive a state space model for the *LCR* circuit of Example 4.6. Use the current through the capacitor, i, and the voltage across the capacitor, v_C, as the state space variables. The system input is v_i and the system output is v_o.

13.3 Solving a state space model

The most common way of solving a state space model is to use a digital computer. This is really an extension of the approach discussed in Section 7.4. Simulation packages are available that can accept the state space matrices A, B, C and D, together with the system inputs, and produce a simulation of the system outputs. For a system of any complexity this is the only way forward. However, for low order systems subject to simple inputs it is possible to derive an analytical solution to the state space model.

It is worthwhile studying such examples as they can provide insight into computer solutions, thus preventing a simulation package being used in a state of ignorance.

Recall Eqn. [13.7] which gives the standard form of the state equation written in compact vector/matrix form. This is

$$\dot{\boldsymbol{x}} = A\boldsymbol{x} + B\boldsymbol{u}$$

This is a set of coupled first order differential equations. This becomes clearer if we expand this compact form. We have

$$\begin{aligned}
\dot{x}_1 &= a_{11}x_1 + a_{12}x_2 + \cdots + a_{1n}x_n + b_{11}u_1 + b_{12}u_2 + \cdots + b_{1n}u_n \\
\dot{x}_2 &= a_{21}x_1 + a_{22}x_2 + \cdots + a_{2n}x_n + b_{21}u_1 + b_{22}u_2 + \cdots + b_{2n}u_n \\
&\vdots \\
\dot{x}_n &= a_{n1}x_1 + a_{n2}x_2 + \cdots + a_{nn}x_n + b_{n1}u_1 + b_{n2}u_2 + \cdots + b_{nn}u_n
\end{aligned}$$

We can solve these equations using the Laplace transform. However, as we require a full solution of the equations we cannot assume zero initial conditions as we did when we wished to form a transfer function. Recall that the Laplace transform of a time derivative is given by

$$\mathscr{L}\{\dot{f}\} = sF(s) - f(0)$$

where $F(s) = \mathscr{L}\{f\}$. So when operated on by the Laplace transform the first equation of this set becomes

$$sX_1(s) - x_1(0) = a_{11}X_1(s) + a_{12}X_2(s) + \cdots + a_{1n}X_n(s) + b_{11}U_1(s) + b_{12}U_2(s) + \cdots + b_{1n}U_n(s)$$

This can be carried out for all the equations. However, it is possible to do this and retain the compact vector/matrix notation. We then have

$$s\boldsymbol{X}(s) - \boldsymbol{x}(0) = A\boldsymbol{X}(s) + B\boldsymbol{U}(s) \qquad \text{Eqn. [13.19]}$$

where $\boldsymbol{X}(s) = \mathscr{L}\{\boldsymbol{x}(t)\}$, $\boldsymbol{U}(s) = \mathscr{L}\{\boldsymbol{u}(t)\}$ and $\boldsymbol{x}(0)$ is the initial condition of the state vector at time $t=0$.

Eqn. [13.19] can be rearranged to find a solution of the state space model. We have

$$s\boldsymbol{X}(s) - A\boldsymbol{X}(s) = \boldsymbol{x}(0) + B\boldsymbol{U}(s)$$

When manipulating matrix equations, we need to be careful to maintain the correct order of matrix multiplication as it is not reversible. We can factorise $\boldsymbol{X}(s)$ out of the left-hand side of this equation but we need to introduce the identity matrix I to ensure that similar quantities are being subtracted as s is a scalar and A is a matrix. This is acceptable because when a vector is multiplied by the identity matrix it remains unchanged. So we have

$$sI\boldsymbol{X}(s) - A\boldsymbol{X}(s) = \boldsymbol{x}(0) + B\boldsymbol{U}(s)$$

Now we can factorise $\boldsymbol{X}(s)$ out of the left-hand side of this equation. This gives

$$(sI - A)\boldsymbol{X}(s) = \boldsymbol{x}(0) + B\boldsymbol{U}(s)$$

Note that $\boldsymbol{X}(s)$ has been brought out on the right side of the bracket so that the order of multiplication is preserved.

In order to make $\boldsymbol{X}(s)$ the subject of the equation we need to pre-multiply the equation by $(sI - A)^{-1}$. This gives

$$(sI - A)^{-1}(sI - A)\boldsymbol{X}(s) = (sI - A)^{-1}\boldsymbol{x}(0) + (sI - A)^{-1}B\boldsymbol{U}(s)$$

Finally, noting that $(sI - A)^{-1}(sI - A) = I$ we have

KEY POINT

$$\boldsymbol{X}(s) = (sI - A)^{-1}\boldsymbol{x}(0) + (sI - A)^{-1}B\boldsymbol{U}(s) \qquad \text{Eqn. [13.20]}$$

This is a solution to the state equations and given the initial value of the state vector, $\boldsymbol{x}(0)$, and knowledge of the inputs to the system, $\boldsymbol{U}(s)$, we can calculate the value of the state vector for all $t > 0$. Note that Eqn. [13.20] has been left in the s domain. It is possible to express this equation in the time domain but it is more complicated. A better approach is to evaluate $\boldsymbol{X}(s)$ for a particular system and then use the inverse Laplace transform to obtain $\boldsymbol{x}(t)$. It is a simple matter to obtain an expression for the system outputs once an expression for Eqn. [13.20] has been obtained.

We now consider an example of how to obtain the solution of a state space model.

Example

13.4 Recall the spring–mass–damper system of Example 13.2. Given $M=2$ kg, $K=6$ N m^{-1}, $B=4$ N s m^{-1} obtain a solution for the position of the mass when a step input force of magnitude 1 N is applied at $t=0$ s. Assume the initial position of the mass is $x=1$ m and the initial velocity of the mass is $v=1$ m s^{-1}.

Solution The state space equations are

$$\begin{pmatrix} \dot{x} \\ \dot{v} \end{pmatrix} = \begin{pmatrix} 0 & 1 \\ -\dfrac{K}{M} & -\dfrac{B}{M} \end{pmatrix} \begin{pmatrix} x \\ v \end{pmatrix} + \begin{pmatrix} 0 \\ \dfrac{1}{M} \end{pmatrix} f$$

$$\boldsymbol{x} = (1 \quad 0) \begin{pmatrix} x \\ v \end{pmatrix}$$

Putting in the given values we obtain

$$A = \begin{pmatrix} 0 & 1 \\ -\frac{6}{2} & -\frac{4}{2} \end{pmatrix} = \begin{pmatrix} 0 & 1 \\ -3 & -2 \end{pmatrix}$$

Also,

$$B = \begin{pmatrix} 0 \\ \frac{1}{2} \end{pmatrix} = \begin{pmatrix} 0 \\ 0.5 \end{pmatrix}$$

We next need to calculate $(sI - A)^{-1}$. First we have

$$sI - A = s\begin{pmatrix} 1 & 0 \\ 0 & 1 \end{pmatrix} - \begin{pmatrix} 0 & 1 \\ -3 & -2 \end{pmatrix}$$

$$= \begin{pmatrix} s & 0 \\ 0 & s \end{pmatrix} - \begin{pmatrix} 0 & 1 \\ -3 & -2 \end{pmatrix}$$

$$= \begin{pmatrix} s & -1 \\ 3 & s+2 \end{pmatrix}$$

The inverse of a 2 × 2 matrix is given by

$$\begin{pmatrix} a & b \\ c & d \end{pmatrix}^{-1} = \frac{1}{ad - bc} \begin{pmatrix} d & -b \\ -c & a \end{pmatrix}$$

Calculating the inverse of a matrix can be quite messy. Fortunately for a 2 × 2 matrix we can use a simple formula.

Using this formula we have

$$(sI - A)^{-1} = \frac{1}{s(s+2) - (-1)(3)} \begin{pmatrix} s+2 & 1 \\ -3 & s \end{pmatrix}$$

$$= \frac{1}{s^2 + 2s + 3} \begin{pmatrix} s+2 & 1 \\ -3 & s \end{pmatrix}$$

This system has a single input and so $U(s)$ is one dimensional. For a unit step input $U(s) = 1/s$. So,

$$(sI - A)^{-1}\boldsymbol{B}\boldsymbol{U}(s) = \frac{1}{s^2 + 2s + 3}\begin{pmatrix} s+2 & 1 \\ -3 & s \end{pmatrix}\begin{pmatrix} 0 \\ 0.5 \end{pmatrix}\frac{1}{s}$$

$$= \frac{1}{s^2 + 2s + 3}\begin{pmatrix} 0.5 \\ 0.5s \end{pmatrix}\frac{1}{s}$$

$$= \frac{1}{s^2 + 2s + 3}\begin{pmatrix} \dfrac{0.5}{s} \\ 0.5 \end{pmatrix}$$

$$= \begin{pmatrix} \dfrac{0.5}{s(s^2 + 2s + 3)} \\ \dfrac{0.5}{s^2 + 2s + 3} \end{pmatrix}$$

We are given

$$\begin{pmatrix} x(0) \\ v(0) \end{pmatrix} = \begin{pmatrix} 1 \\ 1 \end{pmatrix}$$

and so the solution of the state equations in the s domain is

$$\boldsymbol{X}(s) = (sI - A)^{-1}\boldsymbol{x}(0) + (sI - A)^{-1}\boldsymbol{B}\boldsymbol{U}(s)$$

$$\boldsymbol{X}(s) = \frac{1}{s^2 + 2s + 3}\begin{pmatrix} s+2 & 1 \\ -3 & s \end{pmatrix}\begin{pmatrix} 1 \\ 1 \end{pmatrix} + \begin{pmatrix} \dfrac{0.5}{s(s^2 + 2s + 3)} \\ \dfrac{0.5}{s^2 + 2s + 3} \end{pmatrix}$$

$$= \begin{pmatrix} \dfrac{s+2+1}{s^2 + 2s + 3} \\ \dfrac{-3+s}{s^2 + 2s + 3} \end{pmatrix} + \begin{pmatrix} \dfrac{0.5}{s(s^2 + 2s + 3)} \\ \dfrac{0.5}{s^2 + 2s + 3} \end{pmatrix}$$

$$= \begin{pmatrix} \dfrac{s(s+3) + 0.5}{s(s^2 + 2s + 3)} \\ \dfrac{s - 3 + 0.5}{s^2 + 2s + 3} \end{pmatrix}$$

$$= \begin{pmatrix} \dfrac{s^2 + 3s + 0.5}{s(s^2 + 2s + 3)} \\ \dfrac{s - 2.5}{s^2 + 2s + 3} \end{pmatrix}$$

We can obtain the time domain solution of the state equations by inverting these expressions. This gives

$$x(t) = \mathscr{L}^{-1}\left\{\frac{s^2 + 3s + 0.5}{s(s^2 + 2s + 3)}\right\}$$

and

$$v(t) = \mathscr{L}^{-1}\left\{\frac{s - 2.5}{s^2 + 2s + 3}\right\}$$

In order to obtain the final expressions for $x(t)$ and $v(t)$ it is necessary to split these expressions into partial fractions and then invert them using a table of Laplace transforms. This is left as an exercise for the reader.

We see that obtaining analytical solutions of the state equations for even very simple systems is long and complicated. This is the reason why, in practice, solutions are obtained using a digital computer.

Self-assessment questions 13.3

1. Describe the steps needed to obtain an analytical solution of the state equations of a system.
2. Why are computers usually used to obtain solutions of the state equations of a system?

Exercises 13.3

1. Split the expressions obtained in Example 13.4 into partial fractions. Hence invert them in order to form analytical expressions for $x(t)$ and $v(t)$.
2. A system has the following state space matrices:

$$A = \begin{pmatrix} 0 & 1 \\ 2 & 1 \end{pmatrix} \qquad B = \begin{pmatrix} 1 \\ 1 \end{pmatrix} \qquad C = (1 \quad 0) \qquad D = 0$$

Derive an analytical solution for this system when it is subject to a unit step input. Assume the initial conditions vector is

$$\mathbf{x}(0) = \begin{pmatrix} 1 \\ 1 \end{pmatrix}$$

Test and assignment exercises 13

1. Obtain a state space model for the coupled mass system of Example 2.6. This is a fourth order system. Use the displacement and velocity of each of the masses as the state space variables. The system input is the applied force f and the system output is the displacement of mass 2, y.

2. Derive a state space model for the armature-controlled d.c. motor of Example 4.8. Use the angular position θ of the motor, the angular velocity ω of the motor and the armature current i_a as the state variables. The system input is the armature voltage v_a and the system output is θ.
3. A system has the following state space matrices:

$$A = \begin{pmatrix} 1 & 2 \\ 3 & 4 \end{pmatrix} \qquad B = \begin{pmatrix} 2 \\ 1 \end{pmatrix} \qquad C = (1 \quad 1) \qquad D = 0$$

   Derive an analytical solution for this system when it is subjected to a unit step input.
4. A system has the following state space matrices:

$$A = \begin{pmatrix} 4 & 8 \\ 5 & 6 \end{pmatrix} \qquad B = \begin{pmatrix} 2 \\ 4 \end{pmatrix} \qquad C = (0 \quad 1) \qquad D = 0$$

   Derive an analytical solution for this system when it is subjected to a unit step input.

Appendices

A.1 Laws of indices

Sometimes it is necessary to multiply a number by itself several times, for example, $a \times a \times a \times a \times \ldots \times a$. A shorthand notation exists to specify this. If the number a is multiplied by itself p times then we write the result as a^p. The number p is called a **power** or **index** and the number a is called the **base**. The following laws hold when manipulating indices:

KEY POINT

$$a^m \times a^n = a^{m+n}$$

KEY POINT

$$\frac{a^m}{a^n} = a^{m-n}$$

KEY POINT

$$(a^m)^n = a^{mn}$$

KEY POINT

$$a^{-m} = \frac{1}{a^m}$$

There are two special cases worth noting. These are

KEY POINT

$$a^0 = 1$$

That is, any number raised to the power 0 is 1. Also,

KEY POINT

$$a^1 = a$$

That is, any number raised to the power 1 is itself.

A.2 Partial fractions

Given several fractions it is possible to combine them together to form a single fraction. For example,

$$\frac{1}{x+1} + \frac{1}{x+2} = \frac{x+2+x+1}{(x+1)(x+2)} = \frac{2x+3}{x^2+3x+2}$$

This process can be reversed. So given a single fraction it is possible to split this fraction into a sum of several simpler fractions. These simpler fractions are known as **partial fractions**. For example,

$$\frac{2x+3}{x^2+3x+2}$$

has partial fractions

$$\frac{1}{x+1} \qquad \frac{1}{x+2}$$

When splitting partial fractions the form of the partial fractions depends on the nature of the original fraction. The key to deciding which form to use is to factorise the denominator of the original fraction. There are several different cases. Only the ones used in this book are stated here.

KEY POINT

Linear factors $ax+b$ in the denominator give rise to partial fractions of the form

$$\frac{A}{ax+b}$$

where A is a constant to be determined.

KEY POINT

Quadratic factors $ax^2 + bx + c$ in the denominator give rise to partial fractions of the form

$$\frac{Ax + B}{ax^2 + bx + c}$$

where A and B are constants to be determined.

Note that quadratic factors are usually retained when further factorisation would give rise to complex factors.

KEY POINT

Repeated linear factors $(ax + b)^2$ in the denominator give rise to partial fractions of the form

$$\frac{A}{ax + b} + \frac{B}{(ax + b)^2}$$

where A and B are constants to be determined.

The methods of obtaining the coefficients of the partial fractions are illustrated in many examples throughout the book.

Solutions to exercises

Solutions to Chapter 1

Exercises 1.2

1. The fuel delivery system, the engine, the transmission system, the suspension system and the bus body.
2. The keypad, the computer system, the display and the calculator body.
3. The sails, the power transmission and grinding system and the windmill building.
4. Inputs: electricity, gas, water and food.
   Outputs: heat, waste and noise.
5. Inputs: electricity and control signals.
   Outputs: sound and heat.
6. The engine has fuel rate as an input and rotational mechanical power as an output. The rest of the bus has rotational mechanical power as an input and translational motion as an output.
7. The keypad has button depressions as an input and electronic coded signals as an output. The display has electronic coded signals as an input and display information as an output.
8. The sails have wind power as an input and rotational mechanical power as an output. The power transmission and grinding system has rotational power as an input and grinding stone rotation speed as an output.

Exercises 1.3

1. see Figure A1.
2. see Figure A2.

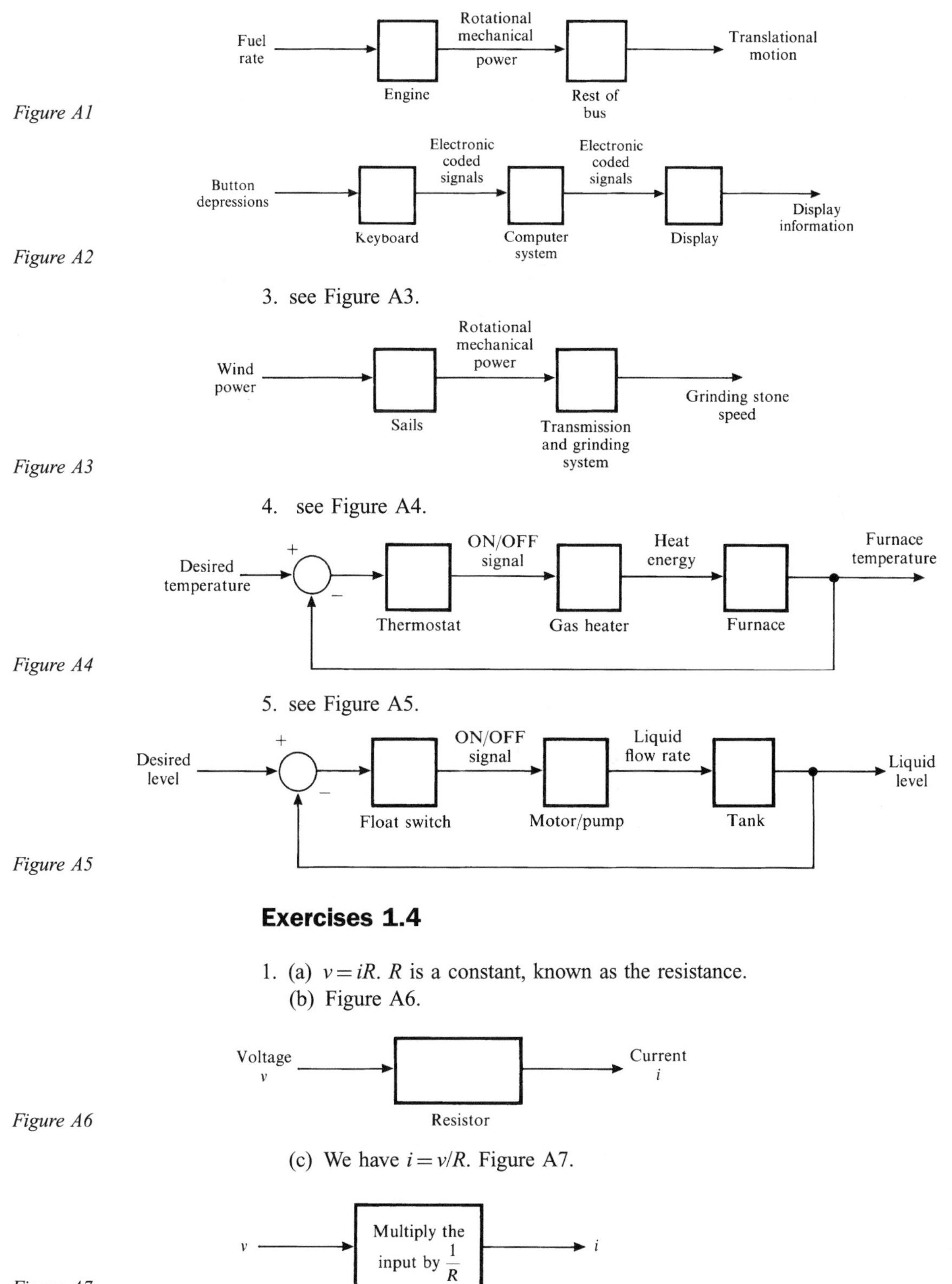

Figure A1

Figure A2

3. see Figure A3.

Figure A3

4. see Figure A4.

Figure A4

5. see Figure A5.

Figure A5

Exercises 1.4

1. (a) $v = iR$. R is a constant, known as the resistance.
   (b) Figure A6.

Figure A6

   (c) We have $i = v/R$. Figure A7.

Figure A7

2. (a) The equation of the tangent to the curve at the operating point is $p = v - 5$.
   (b)

Voltage v	Power p	Power (linear model)
10	5	5
10.01	5.010005	5.01
10.1	5.1005	5.1
20	20	15

Exercises 1.5

1. (a) 0.035 m
   (b) 5.6×10^{-6} m^2
   (c) 6230 Ω
   (d) 0.0256 kg
   (e) 1.003×10^5 V
   (f) 2830 J
   (g) 6.32×10^{-4} m^2
   (h) 8.76×10^{-9} m^3
   (i) 0.0825 m s^{-1}
   (j) 9.853×10^{-3} J s^{-1}
   (k) 5.82×10^4 m s^{-2}

Solutions to Chapter 2

Exercises 2.2

1. Figure A8.

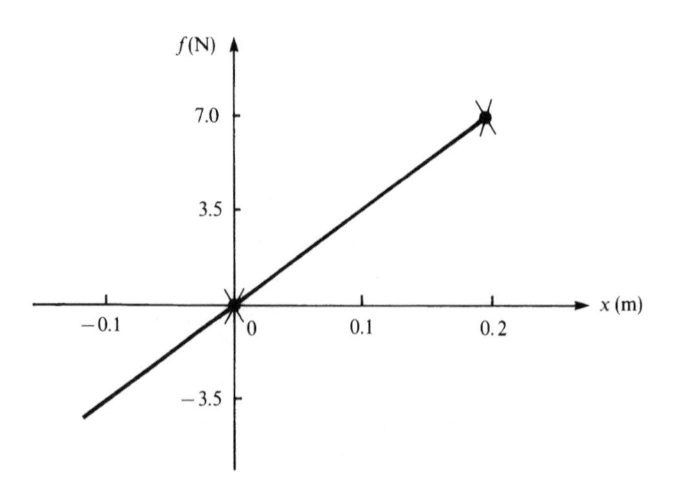

Figure A8

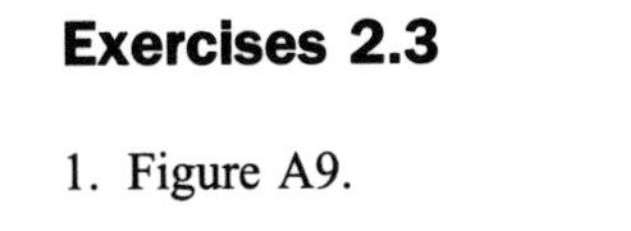

Exercises 2.3

1. Figure A9.

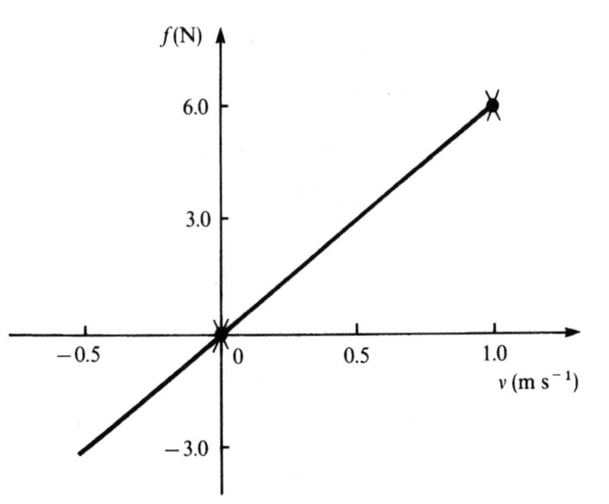

Figure A9

Exercises 2.4

1. $a = \dfrac{f}{M} = \dfrac{4}{M}$

   (a) $a = 0.1739 \text{ m s}^{-2}$
   (b) $a = 0.6349 \text{ m s}^{-2}$
   (c) $a = 1.465 \text{ m s}^{-2}$

Exercises 2.5

1. For the spring $f = K(x - y)$
   For the damper $f = B(\mathrm{d}y/\mathrm{d}t)$

   System differential equation is $Kx = B\dfrac{\mathrm{d}y}{\mathrm{d}t} + Ky$

2. We assume that a force f is applied to spring 1 to displace it. Let r be the displacement of the right side of spring 1.
   For spring 1 $f = K_1(x - r)$

   For the damper $f = B\left(\dfrac{\mathrm{d}r}{\mathrm{d}t} - \dfrac{\mathrm{d}y}{\mathrm{d}t}\right)$

   For spring 2 $f = K_2 y$
   System differential equation is

$$BK_1\frac{\mathrm{d}x}{\mathrm{d}t} = B(K_1 + K_2)\frac{\mathrm{d}y}{\mathrm{d}t} + K_1K_2y$$

3. We denote the force in spring 1 by f_{s1}, the force in spring 2 by f_{s2} and the force in the damper by f_{d}.

For spring 1 we have $f_{s1} = K_1(x - y)$
For spring 2 we have $f_{s2} = K_2 y$

For the damper we have $f_d = B\dfrac{dy}{dt}$

The free-body diagram for the mass is given in Figure A10.

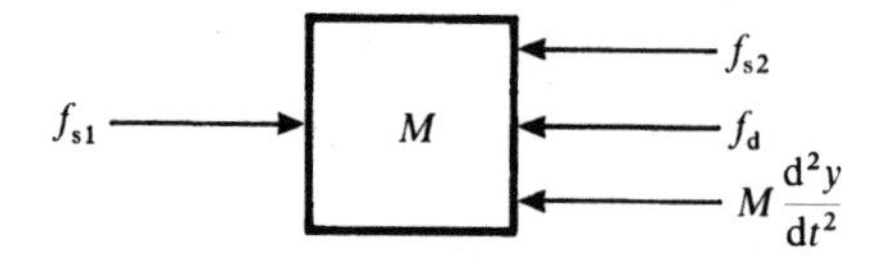

Figure A10

The system differential equation is

$$K_1 x = M\frac{d^2y}{dt^2} + B\frac{dy}{dt} + (K_1 + K_2)y$$

4. We denote the force in the damper by f_d and the force in the spring by f_s.

For the damper $f_d = B\left(\dfrac{dx}{dt} - \dfrac{dy}{dt}\right)$

For the spring $f_s = Ky$
The free-body diagram for mass 1 is given in Figure A11.

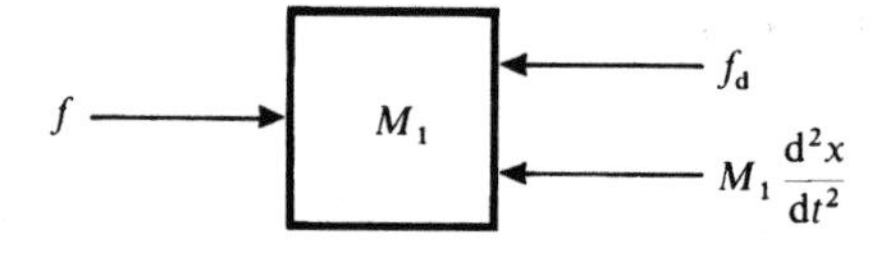

Figure A11

Equating the forces to zero gives $f - f_d - M_1\dfrac{d^2x}{dt^2} = 0$

The free-body diagram for mass 2 is give in Figure A12.

Figure A12

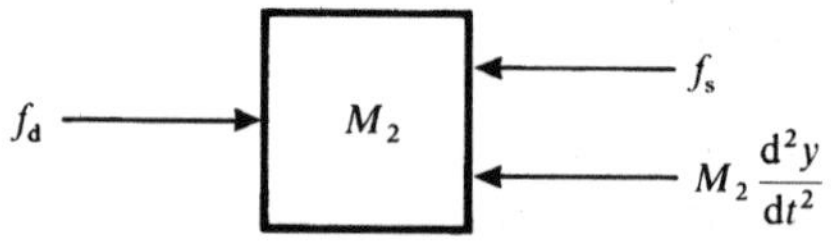

Equating the forces to zero gives $f_d - f_s - M_2\dfrac{d^2y}{dt^2} = 0$

The system differential equation is

$$Bf = M_1M_2\frac{d^3y}{dt^3} + B(M_1 + M_2)\frac{d^2y}{dt^2} + M_1K\frac{dy}{dt} + BKy$$

Solutions to Chapter 3

Exercises 3.3

1. Figure A13.

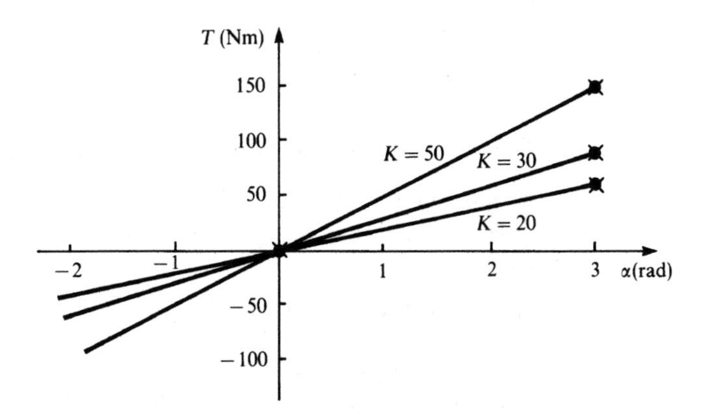

Figure A13

Exercises 3.4

1. Figure A14.

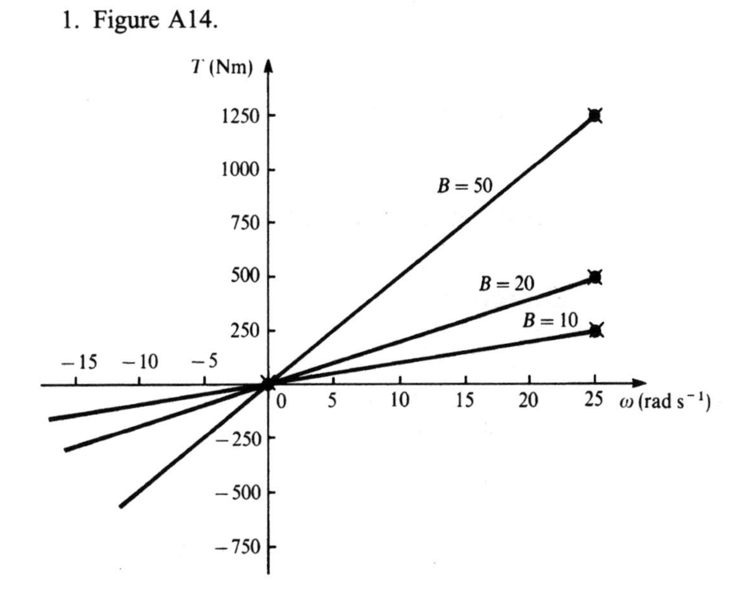

Figure A14

Exercises 3.5

1. (a) $\alpha = 0.3889$ rad s^{-2}
   (b) $\alpha = 0.7511$ rad s^{-2}
   (c) $\alpha = 4.3210$ rad s^{-2}

Exercises 3.7

1. Assume a torque T is applied to the left-hand side of the spring in order to make it rotate.

For the spring $T = K(\theta_i - \theta_o)$

For the damper $T = B\dfrac{d\theta_o}{dt}$

The system differential equation is $K\theta_i = B\dfrac{d\theta_o}{dt} + K\theta_o$

2. Assume the left-hand side of the damper is displaced by θ_i. Assume the torque in the spring is T_s. The torque in the damper is T by Newton's third law.

   For the damper $T = B\left(\dfrac{d\theta_i}{dt} - \dfrac{d\theta}{dt}\right)$

   For the spring we have $T_s = K\theta$

   The free-body diagram for the mass is given in Figure A15.

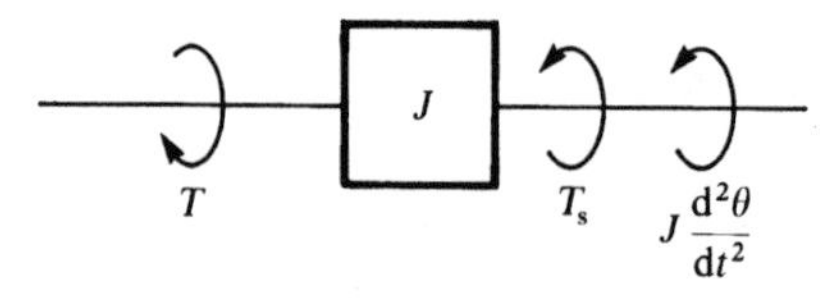

Figure A15

The system differential equation is $T = J\dfrac{d^2\theta}{dt^2} + K\theta$

Solutions to Chapter 4

Exercises 4.2

1. $v = 15.79$ V
2. $i = 0.5111$ A
3. Figure A16.

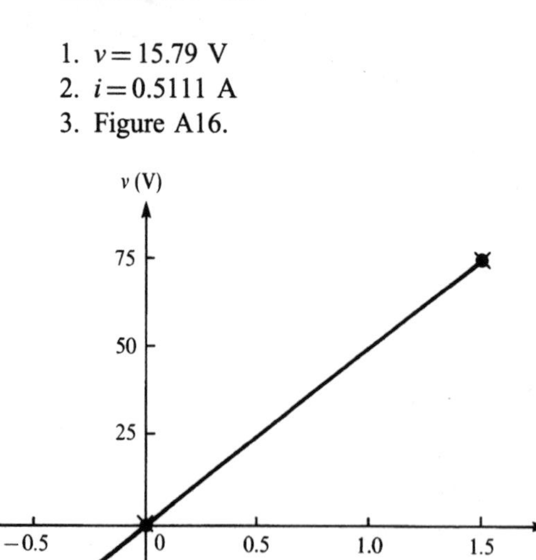

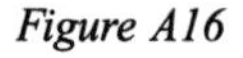

Figure A16

Exercises 4.3

1. $i = 2.76 \times 10^{-3}$ A
2. $\dfrac{\mathrm{d}v}{\mathrm{d}t} = 300 \text{ V s}^{-1}$

Exercises 4.4

1. $v = 0.05$ V
2. $\dfrac{\mathrm{d}i}{\mathrm{d}t} = 480 \text{ A s}^{-1}$

Exercises 4.5

1. Let v_C be the voltage across the capacitor. Using Kirchhoff's voltage law we have

   $$v_i = v_C + v_o$$

   For the capacitor we have $i = C(\mathrm{d}v_C/\mathrm{d}t)$ where i is the current in the circuit. For the resistor $v_o = iR$

   The system differential equation is $RC\dfrac{\mathrm{d}v_i}{\mathrm{d}t} = RC\dfrac{\mathrm{d}v_o}{\mathrm{d}t} + v_o$

2. Let v_C be the voltage across the capacitor, v_R the voltage across the resistor and i the current in the circuit. Using Kirchhoff's voltage law gives

   $$v_i = v_R + v_C + v_o$$

   For the resistor $v_R = iR$

   For the capacitor $i = C\dfrac{\mathrm{d}v_C}{\mathrm{d}t}$

   For the inductor $v_o = L\dfrac{\mathrm{d}i}{\mathrm{d}t}$

   The system differential equation is $LC\dfrac{\mathrm{d}^2 v_i}{\mathrm{d}t^2} = LC\dfrac{\mathrm{d}^2 v_o}{\mathrm{d}t^2} + RC\dfrac{\mathrm{d}v_o}{\mathrm{d}t} + v_o$

Exercises 4.6

1. Assume the armature current is i_a and the back e.m.f. is e_b. For the armature circuit we have

   $$v_a = i_a R_a + e_b$$

   For the motor $e_b = K_e\omega$ and $T = K_T i_a$ where T is the motor torque.
   For the damper we have $T_d = B\omega$ where T_d is the torque in the damper.
   The free-body diagram for the mass is given in Figure A17.

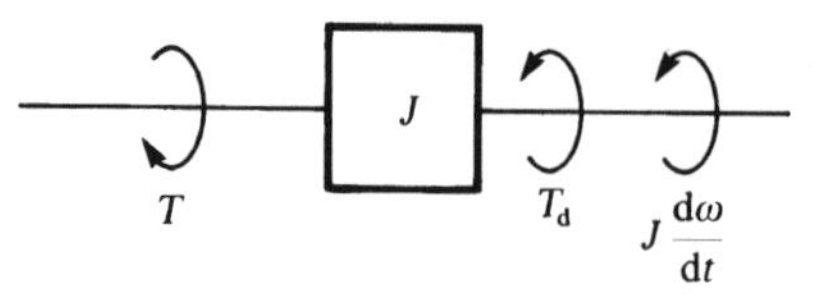

Figure A17

The system differential equation is

$$K_T v_a = JR_a \frac{d\omega}{dt} + (BR_a + K_e K_T)\omega$$

2. Assume the armature current is i_a and the back e.m.f. is e_b. For the armature circuit we have

$$v_a = i_a R_a + L_a \frac{di_a}{dt} + e_b$$

For the motor $e_b = K_e\omega$ and $T = K_T i_a$
For damper 1 we have $T = B_1(\omega - \omega_o)$
For damper 2 we have $T_d = B_2\omega_o$ where T_d is the torque in the damper.
The free-body diagram for the mass is given in Figure A18.

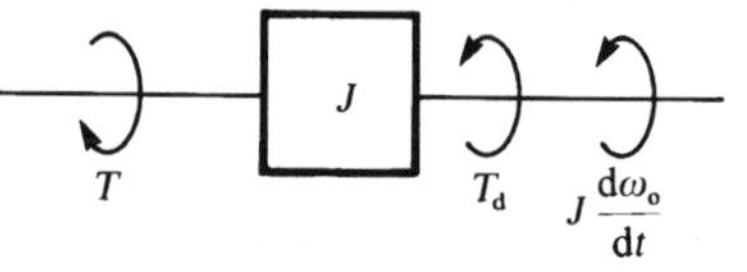

Figure A18

The system differential equation is

$$B_1 K_T v_a = B_1 J L_a \frac{d^2\omega_o}{dt^2} + (B_1 B_2 L_a + B_1 R_a J + K_e K_T J)\frac{d\omega_o}{dt} + (B_1 B_2 R_a + B_2 K_e K_T + K_e B_1 K_T)\omega_o$$

Solutions to Chapter 5

Exercises 5.4

1. Let the intermediate flow between the tanks be q. For tank 1 the conservation of mass gives

$$q_i - q = A_1 \frac{dh_1}{dt}$$

For the outlet valve $\rho g h_1 = qR$

For tank 2 conservation of mass gives $q - q_o = A_2 \frac{dh_2}{dt}$

For the outlet valve $\rho g h_2 = q_o R$
The system differential equation is

$$q_i = \frac{A_1 A_2 R^2}{(\rho g)^2}\frac{d^2 q_o}{dt^2} + \frac{(A_1 + A_2)R}{\rho g}\frac{dq_o}{dt} + q_o$$

Solutions to Chapter 6

Exercises 6.2

1. (a) $R = 2.488\ \text{K W}^{-1}$
   (b) $R = 6.25 \times 10^{-3}\ \text{K W}^{-1}$
2. (a) $q = 7200\ \text{W}$
   (b) $q = 2000\ \text{W}$

Exercises 6.3

1. (a) $C = 5752\ \text{J K}^{-1}$
   (b) $C = 3696\ \text{J K}^{-1}$
2. (a) $q = 177.6\ \text{W}$
   (b) $q = 53.6\ \text{W}$

Exercises 6.4

1. For convenience assume the liquid is warmer than the probe and the energy flow rate from liquid to probe is q. For the probe we have

$$q = C\frac{dT_p}{dt}$$

and $T_f - T_p = qR$
The system differential equation is

$$T_f = RC\frac{dT_p}{dt} + T_p$$

2. Let the energy flow rate between room 1 and room 2 be q_a, the energy flow rate between room 1 and the outside be q_b and that between room 2 and the outside be q_c. For room 1 we have

$$q_1 - q_a - q_b = C_1\frac{dT_1}{dt}$$

and $T_1 - T_2 = q_aR$ and $T_1 - T_e = q_bR_1$
For room 2 we have

$$q_2 + q_a - q_c = C_2\frac{dT_2}{dt}$$

and $T_2 - T_e = q_eR_2$
The system differential equation is

$$R_1RC_2\frac{dq_1}{dt} + R_1^2(R + R_2)q_1 + RR_1R_2q_2 + (RR_1 + R^2 + R_1R_2)T_e$$
$$= C_1C_2\frac{d^2T_1}{dt^2} + (2R + R_1 + R_2)\frac{dT_1}{dt} + (R^2 + RR_1 + RR_2)T_1$$

Solutions to Chapter 7

Exercises 7.2

1. Figure A19.

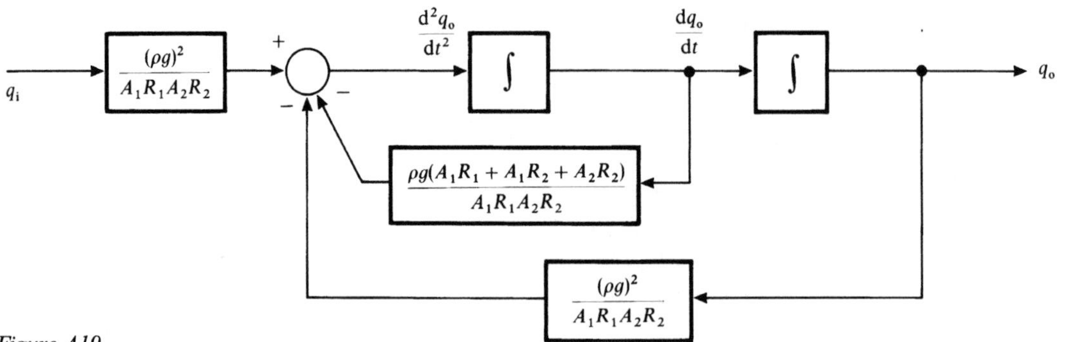

Figure A19

2. Figure A20.

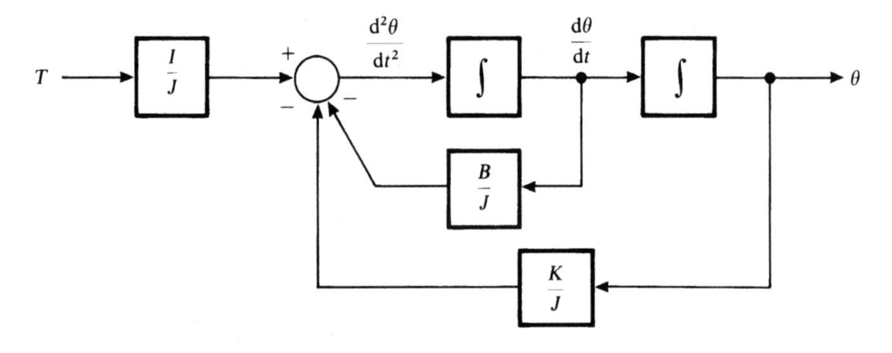

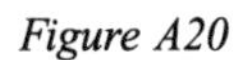
Figure A20

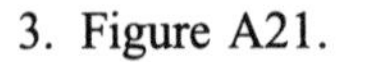
3. Figure A21.

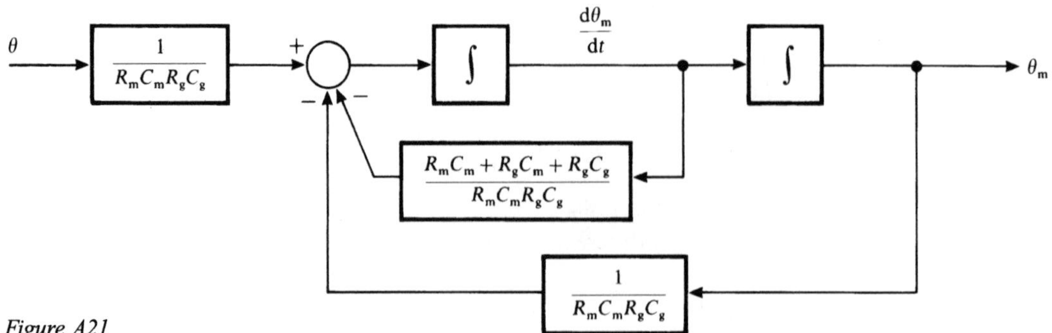

Figure A21

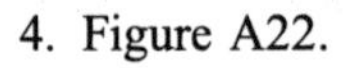
4. Figure A22.

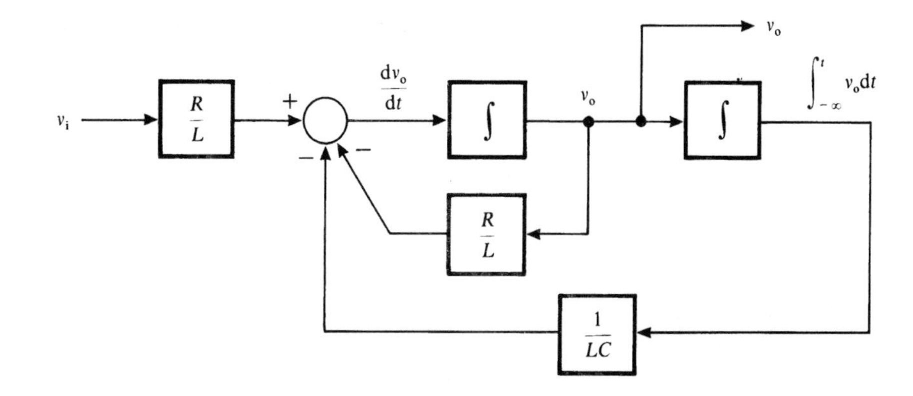

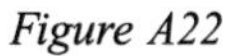
Figure A22

5. Figure A23.

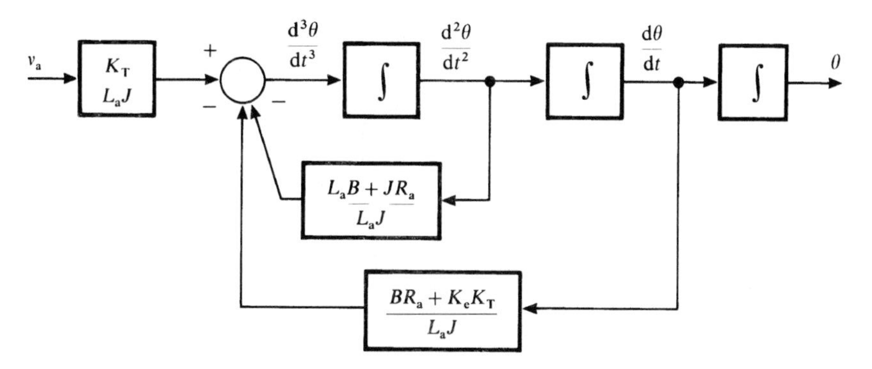

Figure A23

Exercises 7.3

1. Let $K_1 = \dfrac{(\rho g)^2}{A_1 R_1 A_2 R_2}$ and $K_2 = \dfrac{\rho g(A_1 R_1 + A_1 R_2 + A_2 R_2)}{A_1 R_1 A_2 R_2}$
   Figure A24.

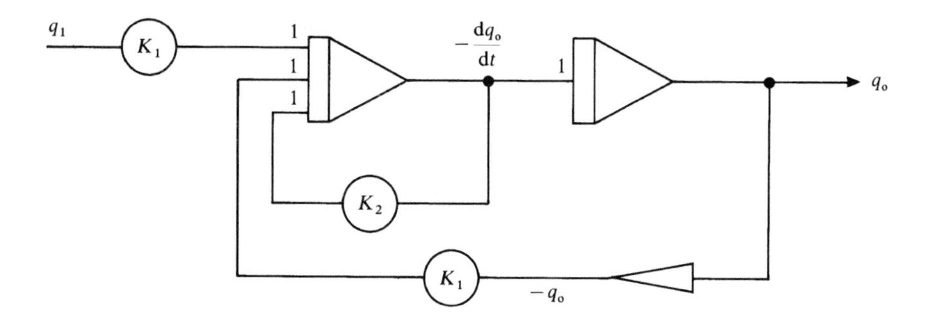

Figure A24

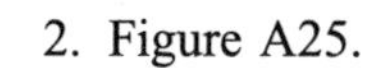
2. Figure A25.

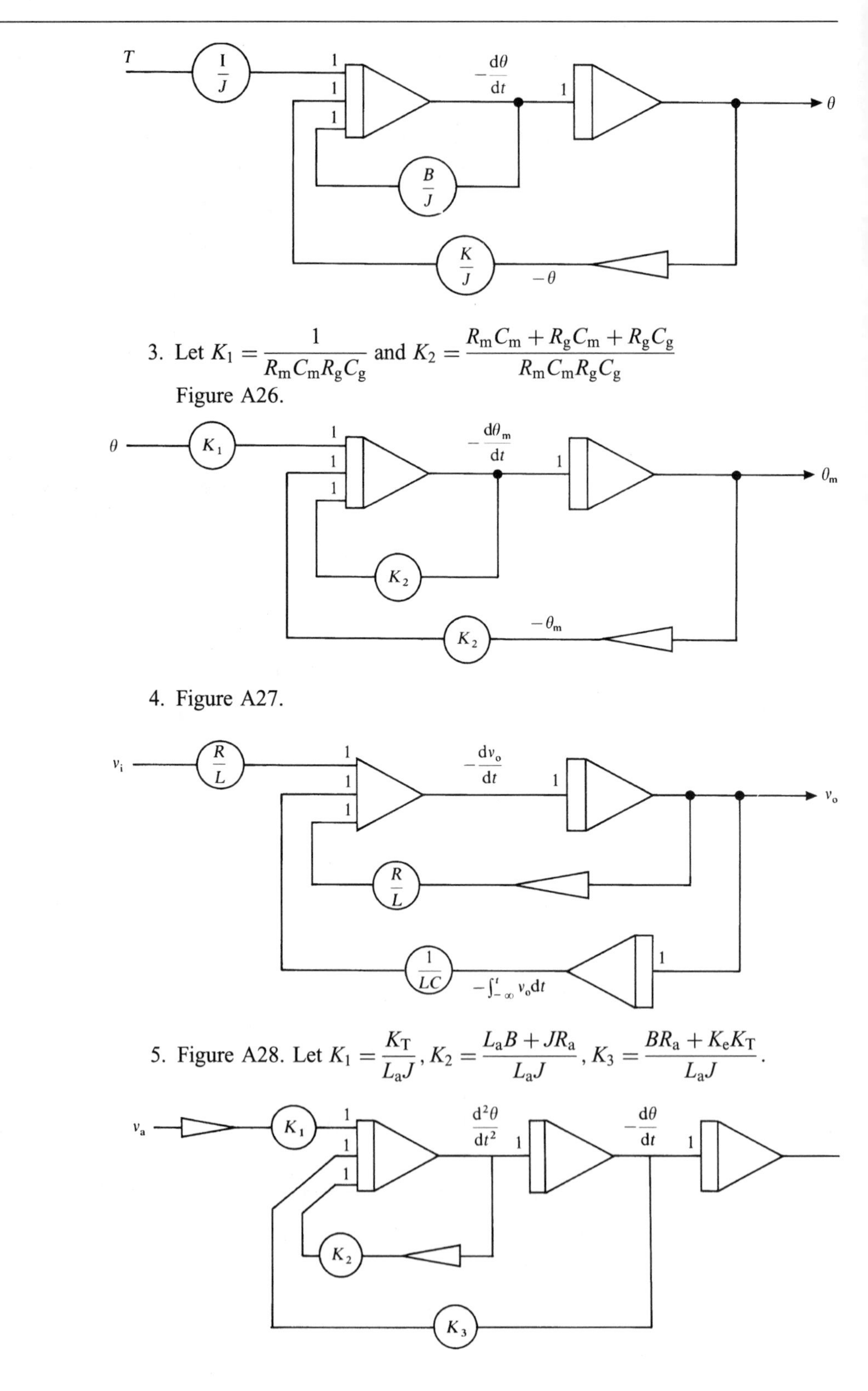

Figure A25

3. Let $K_1 = \dfrac{1}{R_m C_m R_g C_g}$ and $K_2 = \dfrac{R_m C_m + R_g C_m + R_g C_g}{R_m C_m R_g C_g}$
Figure A26.

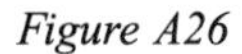

Figure A26

4. Figure A27.

Figure A27

5. Figure A28. Let $K_1 = \dfrac{K_T}{L_a J}$, $K_2 = \dfrac{L_a B + J R_a}{L_a J}$, $K_3 = \dfrac{B R_a + K_e K_T}{L_a J}$.

Figure A28

Solutions to Chapter 8

Exercises 8.2

1. (a) $\frac{2}{s^3}$

   (b) $\frac{6}{s^2+36}$

   (c) $\frac{s}{s^2+16}$

   (d) $\frac{1}{s-5}$

   (e) $\frac{e^{-3s}}{s}$

   (f) $\frac{2}{(s+3)^3}$

   (g) $\frac{s^2-4}{(s+4)^2}$

   (h) $\frac{16s}{(s^2+64)^2}$

2. (a) $u(t)$

   (b) t

   (c) t^3

   (d) $\sin(4t)$

   (e) $t^4 e^{-6t}$

Exercises 8.3

1. (a) $\frac{3}{s+1}+\frac{5}{s+5}$

   (b) $\frac{4}{s^2+1}+\frac{2s}{s^2+4}$

   (c) $\frac{8}{(s+2)^2}-\frac{10}{s^3}$

   (d) $\frac{12}{(s+4)^2+4}+\frac{5(s+2)}{(s+2)^2+16}$

2. (a) $F(s)(6s+2)-72$

   (b) $F(s)(8s^2+3s)-24s-5$

   (c) $F(s)\left(4s^2+\frac{8}{s}\right)-8s-4$

4. $s^3F(s) - s^2f(0) - s\frac{df(0)}{dt} - \frac{d^2f(0)}{dt^2}$ and

$s^4F(s) - s^3f(0) - s^2\frac{df(0)}{dt} - s\frac{d^2f(0)}{dt^2} - \frac{d^3f(0)}{dt^3}$

Exercises 8.4

1. (a) $y(t) = \frac{2}{3} + \frac{1}{3}e^{-3t}$

   (b) $f(t) = \frac{5}{8} + \frac{19}{8}e^{-8t}$

   (c) $c(t) = \frac{1}{3} + \frac{1}{10}e^{-2t} + \frac{77}{30}e^{-12t}$

   (d) $f(t) = te^{-3t} + \frac{1}{2}e^{-3t} + \frac{1}{2}e^{-t}$

   (e) $y(t) = \frac{1}{2}e^{-3t} - 13e^{-2t} + \frac{33}{2}e^{-t}$

Exercises 8.5

1. (a) $\frac{4}{s^2+6}$

   (b) $\frac{6s+2}{s^3+8s^2+7s+1}$

2. (a) Figure A29.

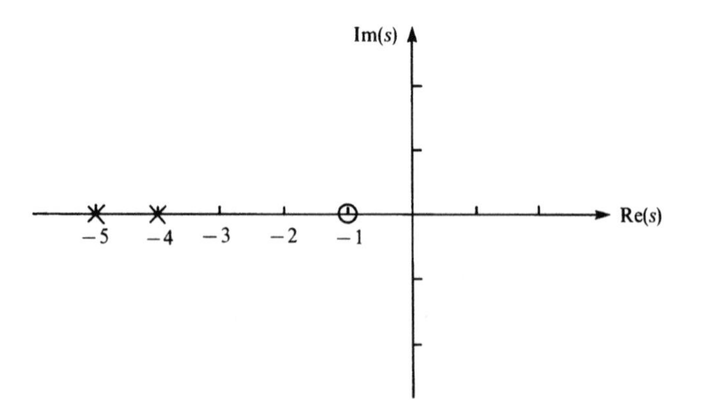

Figure A29

(b) Figure A30.

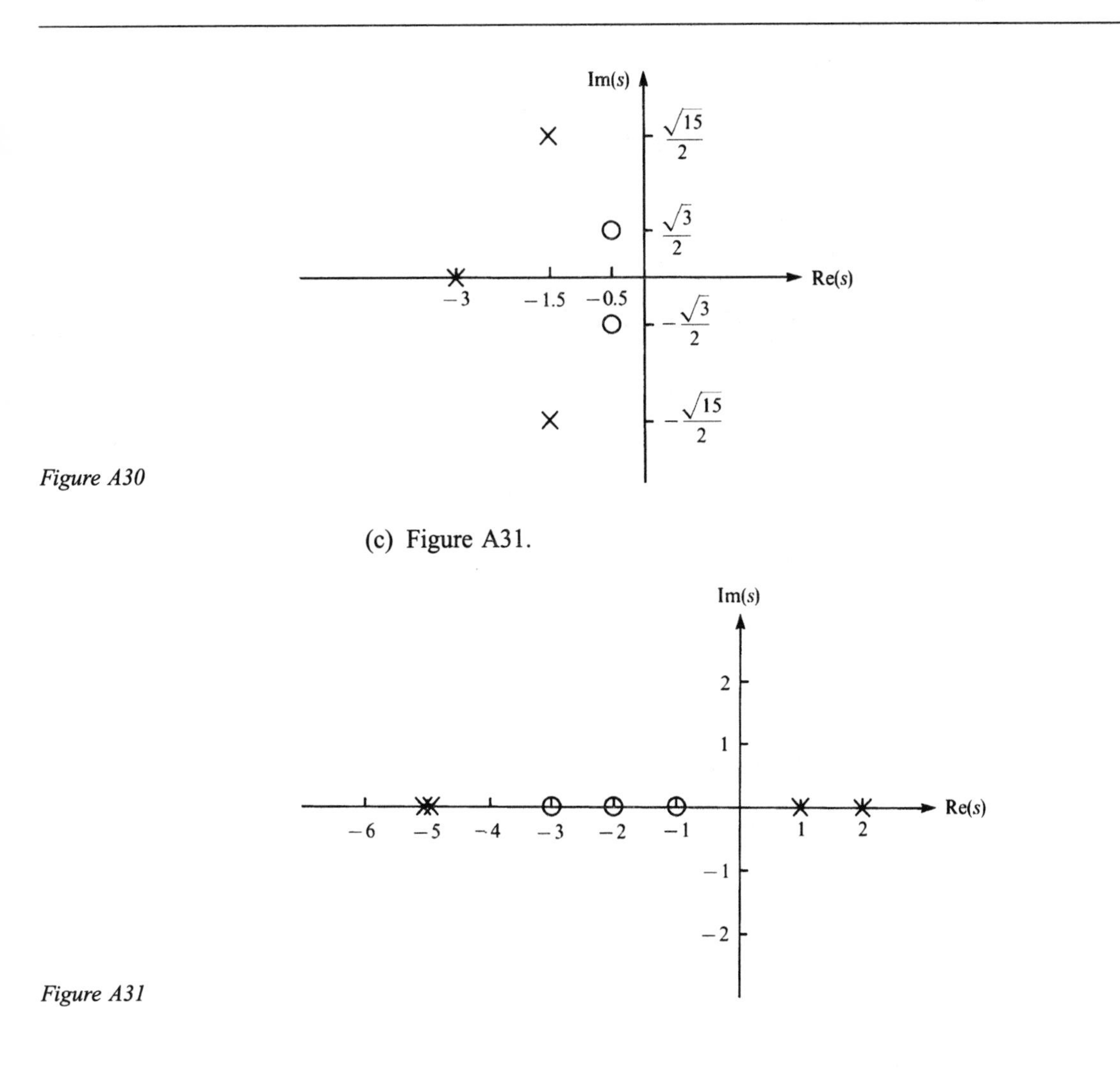

Figure A30

(c) Figure A31.

Figure A31

Solutions to Chapter 9

Exercises 9.3

1. D.c. gain is 1 and is dimensionless. Time constant is CR and has units of seconds.
2. (a) $K=5$ and $\tau=4$
   (b) $K=2$ and $\tau=8/3$
   (c) $K=3.5$ and $\tau=1.5$
   (d) $K=2.56$ and $\tau=1.97$
3. $v_o(t)=10-10e^{-100t}$
4. $q_o(t)=5-5e^{-0.1t}$

Exercises 9.4

1. $v_o(t) = \frac{2}{7} + \frac{5}{7}e^{-7t}$

Solutions to Chapter 10

Exercises 10.2

1. (a) $K=1$, $\omega_n=3.162$, $\zeta=1.581$
   (b) $K=2$, $\omega_n=3.162$, $\zeta=0.7910$
   (c) $K=1.5$, $\omega_n=2$, $\zeta=0.75$
   (d) $K=0.3333$, $\omega_n=1.732$, $\zeta=0.7217$
   (e) $K=0.8890$, $\omega_n=1.732$, $\zeta=0.3848$
   (f) $K=0.9843$, $\omega_n=1.069$, $\zeta=0.2422$

Exercises 10.3

1. (a) $c(t)=2-6e^{-2t}+4e^{-3t}$, settling time 2 s
   (b) $c(t)=1-\frac{9}{8}e^{-t}+\frac{1}{8}e^{-9t}$, settling time 4 s
   (c) $c(t)=\frac{8}{3}-4e^{-t}+\frac{4}{3}e^{-3t}$, settling time 4 s

Exercises 10.4

1. (a) $c(t)=1-3te^{-3t}-e^{-3t}$, settling time 1.333 s
   (b) $c(t)=4-16te^{-4t}-e^{-4t}$, settling time 1 s
   (c) $c(t)=\frac{4}{3}-4e^{-0.5t}-2te^{-0.5t}$, settling time 8 s

Exercises 10.5

1. (a) $c(t)=1-e^{-0.5t}\cos(2.96t)-0.169e^{-0.5t}\sin(2.96t)$, settling time 8 s
   (b) $c(t)=2-2e^{-2t}\cos(4.58t)-0.874e^{-2t}\sin(4.58t)$, settling time 2 s
   (c) $c(t)=2-2e^{-0.5t}\cos(1.94t)-0.512e^{-0.5t}\sin(1.94t)$, settling time 4 s

Exercises 10.6

1. (a) $\omega_n = \sqrt{\dfrac{K}{J}}$, $K' = \dfrac{1}{K}$, $\zeta = \dfrac{B}{2\sqrt{JK}}$

   (b) and (c)

   (i) $B=16$, $c(t)=0.25+0.01934e^{-7.464t} - 0.2693e^{-0.5360t}$

   (ii) $B=8$, $c(t)=0.25 - 0.25e^{-2t} - 0.5te^{-2t}$

   (iii) $B=4$, $c(t)=0.25 - 0.25te^{-t}\cos(\sqrt{2}t) - 0.1768e^{-t}\sin(\sqrt{2}t)$

2. For the armature circuit

$$v_a = i_a R_a + L_a \frac{di_a}{dt} + e_b$$

   For the motor $e_b = K_e\omega$ and $T = K_T i_a$

   For the mechanical system $T - B\omega - J\dfrac{d\omega}{dt} = 0$

$$\omega_n = \sqrt{\frac{BR_a + K_e}{L_a J}}$$

$$K' = \frac{K_T}{BR_a + K_e}$$

$$\zeta = \frac{JR_a + L_a B}{2\sqrt{L_a J(BR_a + K_e)}}$$

Solutions to Chapter 11

Exercises 11.4

1. (a) Maximum percentage overshoot 72%

   (b) Settling time 4.3 s

   (c) Rise time 0.25 s

   (d) Number of oscillations before the output signal remains within the settling band is 3

   (e) Steady state error 10%

2. (a) Maximum percentage overshoot 79%

   (b) Settling time 5.6 s

   (c) Rise time 0.2 s

   (d) Number of oscillations before the output signal remains within the settling band is 6

   (e) Steady state error 2%

Exercises 11.5

1. $G(s) = \dfrac{10}{1 + 3s}$
2. $G(s) = \dfrac{3.84}{s^2 + 0.915s + 1.92}$

Solutions to Chapter 12

Exercises 12.2

1. Figure A32.

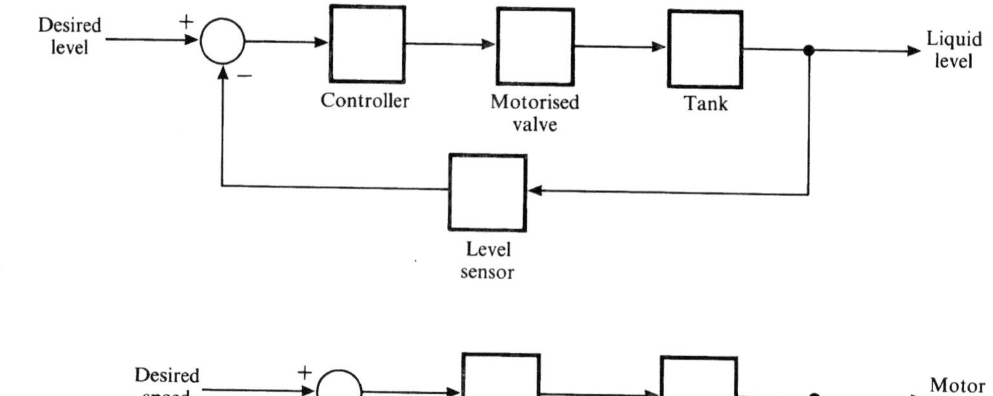

Figure A32

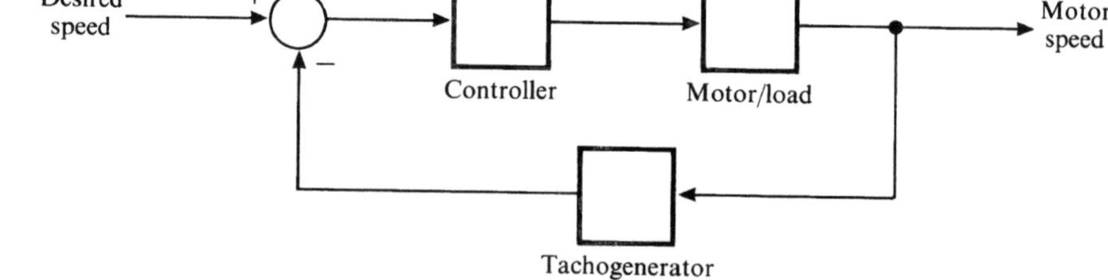

Figure A33

2. Figure A33.

Exercises 12.3

1. (a) $\dfrac{18}{4s + 37}$

   (b) $\dfrac{3(1 + s)}{s(1 + 2s)(s + 3)(s + 4) + 18(1 + s)}$

Exercises 12.4

1. $M = 0.025$
2. (a) $s = -0.586, -3.414$, overdamped response
   (b) $s = -2$ (twice), critically damped response

(c) $s = -2 \pm 2.45j$, underdamped response

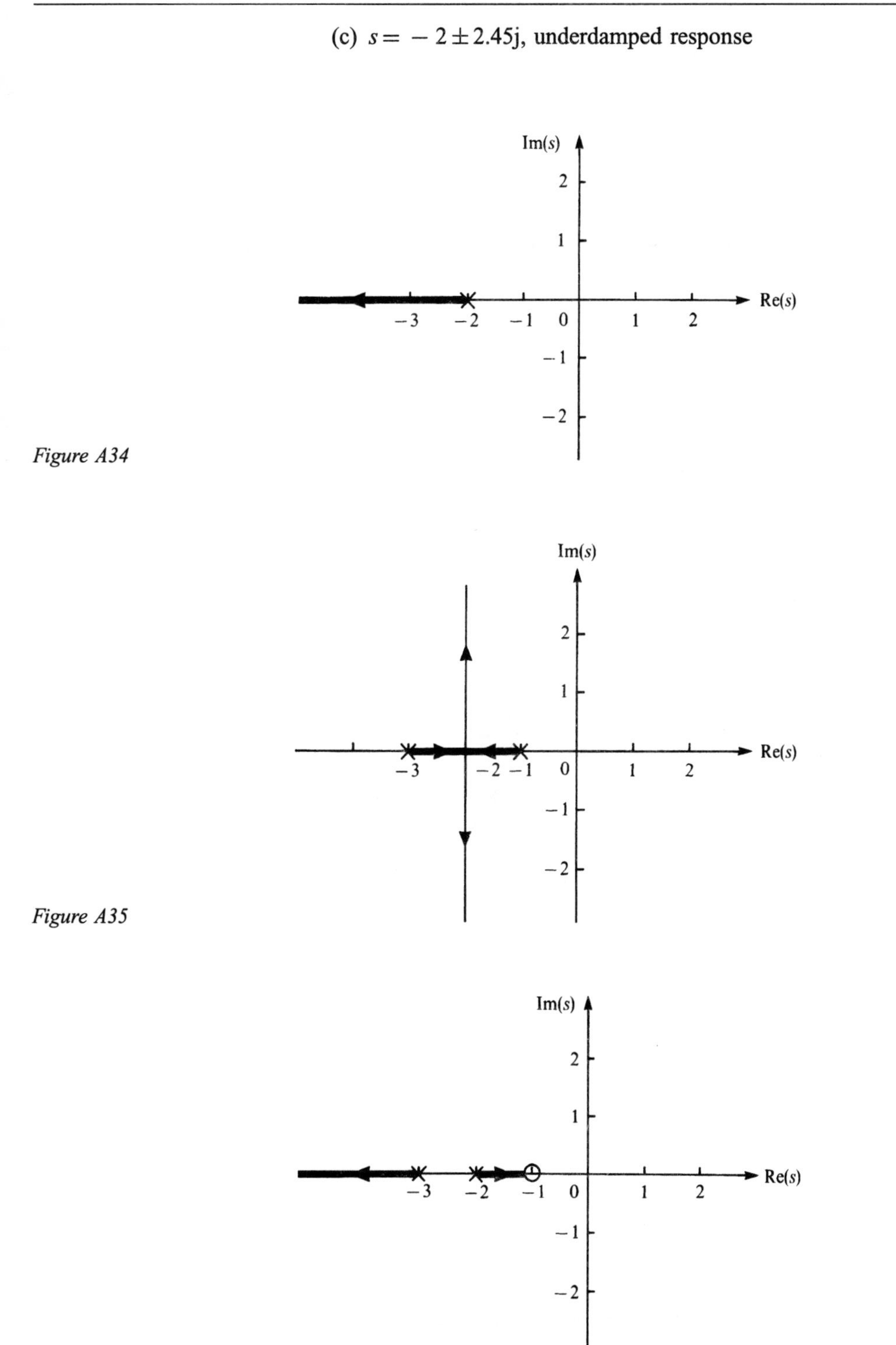

Figure A34

Figure A35

Figure A36

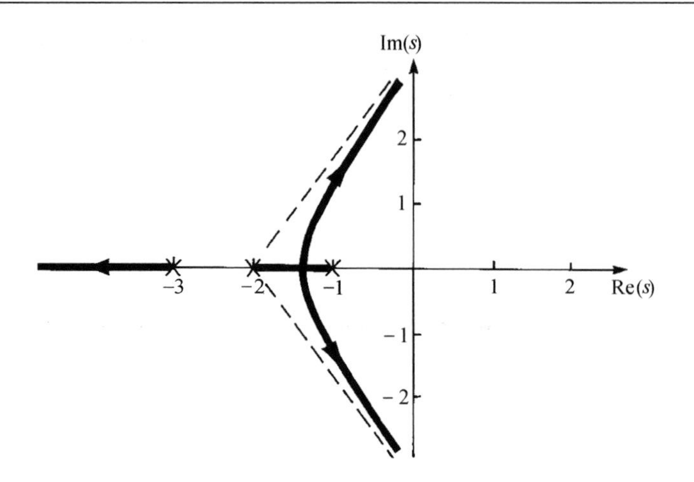

Figure A37

Exercises 12.5

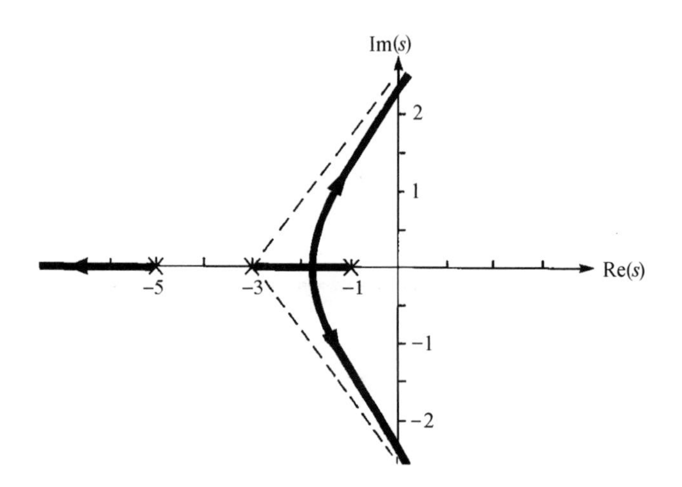

Figure A38

1. (a) Figure A34.
   (b) Figure A35.
   (c) Figure A36.
   (d) Figure A37.
2. Figure A38.

Solutions to Chapter 13

Exercises 13.2

2. $$\begin{pmatrix} \dot{v}_C \\ \dot{i} \end{pmatrix} = \begin{pmatrix} 0 & \dfrac{1}{C} \\ -\dfrac{1}{L} & -\dfrac{R}{L} \end{pmatrix} \begin{pmatrix} v_C \\ i \end{pmatrix} + \begin{pmatrix} 0 \\ \dfrac{1}{L} \end{pmatrix} v_i$$

$$v_o = (0 \quad R) \begin{pmatrix} v_C \\ i \end{pmatrix}$$

Exercises 13.3

1. $x(t) = \dfrac{1}{6} + \dfrac{1}{6}[5e^{-t}\cos(\sqrt{2}t) + 7.78e^{-t}\sin(\sqrt{2}t)]$

   $v(t) = e^{-t}\cos(\sqrt{2}t) - 2.475e^{-t}\sin(\sqrt{2}t)$

2. $\mathbf{x}(t) = \begin{pmatrix} e^{2t} \\ 2e^{2t} - 1 \end{pmatrix}$

Index